MMIC Design
GaAs FETs and HEMTs

Peter H. Ladbrooke

The Artech House Microwave Library

Analysis, Design, and Applications of Fin Lines, Bharathi Bhat and Shiban K. Koul

E-Plane Integrated Circuits, P. Bhartia and P. Pramanick, eds.

Filters with Helical and Folded Helical Resonators, Peter Vizmuller

GaAs MESFET Circuit Design, Robert A. Soares, ed.

Gallium Arsenide Processing Techniques, Ralph Williams

Handbook of Microwave Integrated Circuits, Reinmut K. Hoffmann

Handbook for the Mechanical Tolerancing of Waveguide Components, W.B.W. Alison

Handbook of Microwave Testing, Thomas S. Laverghetta

High Power Microwave Sources, Victor Granatstien and Igor Alexeff, eds.

Introduction to Microwaves, Fred E. Gardiol

LOSLIN: Lossy Line Calculation Software and User's Manual, Fred. E. Gardiol

Lossy Transmission Lines, Fred E. Gardiol

Materials Handbook for Hybrid Microelectronics, J.A. King, ed.

Microstrip Antenna Design, K.C. Gupta and A. Benalla, eds.

Microstrip Lines and Slotlines, K.C. Gupta, R. Garg, and I.J. Bahl

Microwave Engineer's Handbook: 2 volume set, Theodore Saad, ed.

Microwave Filters, Impedance Matching Networks, and Coupling Structures, G.L. Matthaei, L. Young and E.M.T. Jones

Microwave Integrated Circuits, Jeffrey Frey and Kul Bhasin, eds.

Microwaves Made Simple: Principles and Applications, Stephen W. Cheung, Frederick H. Levien, et al.

Microwave and Millimeter Wave Heterostructure Transistors and Applications, F. Ali, ed.

Microwave Mixers, Stephen A. Maas

Microwave Transition Design, Jamal S. Izadian and Shahin M. Izadian

Microwave Transmission Line Filters, J.A.G. Malherbe

Microwave Transmission Line Couplers, J.A.G. Malherbe

Microwave Tubes, A.S. Gilmour, Jr.

MMIC Design: GaAs FETs and HEMTs, Peter H. Ladbrooke

Modern Spectrum Analyzer Theory and Applications, Morris Engelson

Monolithic Microwave Integrated Circuits: Technology and Design, Ravender Goyal, et al.

Nonlinear Microwave Circuits, Stephen A. Maas

Terrestrial Digital Microwave Communications, Ferdo Ivanek, et al.

MMIC Design GaAs FETs and HEMTs

Peter H. Ladbrooke
Director, GaAs Code Ltd

Artech House
Boston and London

Library of Congress Cataloging-in-Publication Data

Ladbrooke, Peter H.
MMIC design.

1. Microwave integrated circuits—Design and construction. 2. Transistor circuits—Design and construction. 3. Gallium arsenide semiconductors. I. Title.
TK7876.L33 1989 621.381′73 88-34959
ISBN 0-89006-314-1

British Library Cataloging-in-Publication Data

Ladbrooke, Peter, H. *1943-*
MMIC design: GaAs FETs and HEMTs
1. Electronic equipment. Monolithic microwave integrated circuits. Design
I. Title
621.381′73

ISBN 0-89006-314-1

ARTECH HOUSE, INC.
685 Canton Street
Norwood, MA 02062

International Standard Book Number: 0-89006-314-1
Library of Congress Catalog Card Number: 88-34959

10 9 8 7 6 5 4 3 2

To Jo and Abby

Contents

Foreword

Circuit design is always a compromise if the device parameters are fixed. In many current processes the fabrication company concerned prefers to fix the device geometry and material parameters giving the circuit designer few degrees of freedom when carrying out his design. Generally the device used is a compromise between ultra-low noise, broad bandwidth and medium output power.

In this book, Peter Ladbrooke suggests that this practice is outdated and that far better circuit performance could be achieved if each device was tailored to meet a specific circuit application. He says this is now possible with recent advances in understanding the factors affecting technology and a knowledge of the relationship between material properties, device geometries and device performance. The author derives quantitative expressions for the elements of the equivalent circuit for the transistors and other components from which S parameters can be obtained and from which a circuit design can be created.

While circuit designers might welcome this idea, fabrication technologists and production engineers may have other views! Imagine a production process in which almost all geometries and technologies are variable from mask set to mask set and process run to process run. I can see production controllers blanching at the prospect. However if performance is at a premium it may be the only viable alternative.

The book covers most MMIC functions giving numerical relationships between key device and processing parameters. The chapters are littered with worked examples of his technique for HEMT and MESFET based technologies. In Chapter 8 the reader is introduced to the concept of reverse modeling and a number of examples are given in which the major fabrication parameters are derived from device electrical characteristics.

The arguments put forward by the author for his approach are compelling but it will remain to be seen whether or not his revolutionary approach will be accepted, particularly as it cuts across the presently accepted fixed process philosophy adopted almost universally both in GaAs and silicon technologies. However, careful choice of the numerous equations which follow in the text will certainly allow the reader to 'reverse model' his competitors' process and this alone may be enough justification for purchasing this well written and often controversial book.

Jim Turner.

James Turner, MBE
Plessey Research Caswell Ltd.
Caswell, November 1988

Acknowledgments

MMIC design: GaAs FETs and HEMTs is a compilation of many hitherto unpublished results arising from carefully executed FET research programs starting in 1974. Most of this work was done in the University of New South Wales, Sydney, Australia, or with the sponsorship of that organization in Cornell University, Ithaca, USA and at Plessey Research (Caswell), England. The support of these organizations, the Radio Research Board of Australia, and the Australian Research Grants Committee, must be acknowledged throughout. Advancements in the application of the resultant underlying FET physics has benefitted from the developments undertaken at GEC Research, with the very welcome support of the British Ministry of Defence (Directorate of Components, Valves and Devices).

The existence of this book owes much to the efforts of others, spread over many years. Ruey Huang participated in early transport calculations, and helped with the practical realization of $0.7\mu_m$ FETs at a time when such a capability was far less commonplace than it is now. Trevor Barton wrote the computer simulators which helped structure our thinking around the important physical phenomena occurring in GaAs MESFETs. Continuing financial support for FET physics research through the Radio Research Board of Australia occurred because Bill Williamson, at that time with the Telecom Research Laboratories, believed in it and in its long-term usefulness.

Jim Turner and Mike Cardwell, both of Plessey Research (Caswell), offered the opportunity to work on a study of soft breakdown in GaAs FETs which yielded important clues to the structure dependence of breakdown processes. Lester Eastman provided the environment and resources necessary to carry out a specific experiment at a time when the investigation could go very little further without such an experiment being done. While at Cornell University, Peter Zwicknagl grew most of the layers for the experiment in question and, together with Carolyn Lowe, did most of the device fabrication. During this phase, the added assistance of William Schaff (with the provision of DLTS data), of Lovell Camnitz (with E-beam direct writing), and of Fayez Najjar (with programming and physics theory) was greatly appreciated.

It was at Cornell University that I first met Colin Wood who, together with Daniel McCaughan and Daryl Hooper, was subsequently to support the application of device physics in MMIC engineering within GEC's research laboratories. Steve Ransome, Jonathan Nicholas and Anthony Jacomb-Hood gave excellent practical assistance to a programme of MMIC fault diagnosis and correction when it was needed most. Steve Ransome supplied Figure 3.1.

Mike Allenson and Brian Taylor, both then at the Royal Signals and Radar Establishment, supported the idea of yield studies based on device physics. Ged King, of Marconi Defence Systems, Stanmore, was quick to appreciate the potential of the method and, with the very able assistance of Steve Flynn, helped to develop a perspective of it from the all-important end-users' point of view. Chaz Caddick kept the effort well supplied with excellent S-parameter measurements on any number of devices and circuits which allowed careful diagnostic and proving work to be done. Andy Hughes gave assistance with circuit modeling beyond the call of duty, and helped plan the book in its early stages. Co-operation with Pieter Abrie over software for power amplifier design provided a useful demonstration of designing FETs and passive components concurrently. It is a particular pleasure to acknowledge the work done with Steve Blight on some of the obscure rate-dependent effects exhibited by short-channel MESFETs.

James Bridge and Adrian Hill have contributed a great deal by virtue of their continuing collaboration and friendship. Some of the figures in this book are taken from work done by them. Theirs has also been the task of writing the software involved in the application of device physics to MMICs.

I'm inclined to believe I've guessed correctly who the reviewers are although tradition demands they should remain nameless. Their involvement in providing continuous useful comment as the manuscript evolved is appreciated nonetheless. Elaine Peyton prepared all the drawings; her help has been especially welcome.

Finally, and perhaps most importantly, I wish to thank my wife Jo—not just for the fact that she word-processed the entire manuscript—but for her patience and enduring unselfishness.

Notes on nomenclature, acronyms and useful data

In the text, capital letters have been used to indicate differently doped regions of a semiconductor by ion density rather than free-carrier density. The distinction is useful for hot-electron devices such as FETs where, on the one hand, injection of excess carriers can readily occur and, on the other, depletion of free carriers can equally occur in some parts of the device. In either case, it is most useful to distinguish regions of semiconductor by the way they are doped (requiring upper case letters) rather than by the free carriers (denoted by lower case letters) they support. To quote an example, we shall talk of PIN diodes rather than pin diodes.

A bar over a symbol is used throughout to indicate normalized values, in particular currents or current-related quantities per unit gatewidth, e.g. $\overline{I_D} = 120\ \mathrm{mA\ mm^{-1}}$.

The same symbol, G, is used for power gain and conductance. The intended meaning for each use should be clear.

The symbol $\mathscr{L}$ has been used for inductance to save confusion with length, L, in particular gate length L_G.

The symbol m, rather than the more familiar n, has been adopted for the ideality factor of a Schottky or PIN diode to save confusion with the electron density n.

Acronyms

CAD	computer-aided design
CHEMT	conventional high electron mobility transistor
DBS	direct broadcast by satellite
ESM	electronic support measures
EW	electronic warfare
FET	field effect transistor
GW	guided weapons

HEMT	high electron mobility transistor
IHEMT	inverted high electron mobility transistor
LDOS	local density of state
LEC	liquid encapsulated czochralski
M^3IC	millimeter-wave monolithic integrated circuit
MAG	maximum available gain
MBE	molecular beam epitaxy
MIC	microwave integrated circuit
MMIC	monolithic microwave integrated circuit
MOCVD	metal organic chemical vapor deposition
MODFET	modulation doped field effect transistor
MSG	maximum stable gain
MTG	maximum tunable gain
MUG	maximum unilateral transducer power gain
OMVPE	organo-metallic vapor-phase epitaxy
PBT	permeable base transistor
SATCOMS	satellite communications
SDHT	selectively doped heterjunction transistor
SPDT	single-pole, double throw
TGSM	terminally-guided sub-munition
TEGFET	two-dimensional electron gas FET
2-DEG	two-dimensional electron gas
TEM	transverse electromagnetic
VGF	vertical gradient freeze
VPE	vapor phase epitaxy
VSWR	voltage standing wave ratio

Useful data

c	3×10^8 ms^{-1}
kT_0/q	0.026 V at room temperature
m_0	9.1×10^{-31} kg
q	1.6×10^{-19} C
ε	1.15×10^{-10} CV^{-1}m^{-1}
ε_{A1}	$\varepsilon - \varepsilon_0 \times 3x$ for $Al_\chi Ga_{1-\chi}As$
ε_0	8.854×10^{-12} CV^{-1}m^{-1}
μ_0	$4\pi \times 10^{-7}$Hm^{-1}

List of symbols

a	mean radius of loop inductor
b	number of turns, N, multiplied by the pitch, p
$b(x)$	depth of channel opening
$B(T_e)$	rate of electron energy loss
B_{IN}	imaginary part of input admittance
B_L	imaginary part of load admittance
c	velocity of light in free space
C	capacitance; Linville stability factor
$\overline{C}$	capacitance per unit length
C_D	capacitance per drain-line section
$\overline{C_D}$	total capacitance per unit length of drain line
C_G	capacitance per gate-line section
$\overline{C_G}$	total capacitance per unit length of gate line
C_{dpbp}	geometric capacitance, drain pad to back plane
C_{dsl}	drain-source capacitance involving field in layer
C_{gda}	geometric capacitance, gate-drain, air path
C_{gc}	gate-channel space-charge capacitance
C_{gd}	gate-drain space-charge capacitance
C_{gdgeom}	total gate-drain geometric capacitance
C_{gsa}	geometric capacitance, gate-source, air path
C_{gsgeom}	total gate-source geometric capacitance
C_{gpbp}	geometric capacitance, gate pad to back plane
C_{gpdp}	geometric capacitance, gate pad to drain pad
C_{gps}	geometric capacitance, gate pad to source
C_I	space-charge capacitance of region I
$C_{II\text{-}III}$	space-charge capacitance of regions II-III
C_o	output capacitance of FET or HEMT
C_{OFF}	off-state capacitance of diode
$C_{1,2,3}$	parasitic, or geometric, capacitances

C_2	equivalent series input capacitance of FET or HEMT
d	depletion depth; spacing of capacitor plates
d_{min}	depth of depletion for forward gate voltage
D	spacing between co-planar electrodes
D'	virtual drain
D_s	area density of occupied surface states
E	electric field
$\mathbf{E}$	vector electric field
E_x	longitudinal field in FET channel
E_y	transverse field in FET channel
E_B	breakdown electric field
$\mathscr{E}$	total energy of an electron
$\mathscr{E}_c$	electron energy at the conduction band edge
$\mathscr{E}_v$	electron energy at the valence band edge
$\Delta\mathscr{E}_c$	conduction-band offset at a heterojunction
f	frequency; function or functional dependence
f_B	bandwidth, or upper frequency limit
f_C	cut-off frequency of lumped line
f_{max}	maximum frequency oscillation, frequency at which power gain of FET or HEMT falls to unity
f_{res}	resonant frequency
f_T	transition frequency, frequency at which current gain or FET or HEMT falls to unity
g_f	small-signal value of feedback conductance
g_m	microwave value of transconductance
g_{m0}	intrinsic microwave transconductance
$\overline{g_{m0}}$	value of g_{m0} per millimeter of gatewidth
g_o	microwave output conductance of FET or HEMT
G	conductance; gain
$G_{g\text{-}o}$	power gain from generator to output
G_i	small-signal current gain
G_{IN}	real part of input admittance
G_L	real part of load admittance
G_{LS}	large-signal power gain
G_P	small-signal power gain of FET or HEMT
G_{ss}	small-signal power gain of power amplifier
G_T	transducer power gain
G_{TU}	unilateral transducer power gain
G_v	small-signal voltage gain
G_w	operating power gain
$G_{W\text{-}MAX}$	maximum operating power gain, equal to MAG
h	substrate thickness

H	general height or spacing
$\boldsymbol{H}$	vector magnetic field
i	small-signal current
i_d	small-signal drain current
i_D	total instantaneous value of drain current
$\langle i_{Dn}{}^2 \rangle$	mean-squared value of drain noise current
i_{DTOT}	total small-signal current in dummy load
i_g	small-signal gate current
i_o	small-signal output current
i_{oTOT}	total small-signal output current
i_s	small-signal source current
I	dc current
I_{CH}	channel current
$\overline{I_{CH}}$	channel current per unit gatewidth
I_D	dc value of drain current
I_{DSS}	drain current with gate shorted to source
$\overline{I_{DSS}}$	value of I_{DSS} per unitgatewidth
I_F	drain current for forward gate bias
$\overline{I_F}$	value of I_F per unit gatewidth
I_0	reverse saturation current in an ideal diode
I_{SUBS}	substrate current
$\overline{I_{SUBS}}$	I_{SUBS}/I_{CH}
J	current density
k	Boltzmann's constant
K	constant
K_F	Fukui noise constant
l_d	length of microstrip in a drain-line section
l_g	length of microstrip in a gate-line section
L	length
L_D	Debye length
L_{elec}	effective electrical gate length
L_G	metallurgical gate length
L_{GD}	gate-drain spacing
L_{SG}	source-gate spacing
$\mathscr{L}$	inductance
$\overline{\mathscr{L}}$	inductance per unit length
$\mathscr{L}_d$	drain series inductance; microstrip inductance per unit length in drain artificial line
$\mathscr{L}_D$	total inductance per unit length of drain line
$\mathscr{L}_g$	gate series inductance; microstrip inductance per unit length in gate artificial line
$\mathscr{L}_G$	total inductance per unit length of gate line

$\mathscr{L}_s$	source series inductance
$\mathscr{L}_{TH}$	Thévenin equivalent inductance
m	ideality factor of Schottky diode; "m"th device
m_0	free-space mass of an electron
m^*	electron effective mass
n	free electron density per unit volume
n_s	sheet concentration of electrons in a 2-DEG
n_{s0}	equilibrium value of n_s
N	N-type semiconductor or layer
N	N-type doping density; number of stages or turns
N^2	impedance transformation ratio
N_{cap}	doping level in cap layer of HEMT
N_I	N-doping level in intrinsic region
N_T	deep level density
N_u	equivalent uniform doping density
N^+	heavily doped N-type semiconductor
N^+	doping level in N^+ layer
NF_{min}	minimum noise figure
p	free hole density; pitch of turns in a spiral inductor
P_{AV}	available power
P_{IM}	power input to input matching network
P_{IN}	input power
P_O	output power
$\overline{P_O}$	output power per unit gatewidth
P^+	heavily doped P-type semiconductor
q	electron charge; microstrip filling factor
Q	quality factor
r	incremental (or slope) resistance of diode
r_i	inner radius
r_o	microwave output resistance; outer radius
r_{ON}	microwave on-state resistance of a diode
R	resistance; random number
$R_{contact}$	absolute value of contact resistance
R_d	microwave value of drain series resistance
R_{dc}	dc value of channel resistance
R_D	dc value of drain series resistance
R_{FB}	feedback resistance
R_g	gate resistance at microwave frequencies
R_G	dc resistance of gate strip
R_i	intrinsic, or microwave channel resistance
R_L	load resistance
R_N	resistance of N-layer in Schottky diode

R_s	microwave value of source resistance
R_S	dc value of source resistance
R_{SG}	bulk resistance of source-gate region
R_{TH}	Thévenin equivalent resistance
R_Σ	$R_g + R_i + R_s$
R_2	equivalent series input resistance of FET or HEMT
s_{ij}	scattering parameters
s_{ij}^*	complex conjugate of s_{ij}
s_{ij}^T	total S-parameters of circuit or network
S	mean side of spiral inductor
S_i	innermost side of square spiral
S'	virtual source
T	absolute temperature
T_e	electron temperature
T_0	lattice temperature
v	electron drift velocity; small-signal voltage
v_o	small-signal output voltage
v_{ph}	phase velocity
v_{sat}	saturated value of electron drift velocity
v^+	incident voltage wave
v^-	reflected voltage wave
V_{B0}	zero-bias barrier potential
V_{DG}	dc voltage between actual drain and gate
$V_{D'G}$	dc voltage between virtual drain and gate
V_{DGB}	drain-gate voltage for initial avalanche
V_{DS}	dc voltage between actual drain and source
$V_{D'S'}$	dc voltage between virtual drain and source
V_{DSB}	drain-source voltage for initial avalanche
V_F	forward gate bias or drive
V_{GS}	dc voltage between gate and actual source
$V_{GS'}$	dc voltage between gate and virtual source
V_K	knee voltage, value of V_{DS} required for current saturation
V_p	pinch-off voltage
V_{SG}	$-V_{GS}$
$V_{S'G}$	$-V_{GS'}$
W	channel layer thickness under gate
W^+	thickness of N^+ layer
W_I	width of intrinsic region
W_R	gate recess depth
W_S	surface depletion depth
W_{SiN}	thickness of silicon nitride
x	spatial co-ordinate; aluminium mole fraction

$x_{overshoot}$	distance over which electron drift velocity overshoots v_{sat}
X	extension of space-charge layer into gate-drain space
$X_{\mathscr{L}TH}$	reactance of Thévenin equivalent inductance
X_{OFF}	off-state reactance of diode
y	spatial co-ordinate
y_{ij}	small-signal admittance parameters
$y_{ij}{}^{T}$	total Y-parameters of circuit or network
Y_A	thickness of AlGaAs layer in HEMT
Y_{cap}	thickness of capping layer in HEMT
Y_{ij}	y_{ij} normalized to $Y_0 = 1/Z_0$
Y_{IN}	input admittance
Y_L	load admittance
Y_0	characteristic admittance of transmission line
Y_{SP}	thickness of spacer layer in HEMT
z	spatial co-ordinate
Z_g	generator internal impedance
Z_G	total gatewidth of FET or HEMT
Z_{GU}	unit gatewidth, or width of a single gate strip
Z_{IM}	impedance looking into input matching network
Z_{IN}	input impedance
Z_0	characteristic impedance of transmission line
Z_{0D}	characteristic impedance of drain line
Z_{0G}	characteristic impedance of gate line
Z_{01}	Z_0 of microstrip with air dielectric
α	attenuation per unit distance
α_D	drain line attenuation per unit distance
α_G	gate line attenuation per unit distance
β	phase coefficient
β_D	phase shift per unit length of drain line
β_G	phase shift per unit length of gate line
γ	complex propagation coefficient
Γ	reflection coefficient
Γ_g	generator reflection coefficient
Γ_L	load reflection coefficient
δ	skin depth; tunability
δ	incremental operator
Δ	incremental operator
ε	absolute permittivity of GaAs
ε_{A1}	absolute permittivity of AlGaAs
ε_{eff}	effective permittivity of microstrip
ε_0	permittivity of free space
ε_r	relative permittivity

θ	phase delay or phase shift
λ	mean-free path between scattering events
λ_g	guided wavelength
μ	drift mobility; absolute permeability
μ_0	low-field drift mobility; absolute permeability of free space
μ_n	value of μ_0 for electrons in intrinsic GaAs
μ_p	value of μ_0 for holes in intrinsic GaAs
ρ	resistivity
ρ_c	contact resistivity in Ω mm
σ	conductivity
τ	electron transit time; recombination carrier lifetime
τ_d	dielectric relaxation time
τ_{gm}	time delay in current generator in FET or HEMT equivalent circuit
$\tau_{overshoot}$	time for which the electron velocity overshoots its final steady value
τ_I	charging time of region I in FET channel
ϕ_B	barrier energy at N-layer to buffer- or substrate interface
ϕ_S	surface potential energy
ϕ_{S0}	value of ϕ_S in equilibrium
ω	angular frequency
ω_H	highest angular frequency in band
ω_L	lowest angular frequency in band
ω_{res}	resonant angular frequency

Chapter 1

Introduction

1.1 Statement of the problem

Monolithic microwave integrated circuits (MMICs) are integrated circuits which support components such as transmission lines, resistors, capacitors, inductors, field-effect transistors (FETs) and diodes of various types all on a single chip of GaAs measuring a few millimetres per side at the most. Currently, such chips perform individually the microwave functions of amplification, mixing, switching and phase-shifting, and there is the promise of "multi-function" chips which may combine two or more of these functions on a single chip. Although large production volumes and low cost are often promoted as important parts of the rationale of MMICs, experience to date has shown that the realization of these objectives requires new approaches to circuit design and major improvements in materials and process technology, with a proper appreciation of the balances that must exist between them.

In many ways, MMICs are a logical development of hybrid microwave integrated circuits (MICs) in which active devices and other components, for example, chip capacitors, are mounted on a dielectric substrate such as alumina. The alumina also supports inductive and transmission line elements which have been previously fabricated on the front surface of the substrate by either thick-film or thin-film technology. The MIC approach encouraged the disciplines of circuit design, thin-film technology and active-device technology to evolve somewhat separately. The result of these practices is well known: in MIC engineering there has been an almost traditional necessity to "tune" the circuit by the use of metal or dielectric disks, by adjusting bond wire lengths, or even by the substitution of one FET for another, all while observing the microwave response on a network analyzer. It is manpower intensive and expensive.

There is much to be gained by reflecting for a moment upon the MIC experience to see what lessons might be learned to the benefit of MMIC engineering. First, the ability to replace devices and to "tune" circuits has never encouraged an answer to the question of how tightly toleranced active device characteristics need to be to ensure that a completed circuit performs to specification: if the circuit was out of specification with the first trial device, then others could be substituted. Secondly, therefore, the MIC approach placed too little emphasis upon the relationship between the microwave characteristics of, say, FETs and the material and device processing parameters which go into them. A fairly superficial understanding sufficed. In MMICs these questions must be answered because there is no opportunity for device replacement and very limited scope for circuit tuning to achieve an in-specification performance. Consequently, imposing MIC design practices can (and has) led to very low chip yields (of the order of 1%, even for quite modest circuits) and to correspondingly high chip costs. For MMICs, a complete reappraisal of design and technological methods is necessary.

1.2 Design approaches

In contrast to the "performance-driven" rules of MIC design, MMIC design requires a "yield-driven" approach consistent with meeting specified performance criteria. In turn, a yield-driven methodology requires a quantitative understanding of how the efforts of the materials scientist and fabrication technologist affect the likelihood that the completed chip will function to specification. Not only is it necessary to know the degree of uniformity that can be achieved in N-doped layers, ohmic contact resistance, and all the other technological variables, but it is also fundamental to be able to forecast how these variations will affect the electrical behavior of all components and, hence, the entire circuit. Thus device physics and computer analysis (especially random sampling) assume a new importance in tying together the disciplines of technology and circuit design which were hitherto divorced in the MIC scheme of things.

A desirable resource in developing MMIC designs is the capability to predict, using compact and numerically efficient models, the equivalent circuit and scattering parameters of a FET from a knowledge of the material and structural make-up of the device, without the need to perform time consuming and costly experiments at microwave frequencies. The results contained in this book are derived from compact, physically-based device models developed in over thirty man-years of careful experimental and theoretical effort. Compact computer code makes possible the simulation of overall MMIC behavior and the physical models from which it came allow a depth of insight to be attained which is denied by full computer simulation. To challenge the purely software-oriented view, the way to achieve the most efficient solution algorithm and, at the same time, the best optimizing compiler is to be competent in finding device-physics solutions in the first place. At the risk of repeating an overworked truth, such immediacy and closeness of contact with the real world allow the physically important mechanisms to be rapidly identified and included. To cite just one example, charge exchange with surface states can have a profound effect on MMIC yield through the irreproducibility in equivalent circuit element values it induces.

Whatever efforts are made to achieve process reproducibility, however, finite spreads will always result. In that case, successful realization of MMICs requires that the circuit design should be "process tolerant". Evolution of yield-driven design practices presupposes one of two things: either a vast archive of experimental data which, for each and every circuit element, show how the electrical characteristics of that element will vary with all conceivable processing spreads; or a thorough understanding in physical and mathematical terms of how the electrical behavior of all circuit elements depends upon their material and technological make-up. It is the second which substantially forms the background to the present work.

It is interesting to compare the demands of these two approaches upon practical data. The first requires fabrication and careful microwave characterization of sufficient samples over a long enough period for the archive to be sufficiently representative of that particular transistor or component. The second involves experiments with a whole series of devices wherein the structure is systematically varied, interactively with the development of a theoretical description of the device. Of the two approaches, the second is a far more protracted exercise than the first, and so initially seems unappealing. Once an investment in device physics has been made, however, immediate forecasts may be made of the microwave characteristics of any newly proposed FET without the need for fabrication of test samples. Furthermore, the option then exists to design the active elements (the FETs) and the passive network concurrently using computer-based design centering routines to explore the compromises possible between performance and yield.

The GaAs metal-semiconductor field-effect transistor (MESFET) is almost exclusively employed as the active amplifying and switching device in MMICs where a transistor device is needed. Design of MMICs for acceptable yield requires at least a semi-quantitative understanding of how the FET functions *as a microwave device* in terms of all important material properties and structural factors. The GaAs MESFET has presented a fair challenge in this regard and, although much remains to be done, the achievement to date has made possible a meaningful start on yield-driven design. Many hitherto unpublished expressions describing the microwave characteristics of GaAs MESFETs are given here for the first time. Since it was learned quite early in these endeavors that little useful relationship exists between the dc characteristics of a FET and the detailed way it will perform at microwave frequencies, little coverage is given in the book to dc behavior.

The idea that the GaAs FET is a component which can be designed is very persuasive. FETs should be configured to meet the needs of MMIC designers, rather than circuit designers having to design MMICs to meet the needs of FETs. The possibilities for exploiting the compromises possible between performance and yield have only just begun to become apparent; as we shall see, even the simple expedient of changing the FET cross-section in apparently minor ways, in an otherwise fixed passive network, can improve both the performance and the yield of an MMIC amplifier. Equally, apparently minor changes of an unintended nature can ruin the microwave behavior of an MMIC.

As a general rule, therefore, MMICs are particularly sensitive to variations in FET structure, as may be demonstrated by a simple example.

1.3 A simple example

Figure 1.1 shows the basis of an MMIC intended to give 6 dB gain at 10 GHz. It has been designed around a 0.5 μm FET with a reduced equivalent circuit as in Fig. 1.1(b). The amplifier has the following component values:

FET: $C_{gc} = 0.2$ pF

$g_{m0} = 30$ mS

$C_o = 0.05$ pF

Passive network: $R_s = 50\ \Omega$

$\mathscr{L}_i = 1.4$ nH

$\mathscr{L}_o = 4.6$ nH

$R_o = 70\ \Omega$

During fabrication, the gate length on one particular chip has turned out at 0.6 μm, instead of 0.5 μm, due to limits of lithographic control. Further, the channel doping is 10% higher than intended. The process engineer has attempted to compensate for this error by etching the channel thinner so he realizes exactly the right value of open-channel current, I_F. Finally, in operation, the FET in the MMIC is biased at a channel current of

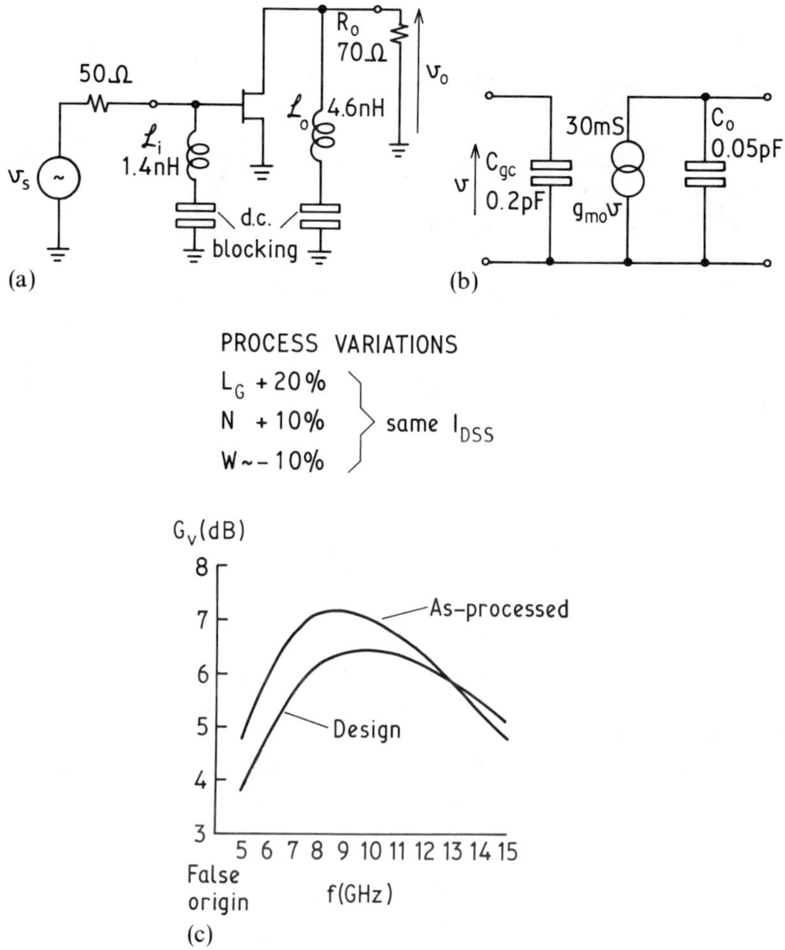

Fig. 1.1 Simple example of MMIC sensitivity to process variations, (a) circuit, (b) 0.5 μm FET, (c) gain responses.

$I_{CH} \approx I_D = I_F/2$, exactly as in the design. The question is, how do the foregoing changes affect the voltage gain of the amplifier, here defined as

$$G_v(\mathrm{dB}) = 20\log_{10}\left|\frac{v_o}{v_s}\right| \tag{1.1}$$

The first step is to solve for G_v(dB) using straightforward methods of circuit theory (indeed, the problem has been chosen so as to make this readily possible):

$G_v(\text{dB}) =$

$$20 \log_{10}\left(\frac{g_{m0}\omega^2 \mathscr{L}_i \mathscr{L}_o R_o}{\{[R_s^2(1-\omega^2\mathscr{L}_i C_{gc})^2 + \omega^2\mathscr{L}_i^2][R_o^2(1-\omega^2\mathscr{L}_o C_o)^2 + \omega^2\mathscr{L}_o^2]\}^{1/2}}\right) \quad (1.2)$$

In this equation, variations in gate length and channel doping principally affect the FET transconductance, g_{m0}, and the device's gate-channel capacitance, C_{gc}. To first order, changes in output capacitance, C_o, will be neglected.

To ascertain the effect on the MMIC, we make use of the following equations from Chapter 6:

$$C_I = \frac{\epsilon L_G Z_G}{d} \quad (6.10)$$

$$I_{CH} = qNv_{sat}(W-d)Z_G \quad (6.11)$$

$$g_{mo} = \frac{\epsilon v_{sat} Z_G}{d} \quad (6.18)$$

C_I is the contribution of the depletion layer in the shadow of the gate to the total gate-channel capacitance, C_{gc}. It is generally the major contribution in a small-signal FET, as will become apparent in Chapter 6. For the present purpose, we shall take $C_{gc} \approx C_I$.

The open-channel current, I_F, is obtained when the gate is forward biased so that the depletion depth, d, is reduced to a value very much less than the channel layer thickness, W. From eqn (6.11),

$$I_F \approx qNv_{sat} W Z_G \quad (1.3)$$

By forming the ratio I_{CH}/I_F and rearranging, we then obtain

$$d = W(1 - I_{CH}/I_F) \quad (1.4)$$

We are now in a position to consider the effects of process variations.

From eqn (1.3), if N is 10% high and I_F is held at the design value, W must be 10% lower than designed. Consequently, from eqns (1.4) and (6.18), g_{m0} will be 10% higher than intended, i.e.

$$g_{m0} = 33 mS$$

Next, if L_G is 0.6 μm instead of 0.5 μm, L_G is 20% high so that, accounting also for the 10% decrease in W, eqns (1.4) and (6.10) give

$$C_{gc} \approx C_I \approx 0.27 \text{ pF}$$

Table 1.1 highlights the changes to the FET brought about by the stated process variations. By substituting all values into eqn (1.2), the gain responses are found to be as in Fig. 1.1(c). We are thus faced with the proposition that apparently small changes in process variables, in this case 10% error in doping density and 0.1 μm error in gate length, could place the MMIC well out of specification. The position becomes much worse if the specification covers the magnitude and phase of the input and output reflection coefficients and the phase of the forward gain, not just the gain magnitude.

Table 1.1 FET errors due to process variations

	As designed	*With* $N+10\%$; $L_G+20\%$
C_{gc}	0.2 pF	0.27 pF
g_{m0}	30 mS	33 mS

1.4 Yield

On the basis of the foregoing simple example, the GaAs FET is evidently a crucial element in determining MMIC performance and yield. We are not here referring to yield determined by mechanical failures such as broken gate strips, poor adhesion of Schottky metal, pinholes in silicon nitride protection and the like: those are issues of process technology which have been well enough canvassed elsewhere (see, for example, Williams, R. E., *Gallium Arsenide Processing Techniques*, Artech House, Dedham, MA, 1984). Yield in the present context means what we shall call "intrinsic" yield, i.e. the fraction of chips that are within microwave specification relative to the number of chips that have been successfully fabricated without "mechanical" failures of the foregoing kind. A useful definition of intrinsic yield is:

$$\text{Intrinsic yield} = \frac{\text{Chips within microwave specification}}{\text{Chips that are dc functional}}$$

We are thus concerned with questions of how spreads in gate length, channel doping, N-layer thickness, nitride thickness, substrate compensation etc. affect the likelihood that an MMIC will perform to specification. Above all, our concern is to establish equations, guidelines and, ultimately, rules which will allow the performance and intrinsic yield of MMICs to be forecast quickly and at low cost using computers, thereby bypassing the time-honored but expensive practice of trying to find answers by technological iteration alone.

1.5 The role of device physics

A major objective in this work has been to develop equations that are compact enough to allow entire analysis and design program packages to run on a reasonably powerful personal computer, i.e. the aim has been to supplant sheer computing power with device physics leading to compact, rapidly-executed code. Experience in developing and using such equations, including the simplified versions given in later chapters, has suggested that the inaccuracies in them are less than the uncertainties evident in contemporary process technology. Thus the methods developed here have pointed the way to desirable changes in technology, most of which have been implemented with a fair degree of success.

It might be asked why it is necessary to take the effects of variations back to the technological level in terms of channel doping, substrate properties, gate length, and so forth. After all, the effects of these variations show up in any case as variations in the microwave scattering parameters, or equivalently the circuit model elements. So, the argument might run, if any opportunity exists for configuring a circuit topology that will be insensitive to the effects of process variations, that fact can be fully determined by designing and testing the MMIC around the electrical variational data.

The answer to this question is fundamentally simple: at the end of the day, the factors over which one has separate control are N, L_G, W . . . , not s_{11}, s_{21} . . . . It is important to address the problem in terms of what the technologist puts into an MMIC. He cannot, for example, tighten up on s_{11} divorced from everything else. He can, however, tighten up on N, for instance. Moreover, a change in any single technological variable results in correlated changes in all the equivalent circuit elements, and the preservation of these correlations is important if a statistical analysis of an MMIC's microwave behavior is to be meaningful. In current practice they are often discarded, or never kept in the first place. A third very important reason for wishing to take the analysis back to the technological level, rather than just the circuit level, is that the engineer thereby gains a facility to design FETs concurrently with the passive network for the purpose of reaching the best possible compromise between performance and yield: that can only be done if we have the ability to predict in detail what mutually consistent values of C_{gc}, C_{gd}, g_{m0}, R_i, g_o, etc. are possible and, once the overall MMIC design is done, what values of N, W, L_G, etc. are needed to realize them.

Out of this background arises the view that the fidelity with which a FET model represents what actually occurs in practice depends more upon the inclusion of all the important physical phenomena than upon the intrinsic accuracy of the mathematical methods of solution. Perhaps it is more important to include in the physical description the effects of transverse fields upon electron transport in the channel of a FET, the effects of charge injected into the buffer or substrate, deep-level charge and surface charge, than it is to accept a less realistic model and then solve it "accurately" using methods of numerical simulation. Maybe the GaAs MESFET is amenable to adequate analytic solution after all, with the many advantages of physical insight that offers. The GaAs FET is primarily a challenge to the inventiveness of semiconductor device physicists in finding approximate solutions to inherently two-dimensional field problems, rather than to the rigid discipline of programming the full physical equations for computer solution.

Experience suggests the uncertainties in device behavior resulting from technological imprecision outweigh the uncertainties due to inaccuracies in approximate analytic solutions. Thus, for the time being, analytic methods are easily adequate to the task. Using them, all the equivalent circuit elements of a FET can be calculated in terms of the material and structural specification, so it is then a short step to compute the overall small-signal response of an MMIC using one of the commercially available microwave network analysis programs. In other words, it is readily possible to forecast the microwave behavior of a complete MMIC directly in terms of the fabrication technology used to realize it. Further progress depends upon gaining improved control over the technology of fabricating MMICs. There is a long way to go before that control is tight enough to make numerical simulation of FETs, and of FETs in MMICs, worthwhile.

The areas where effort needs to be directed are particularly those concerned with the GaAs FET surface or, if the MMIC is protected with a glass-like film (such as silicon nitride), with the interface between the dielectric and the GaAs. Charge exchange with electronic states located at this surface or interface is largely uncontrolled, yet it affects the microwave behavior of a FET, and hence the yield and reproducibility of MMICs. So far as an MMIC designed for small-signal operation is concerned, a lack of control over surface-state charge has an effect roughly equivalent to operating the device at an ill-defined bias point. Its effect tends to be somewhat second order. For power FETs and MMICs the position is different: surface-state occupancy has a dramatic effect on gate-drain breakdown, and on output power density in watts per millimetre of gatewidth, so it is natural to expect power MMICs to exhibit a high degree of irreproducibility in the face of poor control over surface conditions.

For these reasons, this book is an incomplete record of the work that has been done: it emphasizes techniques for the design, analysis and diagnosis of small-signal amplifier MMICs. Whereas design equations are given for all the FET equivalent circuit elements in terms of channel doping, layer thickness, gate length, inter-electrode dimensions, N-layer to buffer–substrate interface energy, and applied gate-source and drain-source voltages, the three factors affecting the net importance of surface charge, namely the surface-state occupancy itself, the gate recess depth, and N^+ cap layer, are not explicitly included in the equations of this book. That is not to say that such equations cannot be written; they certainly can. For the present their value could be in doubt in the light of poor process control and, in any event, their inclusion would require a very much larger book than the present one. Useful work in MMIC engineering can nevertheless be done without them, and the purpose of illustrating the methods of MMIC design which process-independent FET descriptions allow can be fulfilled. Such are the limited objectives of this book.

Chapter 2

MMIC Types and Chip Functions

2.1 Scale of IC fabrication

To help understand the nature and role of MMICs, we shall briefly examine the question of scale, by which we mean the volume in which MMICs should be produced to make it worthwhile investing in the technology in the first place. Analyses of this kind lend balance to the technical issues involved in designing and fabricating GaAs-based circuits, of which the MMIC is a prime example.

As we shall shortly see, the major microwave functions performed by MMICs are amplification, phase shifting of a signal, mixing of two signals and switching of a signal between one or more signal paths. (To this might be added oscillators, although it is rare for the entire component count of an oscillator to be accommodated on a single chip of GaAs.) In combination, these functions allow such sub-systems as receivers and transmitters to be constructed, which in turn form parts of radar systems, radiotelescopes, communications links and so forth. One of the major issues arising in this chain of applications is pricing. How much is a radar systems firm willing to pay for an MMIC chip? What should the selling price of an MMIC chip be to a communications hardware company?

Answers to questions of this sort are difficult to determine in advance; the quality of the forecast depends upon having good marketing intelligence and a sound assessment of the end user's needs. We shall assume the chip selling price has to be around \$25 for "moderate functionality" on a 10 mm^2 die. Then, by estimating the quantity of chips that must be fabricated per annum to justify the whole enterprise, we come to appreciate the nature and type of circuits suitable for integration as MMICs. A fabrication facility with a large enough processing capacity to achieve a chip selling price of \$25 might require in the region of \$50 million to establish. If the ratio of sales turnover to capital employed is, say, 1:1, then \$50 million worth of chips at \$25 each must be sold each year, i.e.

$$2 \times 10^6 \text{ chips/annum}$$

On a 3 in. wafer having a useful area of around 4000 mm^2, there are 400 chip sites of 10 mm^2. If the yield *overall* of RF chips within specification is 25%, then each wafer will yield 100 chips. Therefore, to operate on a scale of 2×10^6 chips per annum

$$\text{Number of wafer starts/annum} = \frac{2 \times 10^6}{10^2} = 20\,000$$

Equivalently, there must be 400 wafer starts per week, with a processing cost per wafer of

$$\frac{\$50 \times 10^6}{2 \times 10^4} = \$2500 \text{ per wafer}$$

For a given process line, the wafer processing cost tends to be fixed regardless of the detailed circuit the wafer supports. Consequently, minimum cost to the end-user follows from any of all of the following:

(a) reducing the chip size for a given functionality,
(b) increasing the functionality for a given chip size,
(c) increasing the yield of chips within RF specification.

Regarding all the foregoing figures as illustrative, we conclude that MMICs are high-volume, low-cost parts. Following the same calculation through with different assumed yields and chip sizes shows just how important both are to competitiveness as a chip producer. Taking a view across the industry as a whole, however, the size, functionality and cost targets are proving hard to meet, and there is a long way to go to achieve chip selling prices appropriate to large volume applications. We consider these applications next, when the types of functions required of MMICs will become apparent.

2.2 Applications

2.2.1 Civil

Reception of transmissions from Direct Broadcast Satellites (DBS) offers a potential market of several hundred million receivers operating at a frequency of about 12 GHz with a bandwidth of 0.5 GHz [2.1]. Figure 2.1 shows a block diagram of an outdoor receiver unit in which the individual functions of RF amplifier and filter, mixer, intermediate frequency (IF) amplifier and local oscillator (with an off-chip dielectric resonator) can be realized as MMICs [2.2]. A single chip receiver, wherein all the functions (apart from the dielectric resonator) are integrated onto a single piece of GaAs measuring 2.5 mm × 2.5 mm, has been described by Kermarrec *et al.* [2.3]. At 12 GHz the conversion gain of the chip was about 25 dB with a noise figure of 4 dB.

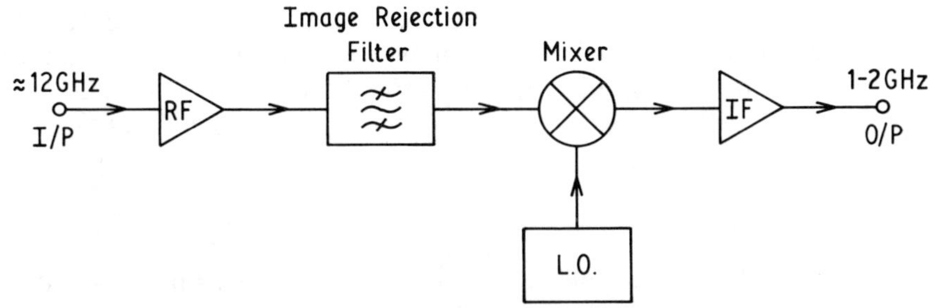

Fig. 2.1 Block diagram of DBS front-end receiver (outdoor unit).

One of the organizations which has applied the principles of yield-driven design most successfully is Pacific Monolithics. Podell and Nelson of Pacific Monolithics have described a single-chip receiver for reception at 3.7–4.2 Ghz [2.4]. From RF input to IF output the gain was greater than 35 dB, with five different circuit types on the chip. Push–pull circuits were used for the oscillator and RF amplifier sections thereby minimizing the size of on-chip bypass capacitors, the effects of grounding inductance and on-chip feedback. The mixer was a double balanced diode type, followed by a three-stage, single ended IF amplifier. In the interests of keeping the chip small, transformer coupling using spiral transformers was used with the FETs located inside the innermost spirals, resulting in an overall chip size of 1 mm × 0.85 mm. This chip is one of the outstanding examples to date of MMIC design.

Other applications in the civil field include cellular radio and radar. Land-mobile telephones operating in the 800–900 MHz band may require transmitters to deliver about 1 W with high efficiency, and local oscillator–mixer MMICs [2.5]. Marine radars for small craft offer a high-volume market for radars operating at about 3 GHz and 10 GHz, and there are applications also at 10 GHz in intruder alarms and right-of access sensors. A possible future application could be in radars for vehicle collision avoidance, with a potential market that is probably the largest of them all.

One of the first cost-effective uses of MMICs has been in instrumentation.

A 2–26.5 GHz, two stage†, dual-gate, distributed amplifier has been described by Orr [2.6]. A gain of about 18 dB and an output power approaching 20 dBm at 1 dB gain compression are obtained from a chip using fourteen FETs, each having gate dimensions of 124 μm × 0.4 μm. The overall chip dimensions are 3.02 mm × 1.62 mm. Generally, MMICs may find application in frequency synthesizers, network analyzers, spectrum analyzers and sampling oscilloscopes. Although the number of chips required for instruments may not even approach those for DBS, instrumentation offers what is often referred to as a "niche" market for MMICs. Such a market renders the chip fabrication facility profitable as part of a so-called vertically integrated company where the added value is derived from the system or, in this case, the instrument which uses the chip.

2.2.2 Military

Radars with electronically-steerable beams are generally known as phased-arrays. A phased-array radar with five facets (or faces), each with 5000 elements, would require 25 000 circuits per system. A block diagram of an experimental receiver module is given in Fig. 2.2 [2.7]. Apart from the addition of a phase shifter, the block diagram is, naturally enough, very similar to that of Fig. 2.1.

In the field of electronic warfare (EW), jammers and decoys typically require integrated circuits covering about one octave bandwidth [2.8]. Decoys, both ship- and aircraft-deployed, are thought to be an important future market with the prospect of large production runs [2.8)]. Size and weight constraints are important in these decoys. Further, a typical electronic support measure (ESM) radar warning receiver must cover a frequency range of about 0.5–18 GHz, with extensions to 40 GHz by 1990 and 100 GHz envisaged by 1995 [2.8].

Anti-radiation missiles make use of what is essentially a simplified ESM receiver [2.8], currently requiring bandwidths of about 10% in the range 3–18 GHz. Although production runs are small, MMICs common to ESM receivers could find application in updates to current systems. Future developments in guided weapons (GW) are likely to involve small phased arrays and multi-channel receivers which could justify MMIC front-end chips [2.8]. The largest volume potential, however, would appear to rest with terminally-guided sub-munitions (TGSMs) operating at 94 GHz. TGSMs are likely to be deployed in very large numbers for anti-tank and anti-ship operations [2.8]. Cost is a sensitive factor. Thus, should it prove possible to

†By "two-stage" we here mean the cascade connection of two distributed amplifiers as in Fig. 10.32, on a single chip.

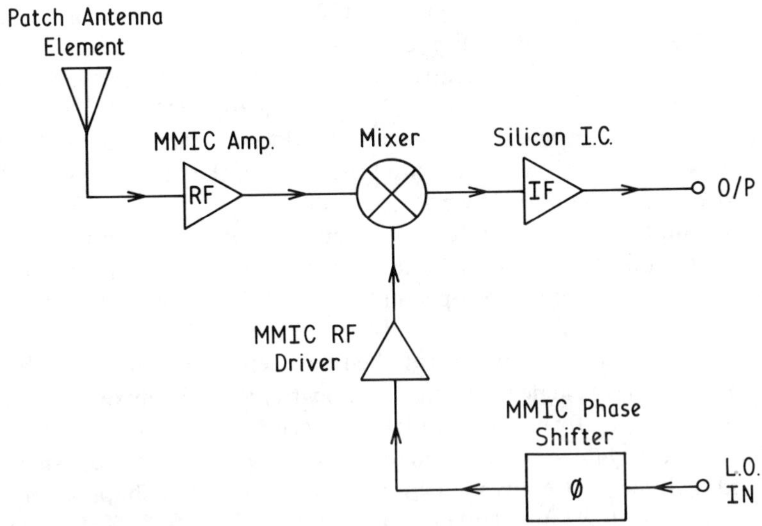

Fig. 2.2 Block diagram of phased-array radar receiver module.

manufacture millimetre-wave monolithic integrated circuits (M^3ICs), TGSMs could provide the market incentive to develop RF amplifiers, mixers and local oscillators in M^3IC form.

2.3 Chip functions

From the foregoing brief survey, the application areas for MMICs involve high production volumes of low-cost chips providing the following functions:

(a) amplifiers (narrow band, broad band, ultra-broad band; low-noise, general gain blocks, and power),
(b) oscillators,
(c) mixers,
(d) phase shifters,
(e) microwave switches.

Chips providing sub-system or system functions on a single piece of GaAs by integrating together some or all of these basic functions is a concept which is attractive. MMICs operating at up to 100 GHz may be envisaged, with applications as summarized in Fig. 2.3 (after references [2.5] and [2.8]).

In this book, we shall concentrate on amplifiers, and in particular upon the role of FETs and high electron mobility transistors (HEMTs) as a factor governing MMIC amplifier performance and cost trade-offs. Only brief mention will be made of the other chip types, and of the diode devices which some of them use. There are several reasons for this.

First, amplifiers are very common components in the systems where MMICs find application. Secondly, the material covered in a work such as this must necessarily be limited in scope, and even MMIC amplifiers alone could occupy many such books (and, no doubt, in time they will). Thirdly, some of the other circuit types use devices switched between the on-state and the off-state, or rely on the existence of some non-linearity (e.g. in a mixer), and these circuits are consequently not as sensitive to device variations as amplifiers are. Amplifiers generally must be designed to absorb the reactances and resistances associated with the FET equivalent circuit into the response-shaping network of the MMIC, so if the equivalent circuit varies from FET to FET, then so too does the response of the overall MMIC. The variation in element values is important because the device capacitances and resistances are comparable with values in the passive part of the network, so they cannot be considered small and cannot be ignored. MMIC amplifiers thus present a major challenge

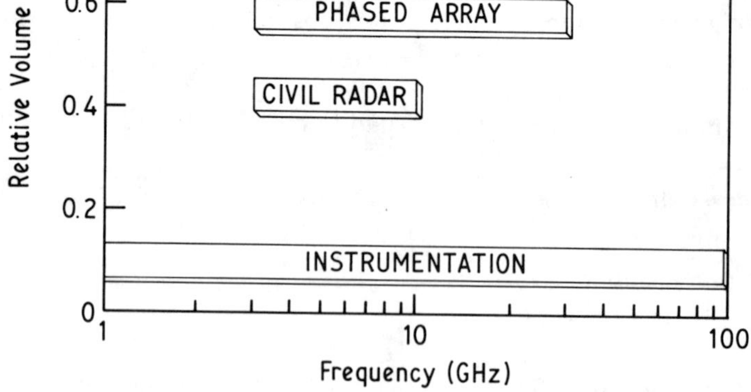

Fig. 2.3 Possible markets for MMICs.

in device physics, technology and engineering, and lead to results which are in any case always applicable to chips performing functions other than amplification. Design equations for amplifiers will be given in Chapter 10.

2.4 Examples of MMICs [2.9]

Figure 2.4 shows an early X-band amplifier MMIC using a single FET with shunt feedback. The photograph illustrates a number of the components generally used in analog integrated circuits, as follows:

Feature A is an integrated GaAs FET having two gate fingers, each of 1 μm × 150 μm. The gate pad is on the left; the drain pad, which is a small triangular area of metal, is on the right. The FET has two sources (top and bottom in Fig. 2.4),

B is a bulk GaAs feedback resistor integrated between the gate pad and the drain (a sketch of this arrangement is to be found in Fig. 11.23),

C is an airbridge connecting one source of the FET to the top plate of the overlay capacitor D. Underneath this airbridge is a bulk GaAs resistor connected between the FET source and ground to provide self-bias for the FET,

D is an overlay capacitor for decoupling the source bias resistor located under the airbridge C,

E is an input bonding pad,

F is a dc blocking overlay capacitor,

G is a high impedance microstripline section for input matching,

H is a dc blocking capacitor at the output terminal,

I is a high impedance microstripline sections for ouptut matching,

J is an overlay capacitor for decoupling the drain bias supply,

K are bond pads for grounding.

No via-hole grounds were used on this chip, which had a design performance of 6 dB gain from 8 to 12 GHz with input and output VSWRs of 2:1 or better. The die area was approximately 1 mm^2.

Figure 2.5 shows a development of the foregoing basic amplifier which still uses the same self-biased FET configuration with shunt feedback. Two stages of amplification are employed, this time using FETs having 0.7 μm gate length, with via-hole grounding of the source resistors and their associated bypass capacitors. The via holes connect the "top hat" metallized pads, at the top and bottom of each FET in Fig. 2.5, through the GaAs substrate to the

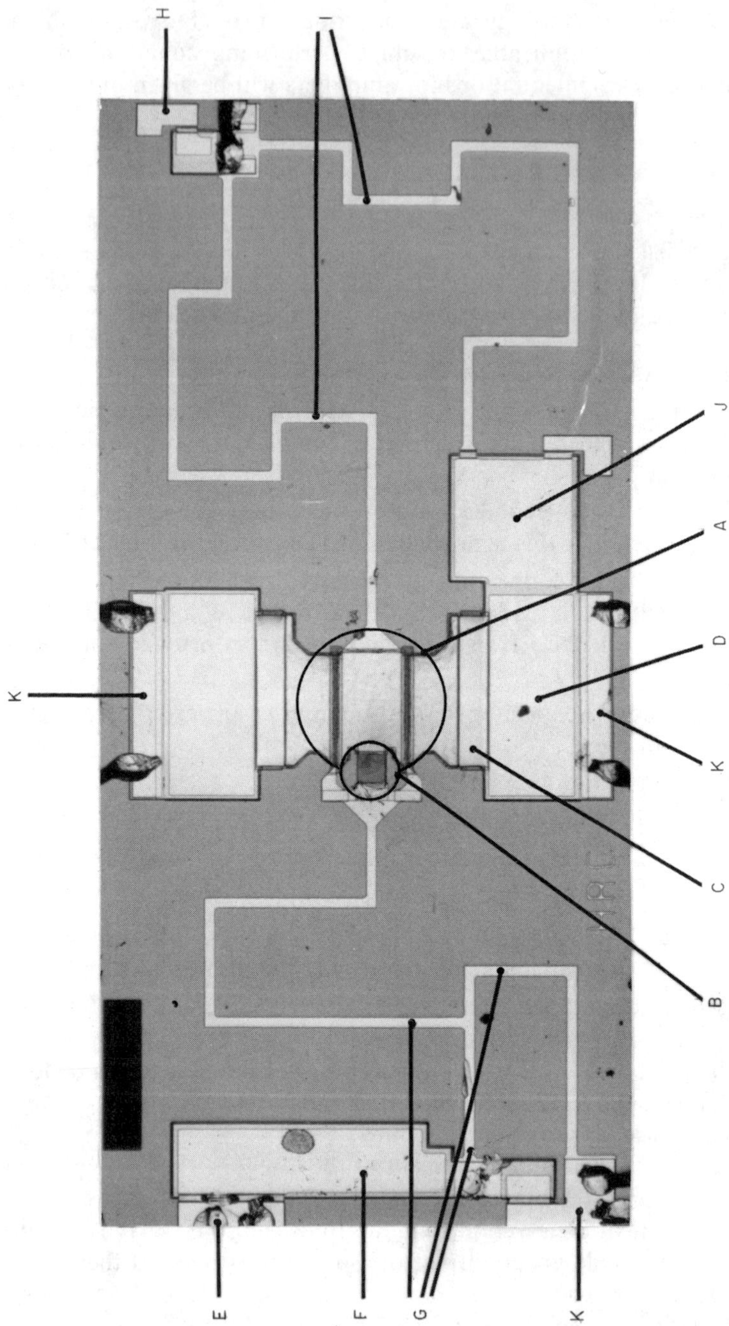

Fig. 2.4 Single-stage X-band amplifier MMIC (from [2.9], with permission).

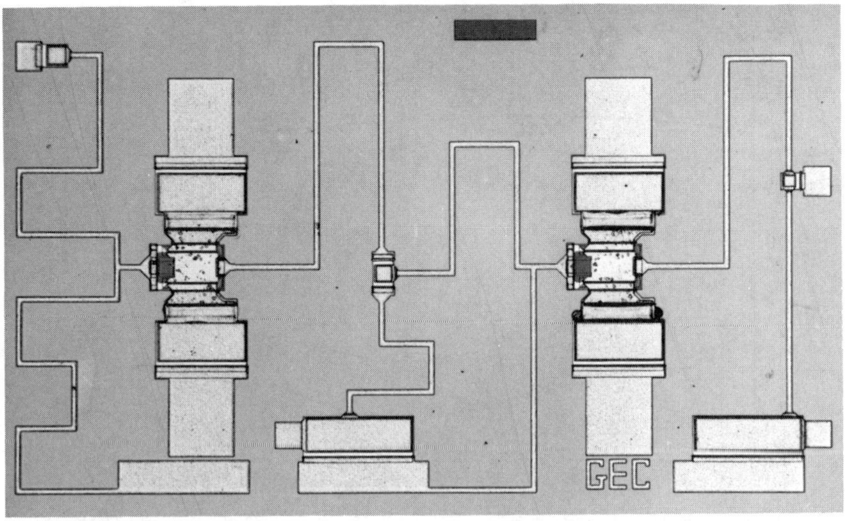

Fig. 2.5 Two-stage X-band amplifier chip (from [2.9], with permission).

metallized backplane on the reverse side of the wafer, which is grounded. A layer of silicon nitride (Si_3N_4) covers the active areas of the FET and integrated feedback resistor. Note how the microstriplines in the input, interstage and output matching networks have been meandered to keep the chip area within reasonable bounds. Meandering introduces inter-component and intra-component coupling which alter the electrical properties of the microstriplines. Such effects require careful characterization of components in isolation before they are designed into a circuit. In addition, the MMIC itself may require several design and fabrication iterations before the performance specification is attained. Figure 2.6 shows the measured gain, and input and output return losses, of an example two-stage X-band amplifier chip (Fig. 2.5).

In Fig. 2.7 is shown an S-band amplifier providing 20 dB gain with a noise figure of 3 dB. Lumped components are used for matching on this chip. The lumped inductors are the coiled elements in the lower half of the photograph. They resemble an "Egyptian frieze" pattern and, unlike multi-turn spiral inductors, are designable with respect to both the inductance and the parasitic capacitance associated with all such structures [2.9]. The input to the chip is the pad located at the bottom end of one of these inductors at the lower left of the chip. Careful examination of Fig. 2.7 will show that the second inductor from the left is in series with the sources of the first FET, the sources themselves being connected by an airbridge passing over the active regions. Series inductance in the source provides quadrature feedback which changes the

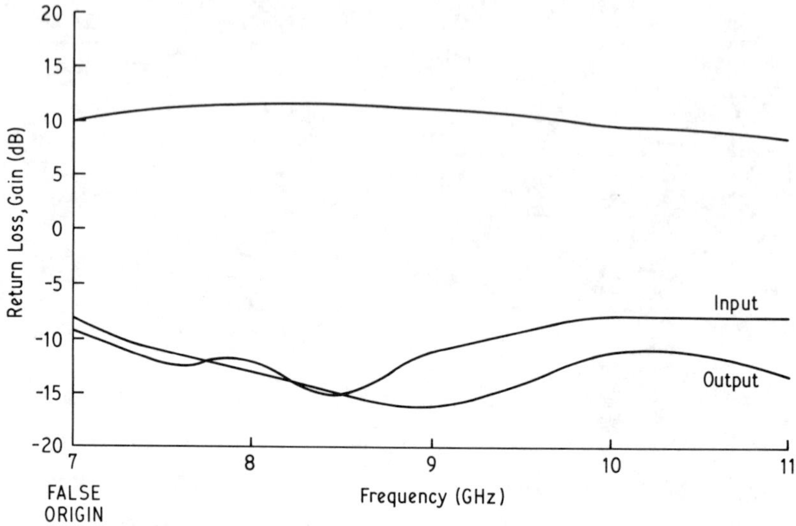

Fig. 2.6 Gain, and input and output return losses, for chip in Fig. 2.5 (from [2.9], with permission).

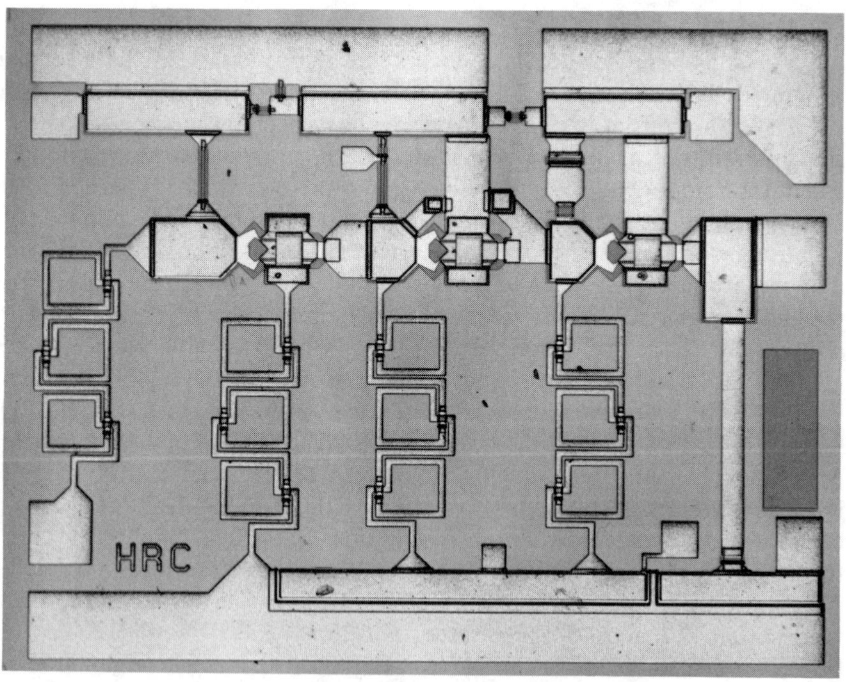

Fig. 2.7 S-band amplifier MMIC (from [2.9], with permission).

impedance required for a low-noise input match to nearer the system impedance (i.e. 50 Ω). A further feature of this particular circuit is that two of the inductors are mirror images of the others, thereby reversing the sense of the stray magnetic coupling without affecting the stray capacitive coupling [2.9]. In this way, unwanted feedback around the amplifier is reduced.

From a technological point of view, the chip in Fig. 2.7 is interesting for the number of airbridges it contains; there are over 30, including bridges over the sharp edges of the Si_3N_4 dielectric in the overlay capacitors to connect to the capacitor top-plates. Such bridges avoid the risk of premature breakdown at points and edges of high electric stress.

In Chapter 10, the theory of distributed amplifiers will be covered. Figure 2.8 is a photograph of an MMIC engineered about five years ago to give 6 dB from dc to 12 GHz. There are four FETs, each of nominal gate

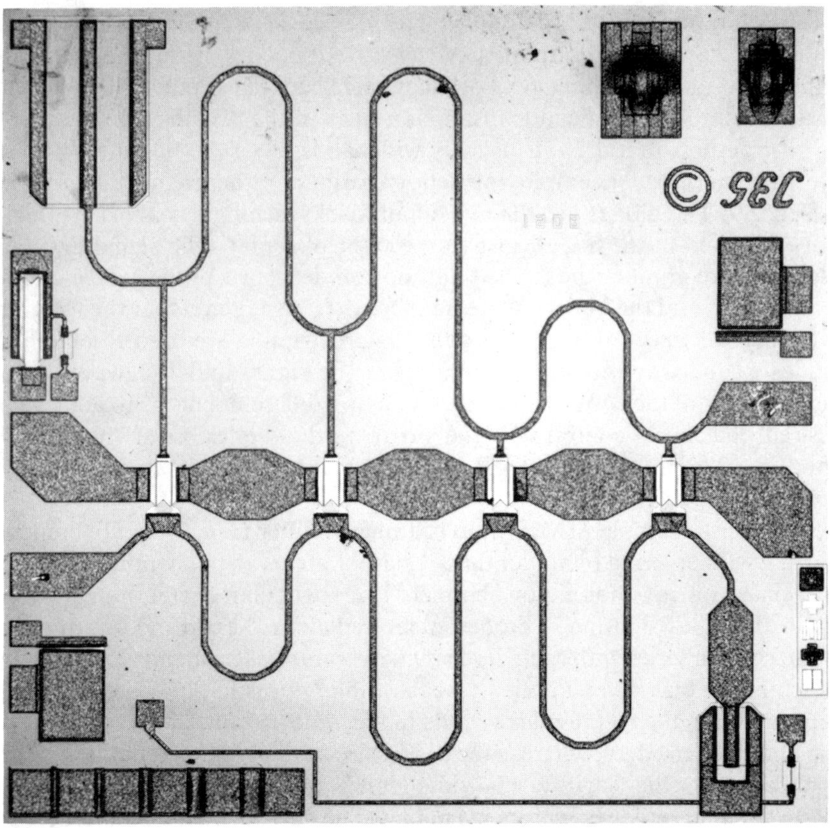

Fig. 2.8 dc-12 GHz distributed amplifier MMIC (from [2.9], with permission).

dimensions 1 μm × 300 μm, with via-grounds connecting the sources to the metallized back surface of the chip (the vias are located beneath the row of trapezoidal metal pads just below the chip center line). The meandered microstriplines between the FETs constitute inductances which, together with the FET input and output capacitances, form artificial transmission lines: one connected to the gates of the FETs, the other to the drains. As will become clear in Chapter 10, signal energy in the form of waves can propagate along these lines, being amplified as it passes from the input to the gate line at the left (below center), through the FETs to the output pad on the right end of the drain line (pad above center). Any signal energy not absorbed by the FETs distributed along the gate line is absorbed in the GaAs bulk resistor which terminates the gate line at the bottom right of the chip. There is a similar, but physically larger, resistor at the top left to dissipate any signals that may propagate backwards along the drain line.

The distributed amplifier does not suffer from the bandwidth limitations of conventional tuned amplifiers. The device capacitance, which in the ordinary way forms a frequency selective network with the generator impedance and matching network connected at the input to the FET, is absorbed into the artificial transmission lines in the distributed amplifier to form structures that are intrinsically wideband.

A balanced mixer employing Schottky diodes rather than FETs is shown in Fig. 2.9. The diodes are the two small, dark, rectangular areas symmetrically disposed in the circuit just to the right of center. The signal and local oscillator are input at the top left and bottom left ports of the Lange coupler respectively, and the intermediate frequency (IF) is taken out at the top right. Meandered lengths of high impedance microstripline are used to match the diodes to the transmission lines carrying the signal and LO waves at the output ports of the coupler, these lines having different lengths to achieve the desired phasing of combined signals driving the diodes. Over the designed band, 8–12 GHz, typically 6 dB conversion loss and 6 dB noise figure are achieved (Fig. 2.10).

The last example MMIC chip is a phase shifter (Fig. 2.11). Reference to Fig. 2.2 will serve as a reminder that a phase shifter is an important component in a phased-array module or channel. The particular circuit shown here is again diode-based, using 32 diodes integrated along the top and bottom edges of the chip. In series with each diode is an overlay capacitor and a bonding pad which allow bias to be applied to each diode individually. The decoupling capacitors are the rectangular islands in the wide metallized planes along the top and bottom edges, whereas the diodes are the smaller rectangular features connecting to the narrow, high-impedance microstripline tracks running along the interior edges of the wide planes. The narrow track sections between each diode constitute inductances which, taken together with the diode off-

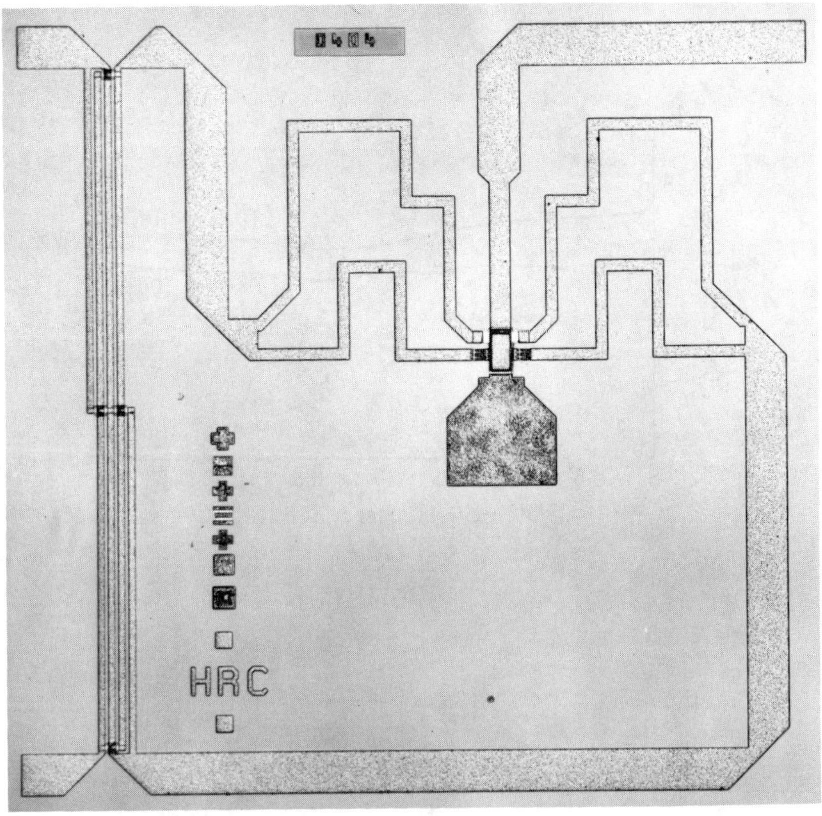

Fig. 2.9 X-band diode mixer (from [2.9], with permission).

state capacitances, again form artificial transmission lines. The circuit functions as follows.

The signal input is at the Lange coupler port at the left, just above center. Half the signal emerges from the coupler at the bottom left port, half at the top right. Signal energy arriving at these ports propagates along the artificial transmission lines from left to right along the lower line, and from right to left along the upper line. If all the diodes are biased in the off-state, the waves propagate the full length of the structure and are absorbed in matched resistors terminating the lines. If, however, a pair of diodes is biased on, say the fourth diode along each line†, signal energy is reflected from these approximate short circuits. The two reflected signals are then recombined in the coupler and appear at the output port of the coupler on the right, just

†That is, the fourth diode from the bottom left, and the fourth diode from the top right.

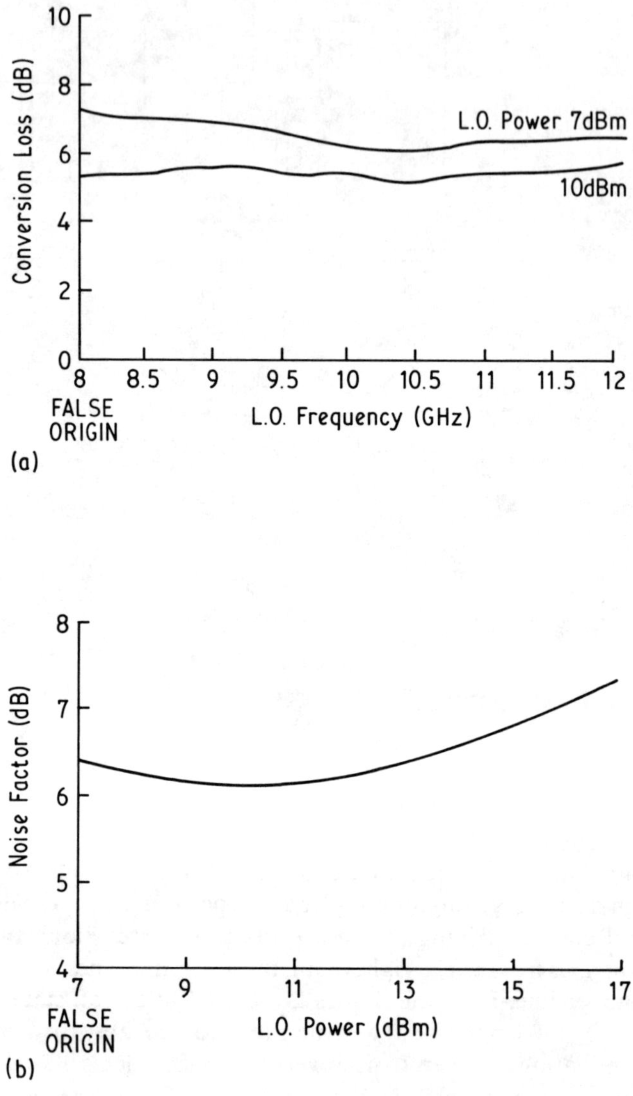

Fig. 2.10 Conversion loss and noise factor of X-band mixer chip (from [2.9], with permission).

Fig. 2.11 X-band diode phase shifter MMIC (4 bit) (from [2.9], with permission).

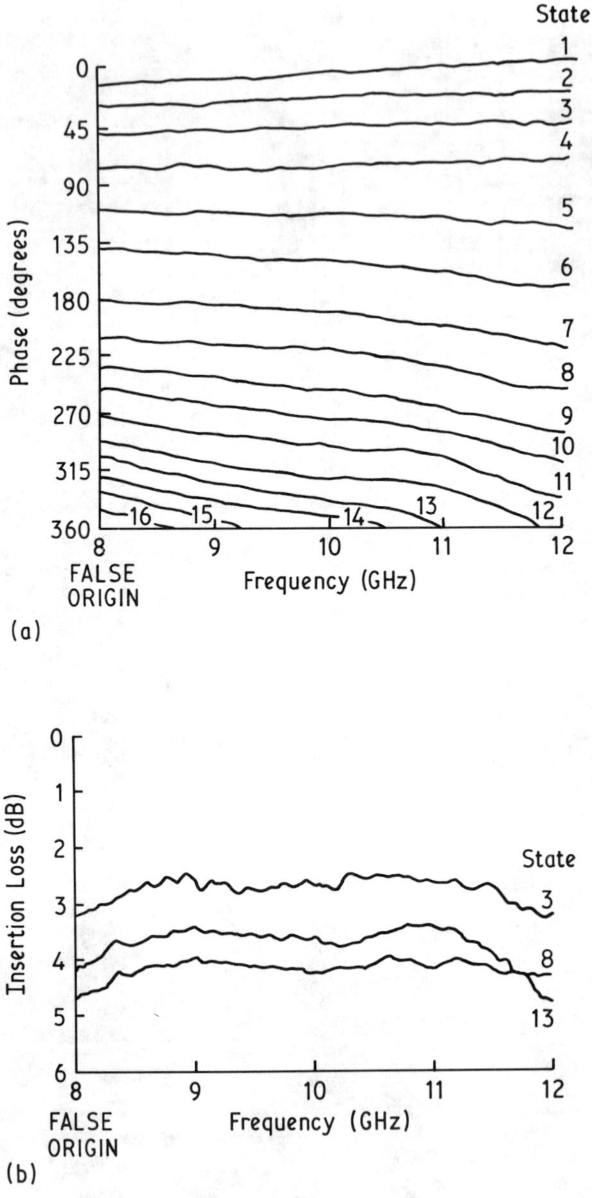

Fig. 2.12 Phase and insertion loss measured for 4-bit phase shifter (from [2.9], with permission).

below center. Figure 2.12(a) shows the phase response, and Fig. 2.12(b) shows the insertion loss. The amount of phase shift is determined by which pair of diodes is turned on. As there are sixteen pairs of diodes, the circuit is described as a four-bit phase shifter. Note also from Fig. 2.12(a) that the phase shift in any given state (say state 7) is not constant but is approximately proportional to frequency. "Phase proportional to frequency" follows from the fact that the circuit is actually a constant group delay circuit rather than a constant phase shifter. Constant group delay is what is required for phased array systems. A full account is given in reference [2.10].

Following this review of some typical MMICs, we will look briefly in Chapter 3 at some of the passive elements that go into them.

References

2.1 Morris, F. W. Potential interference from terrestrial sources into direct broadcast satellite home receiving systems, *IEEE J. Selected Areas in Comms.*, **SAC-3**(1), 155–166, January 1985.

2.2 Shigekazu Hori, Kiyoho Kamei, Kiyoyasu Shibata, Mikio Tatematsu, Katsuhiko Mishima and Susumu Okano, GaAs monolithic MICs for Direct Broadcast Satellite receivers, *IEEE Trans. Electron Devices*, **ED-30**(12), 1867–1874, December 1983.

2.3 Kermarrec, C. *et al.*, The first GaAs fully integrated microwave receiver for DBS applications at 12 GHz, *Proc. 14th European Microwave Conference*, 749–754, Liege, 10–13 September 1984.

2.4 Podell, A. F. and Nelson, W. W. High volume, low cost, MMIC receiver front end, *IEEE Microwave and Millimetre Wave Monolithic Circuits Symp. Dig.*, 57–59, Baltimore, 4–5 June 1986.

2.5 Oxley, T. H. and Corbey, C. D. Systems applications of GaAs devices and ICs, *Microprocessors and Microsystems*, **11**(8), 443–449, October 1987.

2.6 Orr, J. A stable 2–26.5 GHz two-stage dual-gate distributed MMIC amplifier, *IEEE Microwave and Millimetre Wave Monolithic Circuits Symp. Dig.*, 19–22, Baltimore, 4–5 June 1986.

2.7 King, G. and Flynn, S. J. A combined MMIC/MIC 8-channel receive-only phased array demonstrator operating in I/J band, *Military Microwaves '86*, Proceedings, 447–452, Brighton, 24–26 June 1986.

2.8 Oxley, T. H. and Forrest, J. R. The systems need for GaAs devices and integrated circuits, *GEC Journal of Research (Special Issue on Monolithic Microwave Integrated Circuits)*, **4**(2), 66–71, 1986.

2.9 Wilson, K. Monolithic microwave integrated circuits, *GEC Journal of Research (Special Issue on Monolithic Microwave Integrated Circuits)*, **4**(2), 140–146, 1986.

2.10 Wilson, K., Nicholas, J. M. C., McDermott, G. and Burns, J. W. A novel X-band phase shifter, *IEEE Trans. Microwave Theory Tech.*, **MTT-33**(12), 1572–1578, December 1985.

Chapter 3

Overview of Passive Elements

3.1 Introduction

Figure 3.1 is a three-dimensional schematic diagram of an MMIC showing some of the major components that go to make up an integrated circuit. In this chapter we shall cover briefly the passive elements.

Figure 3.2 is a photograph of one section of a diode-based phase shifter MMIC which shows a Schottky diode, an interdigitated capacitor, an overlay,

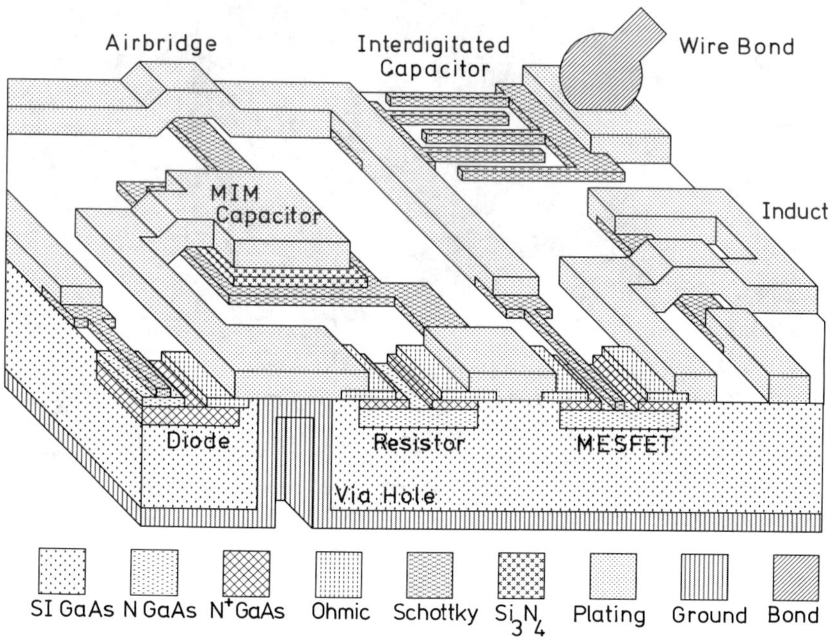

Fig. 3.1 Three-dimensional schematic of basic MMIC construction.

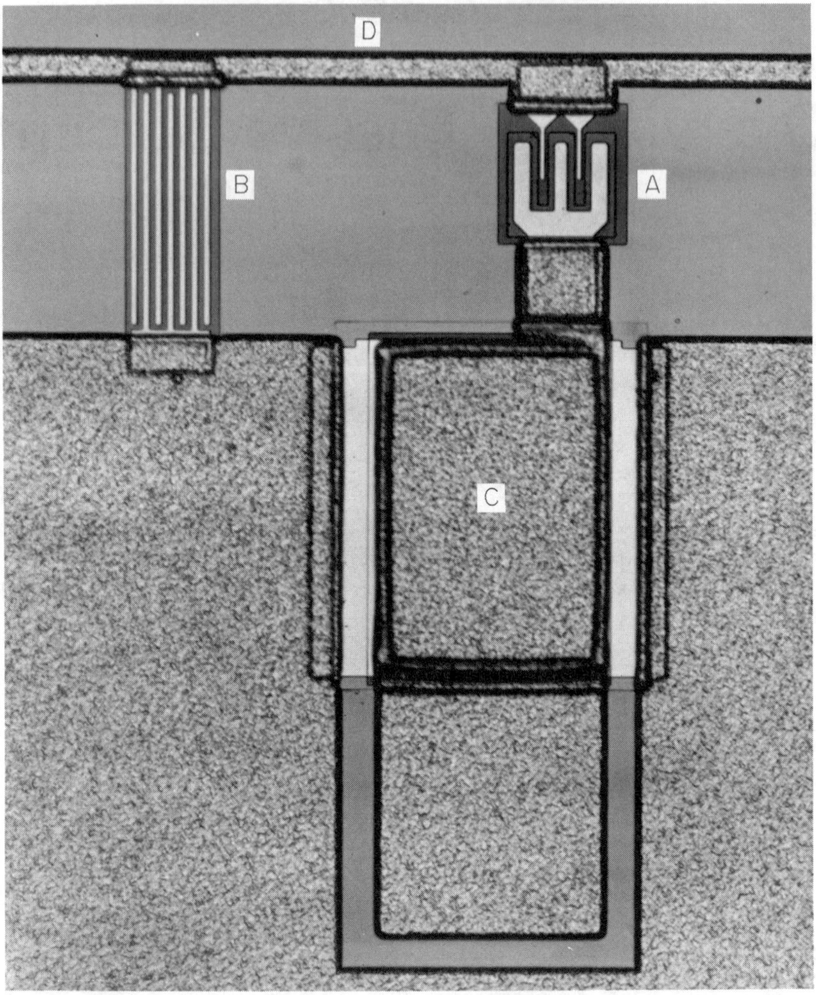

Fig. 3.2 Close-up of diode-based MMIC showing: A—Schottky diode; B—Interdigitated capacitor; C—Overlay capacitor; D—Microstrip (from [3.8], with permission).

or metal–insulator–metal (MIM) capacitor, and a section of microstrip transmission line. The first component we discuss is the microstrip transmission line.

3.2 Microstripline

There are many theories, reviews and accounts of microstripline. Most of the references to original papers can be found in reference [3.1]. Microstripline is a form of miniature waveguide consisting of a narrow strip, of width w, separated from a metal backplane by a dielectric slab of relative permittivity ϵ_r and thickness h (see Fig. 3.3). In the case of an MMIC, of course, the dielectric is the substrate of semi-insulating (SI) GaAs ($\epsilon_r = 13$). Excitation of a microstripline (say at one free end) results in a wave propagating along the strip with most of the energy being carried by the fields in the dielectric. There are, however, fields also in the airspace above the strip which, although weaker, have an important effect upon the wave propagation, and the ease with which we may describe it.

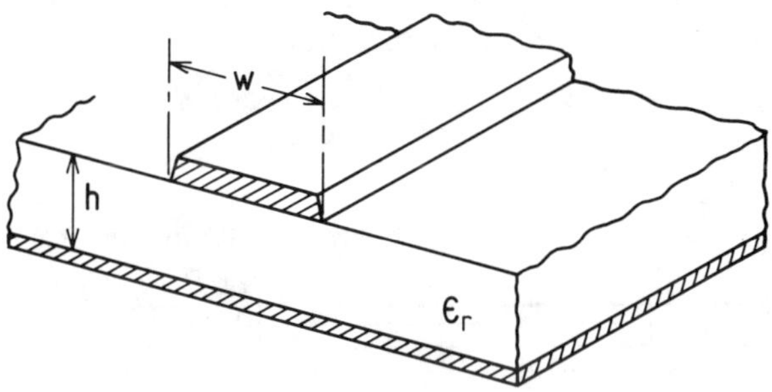

Fig. 3.3 Schematic of a microstrip transmission line.

If we start with the premise that the propagation is transverse electromagnetic (TEM), we immediately suspect that the velocity of propagation in the airspace above the strip should be equal to the velocity of propagation of electromagnetic energy in free-space, c, whereas in the dielectric we would expect the velocity to be more like $c/\epsilon_r^{1/2}$. Consequently, what started as a plane wave soon ceases to be plane, and longitudinal components of field develop to give rise to a hybrid mode of propagation. A good illustration of the fundamental hybrid mode is given in reference [3.1], page 137. The second consequence of the non-uniform dielectric is that the microstrip transmission line is dispersive, i.e. the propagation velocity is a function of frequency. Equivalently, the phase constant or wave number, $\beta = 2\pi/\lambda$ is not proportional to the frequency. For example if f is doubled, β is not exactly doubled, but is rather less than doubled. A readable account is to be found in reference [3.2].

Nowadays, proprietary computer programs for network analysis implement formulae for microstrip accurate to within one or two percent. For present purposes, all we need is an idea of the range of characteristic impedances available from microstrips of technologically convenient dimensions on GaAs. To this end we make use of a low-frequency TEM approximation [3.3], originally due to Presser [3.4]. For the structure sketched in Fig. 3.3, the design curves of Fig. 3.4 give answers which are good to within a few percent at a few gigahertz. The variable q is the microstrip filling fraction, which expresses the idea that, the wider is the strip, w, the more the signal energy tends to propagate in the dielectric; hence the effective dielectric constant increases:

$$\epsilon_{eff} = 1 + q(\epsilon_r - 1) \tag{3.1}$$

In Fig. 3.4, Z_{01} is the characteristic impedance of a microstripline of the same dimensions w and h, but with no dielectric (i.e. $\epsilon_r = 1$). To use the curves at a prescribed value of w/h, Z_{01} and q are determined from the graph; ϵ_{eff} is then calculated from eqn (3.1) and the characteristic impedance of the GaAs-based microstrip is found from the expression

$$Z_0 = \frac{Z_{01}}{\epsilon_{eff}^{1/2}} \tag{3.2}$$

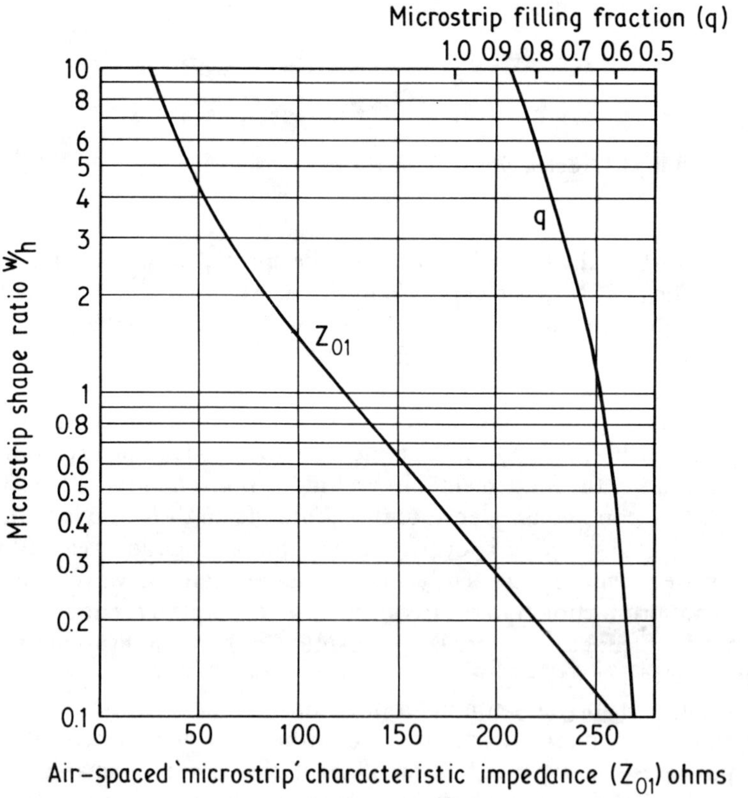

Fig. 3.4 Design curves for microstrip based on quasi-TEM propagation (after [3.2], Fig. 3.7).

The phase velocity of waves propagating along the strip is;

$$v_{\mathrm{ph}} = \frac{c}{\epsilon_{\mathrm{eff}}^{1/2}} \tag{3.3}$$

where c is the velocity of light in free space. The inductance per unit length of the microstripline can then be determined, if required,

$$\bar{\mathscr{L}} = \frac{Z_0}{v_{\mathrm{ph}}} = \frac{Z_{01}}{c}\ \mathrm{H\ m^{-1}} \tag{3.4}$$

Similarly, the capacitance $\bar{C}$ per unit length can be found:

$$\bar{C} = \frac{1}{Z_0 \cdot v_{\mathrm{ph}}} = \frac{\epsilon_{\mathrm{eff}}}{c \cdot Z_{01}}\ \mathrm{F\ m^{-1}} \tag{3.5}$$

To consider the range of values possible, assume that, after processing, the GaAs substrate is lapped to a thickness of $h = 100\ \mu\mathrm{m}$, and that metal tracks down to $10\ \mu\mathrm{m}$ can be realized without too much difficulty. With $w = 10\ \mu\mathrm{m}$, $w/h = 0.1$ and from Fig. 3.4 we find

$$Z_{01} \approx 260\ \Omega \qquad q \approx 0.55$$

From eqns (3.1)–(3.5) we then find, with $\epsilon_r = 13$:

$$\epsilon_{\mathrm{eff}} = 7.6 \qquad Z_0 = 96\ \Omega \qquad v_{\mathrm{ph}} = 1.1 \times 10^8\ \mathrm{m\ s^{-1}}$$

$$\bar{\mathscr{L}} = 0.87\ \mathrm{nH\ mm^{-1}} \qquad \bar{C} = 0.098\ \mathrm{pF\ mm^{-1}}$$

Consider now the other extreme. With $w/h = 2$,

$$\epsilon_{\mathrm{eff}} = 9.28 \qquad Z_0 = 28\ \Omega \qquad v_{\mathrm{ph}} = 0.985 \times 10^8\ \mathrm{m\ s^{-1}}$$

$$\bar{\mathscr{L}} = 0.28\ \mathrm{nH\ mm^{-1}} \qquad \bar{C} = 0.363\ \mathrm{pF\ mm^{-1}}$$

These values give some idea of the range available. Figure 3.4 and eqns (3.1)–(3.5) are useful also for estimating the parasitic inductance and capacitance associated with other elements where the shape is reasonably simple, including GaAs bulk resistors, some capacitors and inductors, and geometric pad capacitance in FETs and HEMTs.

Before leaving the subject of microstripline, two further points should be mentioned. First, the tracks are normally narrow (10–200 μm) and thin (several microns of gold), so the strip has series resistance. The resistance gives rise to loss which cannot be neglected: at 10 GHz for example, a 50 Ω microstrip might have an attenuation of around $10\ \mathrm{dB\ m^{-1}}$. The loss lowers the Q factor of any reactive element we may endeavor to realize on an MMIC. Secondly, if microstrips are meandered as in Fig. 2.8, cross-coupling between the parallel sections reduces both the characteristic impedance and the effective length of these lines. Corrections must be made for both in a practical circuit.

3.3 Inductors

Inductors take one of three forms: a microstripline section, often meandered, a single loop or a spiral inductor. The use of microstriplines to realize inductors has been covered in Section 3.2 (eqn (3.4)).

For a single loop inductor (Fig. 3.5), the inductance is given approximately [3.5] by

$$\mathscr{L} \approx 12.57a[\ln(8\pi a/w) - 2] \tag{3.6}$$

where $\mathscr{L}$ is the inductance in nH, a is the mean radius in cm and w is the strip width in cm.

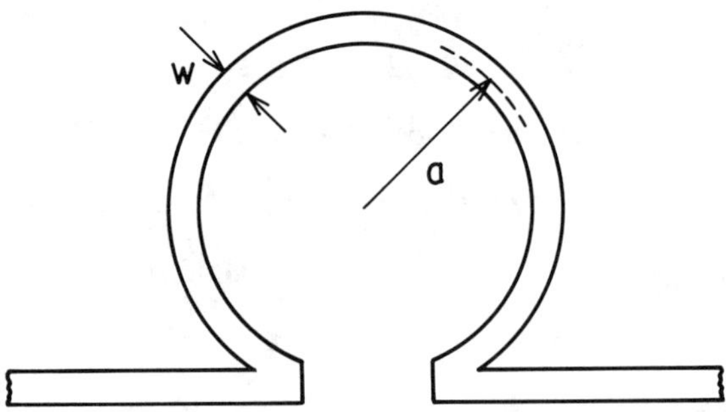

Fig. 3.5 Single loop inductor.

Equation (3.6) applies for a free-standing loop. On an MMIC, where a ground plane is present, it has been argued [3.6] that the inductance is reduced by the factor (for a 200 μm substrate)

$$1 + (5.2 \pm 1.5)a \tag{3.7}$$

Furthermore the capacitance of the loop, formed by the distributed capacitance of the circle of microstrip to the ground plane, gives rise to an approximate equivalent circuit for the loop inductor as in Fig. 3.6, where each capacitor C is half the strip-to-back plane capacitance

$$C \approx \pi a \bar{C} \tag{3.8}$$

When the inductor is used or measured between the terminals 1–1′ in Fig. 3.6,

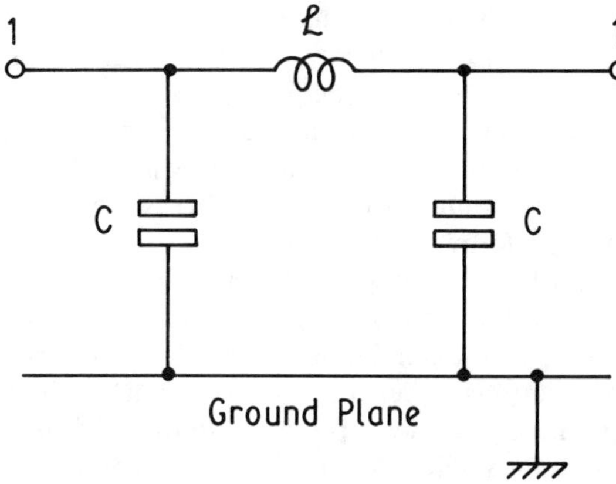

Fig. 3.6 Partial equivalent circuit of inductor, showing capacitance to ground.

as it normally is, the two capacitors C are effectively in series with one another. The shunt capacitance across the inductor is then $C/2$, and the single loop has a self-resonant frequency given approximately by

$$\omega_{\text{res}} \approx \frac{1}{(\mathscr{L}C/2)^{1/2}} \tag{3.9}$$

To obtain an idea of the magnitudes involved, consider a single loop on a 100 μm substrate with

$$a = 200\ \mu\text{m}$$

$$w = 10\ \mu\text{m}$$

Inserting numbers into eqn (3.6) we then find

$$\mathscr{L} \approx 1.06\ \text{nH}$$

Assuming eqn (3.7) approximately applies for 100 μm substrates with the factor $[1+(5.2+1.5)a]$, $\mathscr{L}$ becomes

$$\mathscr{L} \approx 0.93\ \text{nH}$$

From the example calculation in Section 3.2, for $w/h = 0.1$,

$$\bar{C} = 0.098\ \text{pF mm}^{-1}$$

Equation (3.8) now gives $C \approx 0.062$ pF, and eqn (3.9) yields

$$f_{\text{res}} \approx 30\ \text{GHz}$$

Thus, roughly speaking, we can expect to realize single loop inductors up to about a nanohenry as an order of magnitude, but we can expect them to look like inductors only when they are used at frequencies well away from their self-resonant frequency of approximately 30 GHz. Also, it should be made clear that Fig. 3.6 is somewhat inadequate as a circuit model for a loop inductor. To represent series loss in the loop, a resistor should be added in series with $\mathscr{L}$ and, to represent electric field passing from point-to-point across the strip, a capacitance should be connected across the terminals 1–1′.

Figure 3.7 shows a form of spiral inductor where the individual turns are not concentric with one another, thus allowing the loop size to be kept the same from turn to turn. More commonly, the turns are concentric so the turns decrease in size closer to the middle of the spiral. Both circular and square-sided spirals and variations in between (like octagonal inductors) have been constructed. Most theoretical descriptions of spiral inductors are even more approximate than for the single loop, so it is common practice to fabricate and thoroughly characterize a range of inductors using a network analyzer to generate a library of "standard" parts for use in MMICs. The following formulae are, therefore, at best, guidelines only.

For circular spirals, the inductance of a free-standing spiral, with no

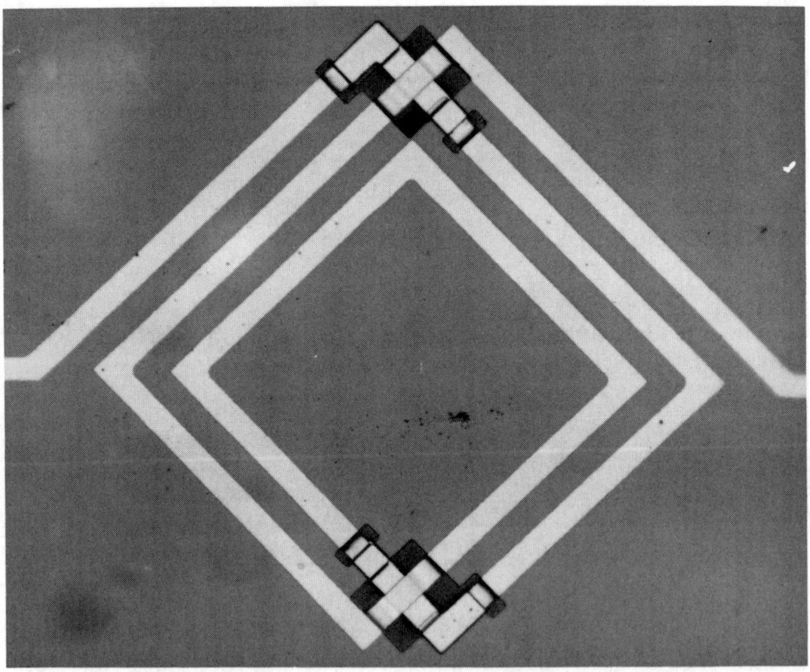

Fig. 3.7 Spiral inductor (from [3.8], with permission).

ground plane in close proximity, may be estimated to within about 20% [3.7] by

$$\mathscr{L} \approx \frac{(r_o + r_i)^2 N^2}{25.4(60r_o - 28r_i)} \tag{3.10}$$

(This formula also seems to have its origins in reference [3.5].)

In eqn (3.10): $\mathscr{L}$ is the inductance in nH, r_i is the inner radius of the spiral in μm, r_o is the outer radius of the spiral in μm and N is the number of turns. For instance, if $r_i = 100$ μm, $r_o = 200$ μm and $N = 3$,

$$\mathscr{L} \approx 3.5 \text{ nH}$$

Again, the inductance value is decreased by the proximity of the back plane, and there is inter-turn capacitance, capacitance from the spiral strip to the backplane, and series loss to be accounted for. A minimum circuit model for an inductive spiral is therefore as in Fig. 3.8. Estimation of the series resistance, R, is relatively straightforward, and so will not be carried out here. We will, however, again examine a method for estimating the shunt capacitances. C_1 and C_2 cannot be expected to be equal because the spiral has no symmetry with respect to its terminals; it is normal for the innermost turn to be brought out to the periphery either by airbridges over the other turns, or by a conducting underpass beneath them.

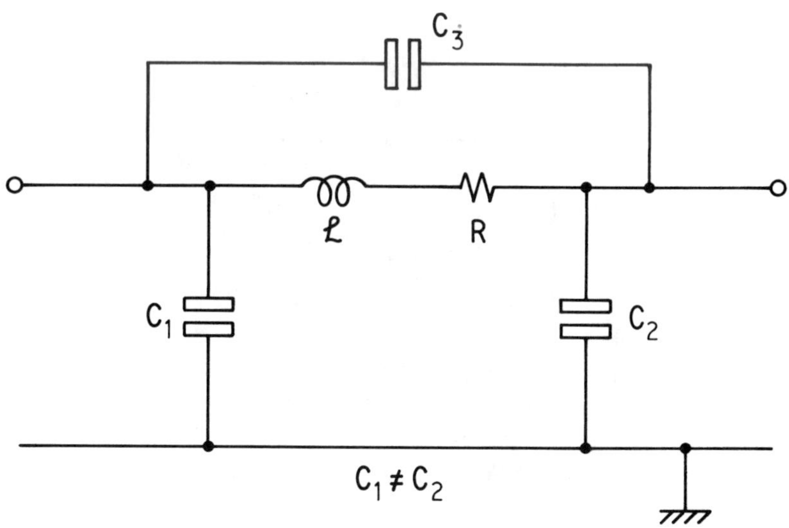

Fig. 3.8 Equivalent circuit for spiral and loop inductors.

For square spirals (Fig. 3.9), an approximate formula† for the inductance [3.6] is

$$\mathscr{L} \approx 0.0008 N^2 S\left[\ln\left(\frac{S}{b}\right) + 0.726 + 0.178\left(\frac{b}{S}\right) + \frac{1}{8}\left(\frac{b}{S}\right)^2 + \cdots\right] \quad (3.11)$$

where $\mathscr{L}$ is the inductance in nH, $b = Np$ (in μm), p is the pitch in μm, N is the number of turns and S is the mean side of the spiral in μm. A reasonable approximation for S is

$$S \approx S_i + Np$$

where S_i is the innermost dimension (Fig. 3.9). For example, if $S_i = 200$ μm, $N = 3$ and $p = 33$ μm, eqn (3.11) gives

$$\mathscr{L} \approx 4.1 \text{ nH}$$

This should be compared with the previous result for a circular spiral for the same inner and outer dimensions (4.1 nH versus 3.5 nH). If inductance were simply proportional to the total length of track, we might expect these two values to be roughly related by the ratio of the periphery of a square to that of an inscribed circle, i.e. in the ratio $8/2\pi = 1.27$. The ratio is in fact 1.17. Once again the presence of the ground plane reduces the value of inductance, typically by around 10% (depending upon the substrate thickness).

It is important to make some kind of estimate of the self-resonant frequency of spiral inductors to help define the range over which they may be

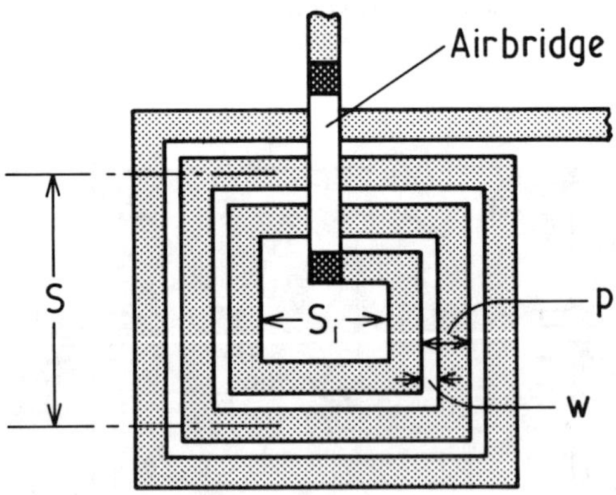

Fig. 3.9 Square-sided spiral inductor.

†Eqn (3.11) neglects an additional term which refines the estimate of $\mathscr{L}$ by taking into account the strip width and inter-turn spacing.

useful. For this purpose we can simplify the equivalent circuit model from Fig. 3.8 to Fig. 3.6, and estimate C from Fig. 3.4 by regarding the inductor as a single loop of microstrip of width equal to $b = Np$ and length equal to the mean periphery. (Because the adjacent turns tend to act as guard rings for each other, it is not valid to estimate the capacitance to ground as the capacitance per metre of the microstrip of width w, multiplied by the total length of the spiral if it were uncoiled: this overestimates C.) For the example square spiral above, the effective value of w is 100 μm, and the periphery (or effective length) is $4S = 1.2$ mm. For the purposes of illustration we shall in this case assume a substrate thickness of $h = 200$ μm, so we have an effective w/h of 0.5. From Fig. 3.4 and eqns (3.2), (3.3) and (3.5), we then find

$$\bar{C} \approx 166 \text{ pF m}^{-1}$$

For a total effective periphery of 1.2 mm, we thus estimate the total capacitance of the spiral to ground to be 0.2 pF. Thus, in Fig. 3.6, each of the capacitors C we estimate as one half of the total capacitance to ground, that is,

$$C \approx 0.1 \text{ pF}$$

Using eqn (3.9) we then estimate f_{res} (with $\mathscr{L} = 4.1$ nH) as:

$$f_{res} \approx 11 \text{ GHz}$$

This is a rough estimate. Nevertheless, we should expect it to be reasonably representative of the self-resonant frequency measured in practice.

The utility of these methods lies not so much in their ability to approximate actual values (often the approximate values are not good enough), but in pointing the way to structural changes that would bring about wanted electrical changes. For instances, suppose the above self-resonant frequency were too low, and we ask the question: "what could we do to raise it?". One manoeuvre we could try would be to decrease the track width, and hence the pitch. Suppose we were to consider

$$S_i = 200 \text{ μm (as before)}$$

$$N = 3 \text{ (as before)}$$

$$p = 15 \text{ μm (decreased from 33 μm)}$$

From eqn (3.11) we then find

$$\mathscr{L} \approx 4.34 \text{ nH}$$

which is a little greater than the previous value (4.1 nH). Further, for the purpose of estimating the capacitance to ground, we now have an effective width of 45 μm, a substrate thickness of 200 μm, and an effective periphery of

$4S = 980$ μm. From Fig. 3.4 and eqns (3.2), (3.3) and (3.5) we now obtain:

$$\bar{C} \approx 120 \text{ pF m}^{-1} \text{ and } C \approx 0.06 \text{ pF}$$

Finally, from eqn (3.9):

$$f_{res} \approx 14 \text{ GHz}$$

The obvious price we have paid for this increase in self-resonant frequency is a decrease in the quality factor Q of the inductor ($Q \approx \omega\mathscr{L}/R$), because we have increased the series resistance R in Fig. 3.8 by making the track narrower. Such are the compromises of integrated circuit engineering.

We have not so far mentioned a method for estimating the capacitance C_3 in Fig. 3.8, which originates from electric field fringing between the turns of the spiral. The problem is complicated by the fact that, in a cross-section through the spiral, there are multiple conductors, all coupled to one another. For the example of the three-turn spiral we have just been considering, a rough estimate can be made using the graph of Fig. 6.49, which relates to the geometry shown in Fig. 6.45. We may think of the central turn as the shaded strip in Fig. 6.45, and the inner and outer turns as the flanking electrodes. By considering only nearest-neighbor fringing in this way, all the electric field coupling, or fringing capacitance, is accounted for by estimating the capacitance of the center turn, of total length $4S$, to the turns on either side. To do this using Fig. 6.49 we need to know ϵ_r, the effective permittivity for use with the graph, and the strip width W. Given W, we can find $D = (p - W)$. (We are using W to conform with Fig. 6.49.)

The permittivity ϵ_r is simply the average of the relative permittivities above and below the strips, being 1 for air (above the strips) and 13 for GaAs (below the strips). Thus $\epsilon_r = 7$. For the purposes of illustration, we shall assume $W/D = 1$. From Fig. 6.49, this gives

$$\frac{C}{\epsilon_r(L + W + D)} \approx 20 \text{ pF m}^{-1}$$

The mean side of the spiral, S, is given by:

$$S = S_i + Np = 200 + (3 \times 15) = 245 \text{ μm}$$

The periphery of the central turn is $L = 4S$ so that, substituting for L, W and D in the above we find, for $C_3 = C$,

$$C_3 \approx 0.14 \text{ pF}$$

Evidently, C_3 may not be neglected.

Finally, it is worthwhile mentioning that one way to keep the self-resonant frequency up in value is to stand the spiral off the surface of the GaAs by mounting it on small metal pillars. In effect, the spiral becomes a sequence of series-connected, long airbridges. Such an expedient reduces the shunt capacitances C_1 and C_2 in Fig. 3.8 and, more importantly, reduces the fringing capacitance C_3. In the example just given, C_3 might decrease to approximately 0.02 pF as ϵ_r reduces from 7 to 1.

In summary, inductance values of a few nanohenries are quite easily realized, but the self-resonant frequency can just as easily be below 20 GHz. As a very rough rule of thumb, the self-resonant frequency of a 1 nH inductor might be around 30 GHz, falling relative to that value as approximately $1/\mathscr{L}^{1/2}$, with $\mathscr{L}$ in nH. Thus for $\mathscr{L}=4$ nH, we could expect a self-resonant frequency of $30\ \text{GHz}/\sqrt{4}=15$ GHz. Measurements confirm this guideline [3.6].

3.4 Capacitors

Two types of capacitor are available: interdigitated and overlay or metal–insulator–metal (MIM). Examples of both were seen in Fig. 3.2.

Interdigitated capacitors (Fig. 3.10) are used where small value, high precision values of capacitance are required. The capacitance results from electric field passing between the two sets of fingers. ***E*** arises from net positive charge on one set of fingers, and terminates on net negative charge on the other set. Such a field is often referred to as a "fringing" field. Calculation of the capacitance can be carried out in terms of elliptic integrals providing end

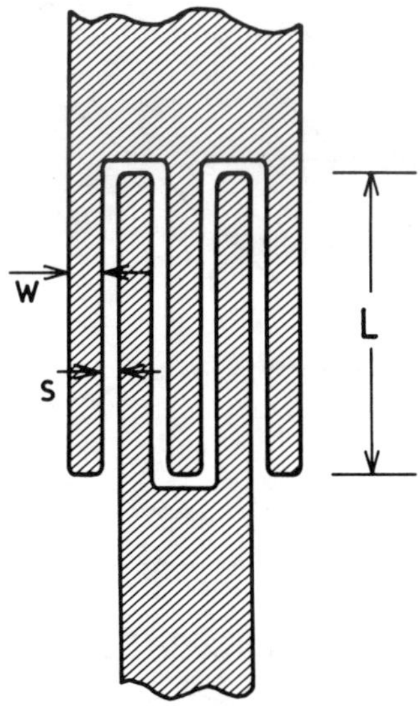

Fig. 3.10 Interdigitated capacitor.

effects are neglected. Wilson [3.8] has published a set of design curves resulting from such calculations, reproduced here as Fig. 3.11, where w is the finger width and s is the spacing between fingers.

To illustrate the use of these curves, suppose we wish to design an interdigitated capacitor having $C = 0.21$ pF (this example follows reference [3.8]). To begin with, we have free choice of finger dimensions, and we set out to determine the number of fingers necessary. A possible choice is:

$$\text{Finger length } L = 60\ \mu\text{m}$$

$$\text{Finger width } w = 2\ \mu\text{m}$$

$$\text{Spacing } s = 2\ \mu\text{m}$$

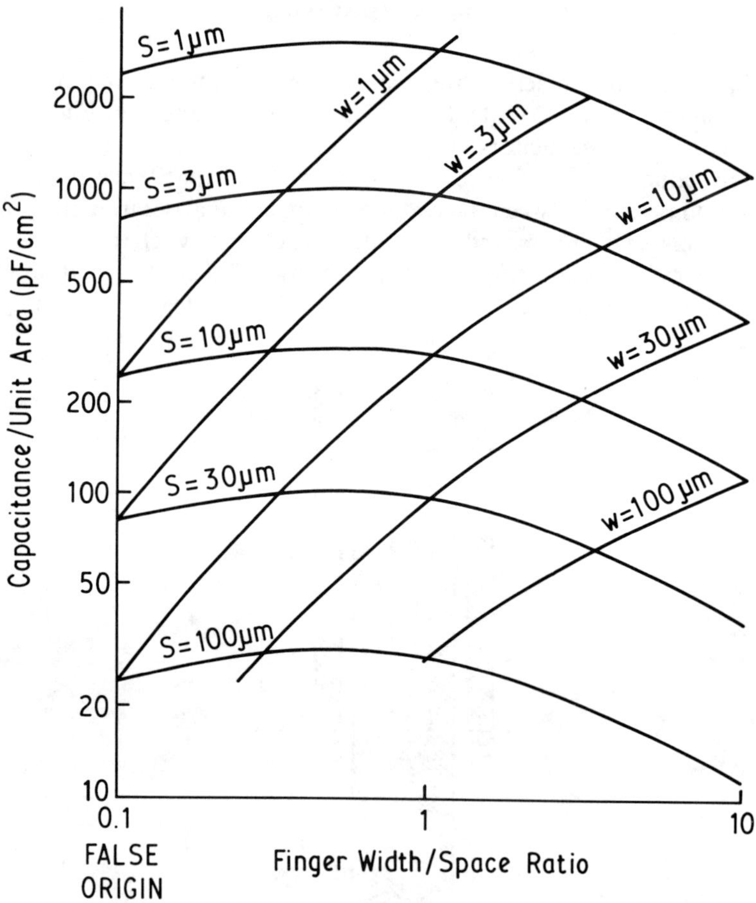

Fig. 3.11 Design curves for interdigitated capacitor (after [3.8], with permission).

From Fig. 3.11, we find the capacitance per unit area at these values of s and w as 1500 pF cm^{-2}. Consequently, to realize 0.21 pF the total area of the interdigitated region must be

$$\frac{0.21\ \text{pF}}{1500\ \text{pF cm}^{-2}} = 1.4 \times 10^{-4}\ \text{cm}^2$$

Hence the overall transverse dimension must be

$$\frac{1.4 \times 10^{-4}\ \text{cm}^2}{60 \times 10^{-4}\ \text{cm}} = 232\ \mu\text{m}$$

As there is one finger every $s + w = 4$ μm, the number of fingers required is thus 58.

The principal parasitic elements associated with an interdigitated capacitor are $\mathscr{L}$, R, C_1 and C_2, as shown in Fig. 3.12. The inductance $\mathscr{L}$ arises from the magnetic field created by the passage of current through the capacitor. The resistance R relates to series loss in the fingers. Capacitors C_1 and C_2 represent the capacitance of the finger array to ground (i.e. the back plane). A more complete circuit model would also allow for shunt loss arising from the fact that semi-insulating GaAs is not a perfect insulator but has a finite density of free electrons which move in sympathy with the fields in the substrate. The energy required to move the electrons in this way is, of course, derived from the field itself. This energy requirement is modeled by resistive elements connected between terminal 1 and ground, and terminal 1′ and ground, in Fig. 3.12. Further, if the fingers are in direct contact with the GaAs surface, the field singularity at the edge where the metal meets the GaAs may

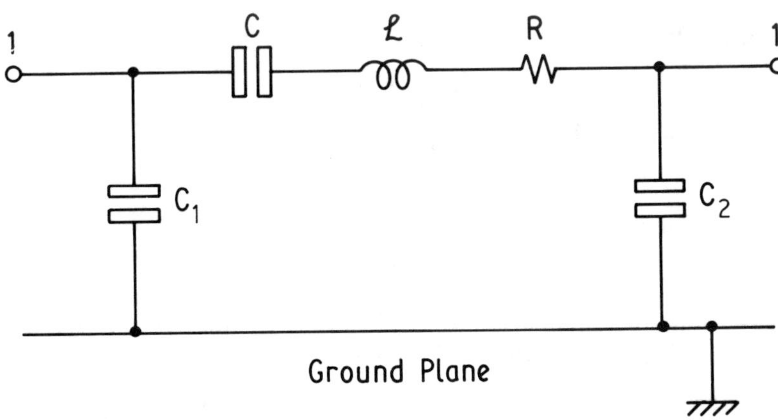

Fig. 3.12 Equivalent circuit of interdigitated and overlay MMIC capacitors.

give rise to thermionic-field emission of electrons, with attendant further loss†. For simplicity, we are ignoring shunt loss throughout.

As with the inductor, the series resistance, R, is relatively straightforward to estimate. Approximate values for the shunt capacitors can be obtained by the same methods as were used for this purpose in the spiral inductor. The inductance $\mathscr{L}$ requires a little more thought.

If we try and generate a picture of what is happening in the interdigitated capacitor by looking at it as a parallel-plate capacitor of modified geometry, we can argue as follows‡. Suppose we were to open-up a parallel plate capacitor from one edge like a box lid, and lay the lid (or capacitor plate), with its interior face upwards, alongside the other plate: the field would now appear as in Fig. 3.13(a). The positive charge giving rise to the $\boldsymbol{E}$ field on the right-hand plate would be supplied by net movement of positive charge into the plate from the bottom right, and the negative charge required to terminate the field on the left-hand plate would come from a net movement of negative charge in from the top left. In both electrodes conventional current flows in the same direction, diagonally from left to right. Just as in the parallel plate capacitor, the current is continuous. Given this simple picture, the magnetic field behaves roughly as in Fig. 3.13(b), i.e. the $\boldsymbol{H}$ field tends to form loops around the cross-section of the entire interdigitated structure, rather than around the individual fingers. On this basis, the parasitic inductance can be estimated by regarding the interdigitated region of length L as a section of microstrip of width equal to the number of fingers multiplied by $s+w$. For the example already under consideration, this dimension is 232 μm. We now proceed as follows (assuming a substrate thickness, h, of 200 μm). From Fig. 3.4, for $w/h = 1.16$,

$$Z_{01} \approx 117\ \Omega \qquad q \approx 0.63$$

From eqn (3.1),

$$\epsilon_{\text{eff}} \approx 8.56$$

From eqns (3.2) to (3.4),

$$\bar{\mathscr{L}} = 3.9 \times 10^{-7}\ \text{H m}^{-1}$$

So, for $L = 60$ μm,

$$\mathscr{L} = 23\ \text{pH}$$

The self-resonant frequency of the 0.21 pF capacitor, which is a series resonance, is thus in the vicinity of 72 GHz.

†This applies to all other passive elements, e.g. microstriplines, inductors, couplers, etc.

‡This argument is based on the reasoning of K. Wilson [3.8].

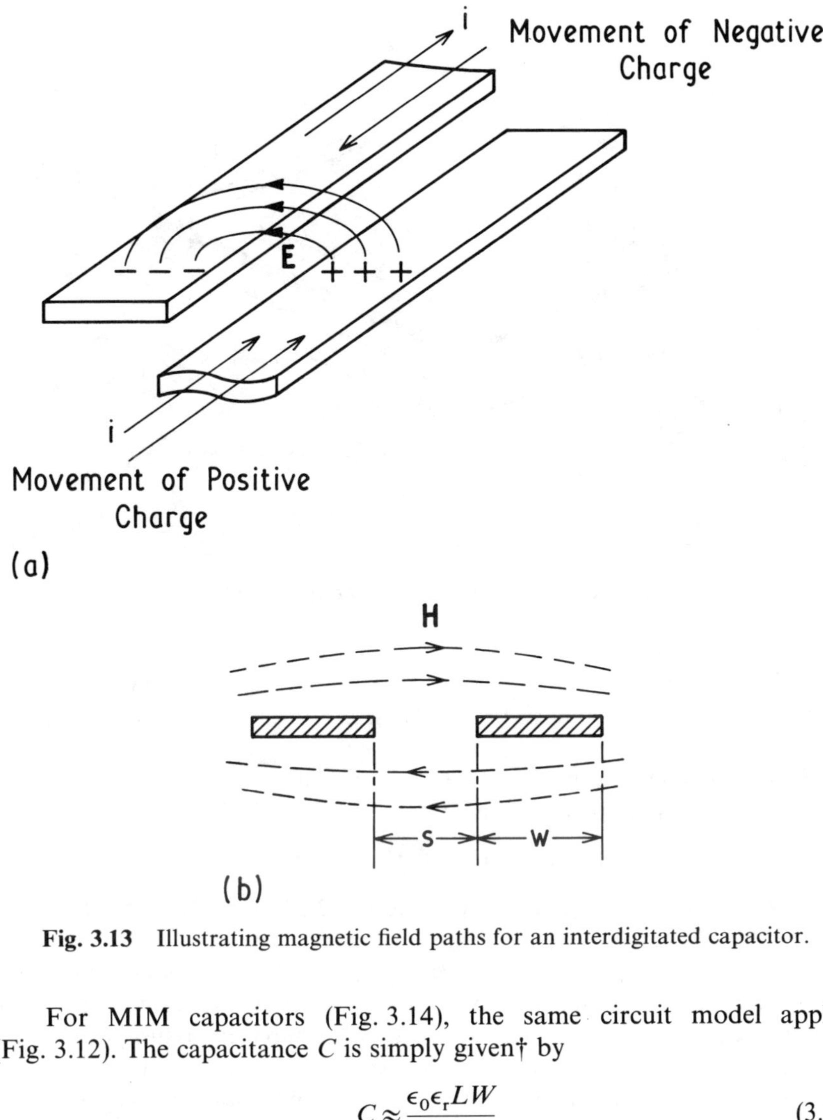

Fig. 3.13 Illustrating magnetic field paths for an interdigitated capacitor.

For MIM capacitors (Fig. 3.14), the same circuit model applies (Fig. 3.12). The capacitance C is simply given† by

$$C \approx \frac{\epsilon_0 \epsilon_r L W}{d} \tag{3.12}$$

where ϵ_r is the relative permittivity of the insulator ($\epsilon_r \approx 6$–7 for Si_3N_4, depending upon deposition conditions). If the way in which the dielectric film is deposited results in the layer having finite conductivity, a resistor should be

†This classical expression omits capacitance due to fringing fields. Corrections can be made using the methods employed for FET pad capacitance in Section 6.10.

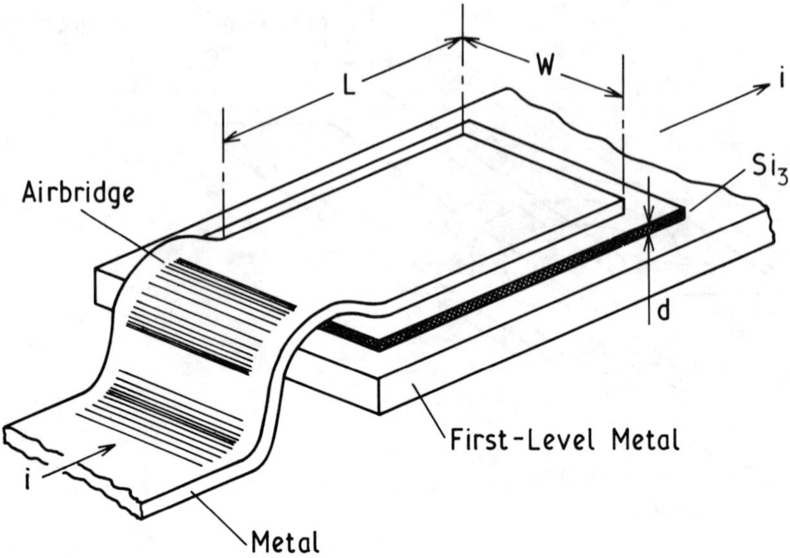

Fig. 3.14 Schematic of an overlay capacitor.

added in parallel with C in Fig. 3.12. The series resistance R mainly results from the bottom plate, which is first-metal level and cannot be plated to reduce its resistance.

Typically, dielectric film thicknesses may be in the range 1000 Å to about 3000 Å, giving the following range for C

$$C \approx 100\text{–}600 \text{ pF mm}^{-2}$$

For example, a 50 μm square top plate on 2000 Å Si_3N_4 would yield a capacitance of about 0.8 pF.

The parasitic inductance may be estimated by again assuming that the magnetic field tends to form loops around the whole structure, rather than around one plate and through the dielectric. An equivalent assumption is that the current i in Fig. 3.14 is continuous and that the return for this current is through the ground plane. Given this rough approximation, we can again employ the same methods as before and look upon the structure as a length L of microstrip having width w. From Fig. 3.4 with $h = 200$ μm for example, we estimate the parasitic inductance for the 50 μm × 50 μm capacitor as

$$\mathscr{L} \approx 35 \text{ pH}$$

Self-resonance of this particular 0.8 pF capacitor therefore occurs at about 30 GHz. Finally, the shunt capacitors C_1 and C_2 can be estimated by the same general method. Assuming C_1 to be equal to C_2, and both to be equal to one half the capacitance of a microstrip section of length L and width w, we find $C_1 = C_2 \approx 0.003$ pF for the example MIM capacitor. Factors unaccounted for in this simple treatment will undoubtedly raise these values somewhat.

3.5 Resistors

Resistors may be constructed from either metal films such as NiCr or from the GaAs *N*-layer itself. Figure 3.1 includes a bulk resistor, practical examples of which are shown in Fig. 3.15.

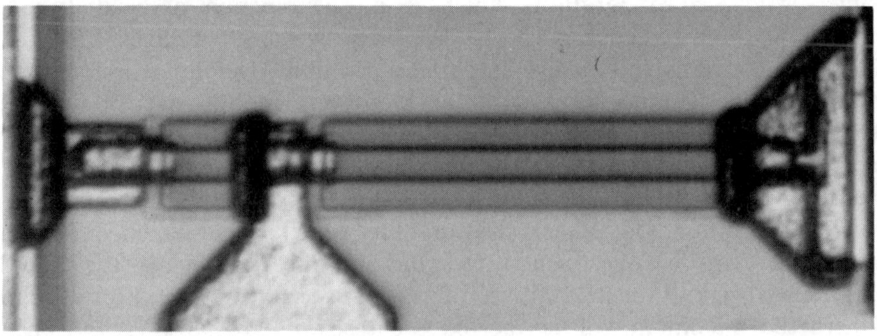

Fig. 3.15 Two GaAs (bulk) resistors on part of an MMIC (from [3.8], with permission).

If the thickness of the resistive film from which the resistor is constructed is W, the distance separating the two ohmic contacts at either end of the resistor is L, and the dimension transverse to the direction of current flow is Z_G, then† the resistor value is given by:

$$R = \frac{\rho L}{W Z_G} + \frac{2\rho_c}{Z_G} \tag{3.13}$$

The second term arises from the contact resistance between the film, of resistivity ρ, and the ohmic contacts. ρ_c has units of (Ω mm), with typical values in the range 0.1–1 Ω mm. For a resistor of width $Z_G = 100$ μm and $\rho_c = 0.5$ Ω mm, the second term amounts to 10 Ω. A contribution to the resistance of this magnitude can rarely be neglected.

Table 3.1 summarizes the salient features of bulk and metal film resistors.

†For ease, the same symbols are used as for the FET in later chapters.

Table 3.1 Properties of MMIC resistors (data are typical only)

Property	*GaAs (or bulk)*	*Metal film (NiCr)*	*Reference*
Resistivity	300–450 Ω per sq	60–600 μΩ cm	[3.9]
	up to 500 Ω per sq	up to 100 Ω per sq	[3.8]
Precision	±5%	±10%	[3.8]
Temperature coefficient	+1500 ppm per deg C	+150 ppm per deg C	[3.8]
	+3000 ppm per deg C	+200 ppm per deg C	[3.9]
Long-term stability	Can be poor (surface)	Good (with annealing)	[3.8]

Because bulk resistors are fabricated from single crystal GaAs they do not suffer from electromigration, which is a phenomenon associated with crystalline grain boundaries [3.8]. They have two disadvantages, however. The first concerns the change in surface potential over long periods; the surface potential may drift over a timescale of several months, even on surfaces that have been dielectric protected. In such a circumstance the depth of surface depletion changes, giving rise to changes in the sheet resistivity governing the resistor value. The second factor to which a bulk resistor is susceptible is current saturation. GaAs has a drift mobility which is field-dependent much above 1 kV cm^{-1}. Consequently, to avoid non-linearity, the resistor should be designed so that the field within it does not exceed this value. For a circuit where the maximum voltage across the resistor can be, say, 3 V, then $L \geqslant 30$ μm will ensure reasonable linearity.

A resistor of either construction has parasitic reactances associated with it. Current flowing through the film gives rise to a magnetic field $\boldsymbol{H}$ which links with the current via the current return in the ground plane. It suggests the resistor should have series inductance associated with it, say $\mathscr{L}$. Likewise, lines of electric field $\boldsymbol{E}$ will pass from the resistor, which generally will be at some instantaneous potential different from ground, through the substrate to the ground plane. Consequently, the resistor possesses capacitance to ground. An equivalent circuit for a resistor is thus as in Fig. 3.16, where we have assumed the layout of the resistor to be of a simple form that allows the total shunt capacitance to be equally divided between the ends.

To consider an example, suppose we wished to fabricate a 50 Ω GaAs resistor from an N-layer having $N = 2 \times 10^{18}$ cm^{-3}, $\mu_0 = 2500$ cm^2 Vs^{-1} and $W = 0.2$ μm (neglecting surface depletion). The contact resistance will be assumed to be $\rho_c = 0.25$ Ω mm. A 10 V working limit is required, so that we must have $L = 100$ μm for the average field to remain at or below 1 kV cm^{-1}. Substituting values into eqn (3.13) we can then find Z_G as

$$Z_G = 135 \text{ μm}$$

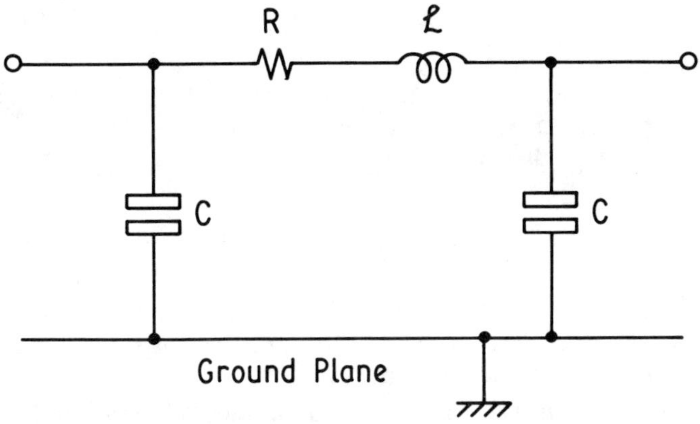

Fig. 3.16 Equivalent circuit of a resistor.

If the substrate thickness is $h = 100\ \mu\text{m}$, for instance, then $\mathscr{L}$ and C are estimated on the basis of a section of microstripline of $w/h = 1.35$. From Fig. 3.4, at $w/h = 1.35$,

$$Z_{01} = 110\ \Omega \qquad q = 0.7$$

From eqn (3.1),

$$\epsilon_{\text{eff}} = 9.4$$

From eqn (3.4),

$$\bar{\mathscr{L}} = 3.7\ \text{H m}^{-1} \quad \text{so} \quad \mathscr{L} = 37\ \text{pH}$$

From eqn (3.5),

$$\bar{C} = 0.28\ \text{pF mm}^{-1} \quad \text{so} \quad C = 0.014\ \text{pF}$$

Wilson [3.8] has given specific design charts for GaAs resistors and NiCr resistors at X band for a 200 μm substrate; these show a preferred range for GaAs resistors covering 50 Ω to 2500 Ω, whereas for NiCr the preferred range is estimated to be only 3 Ω to 20 Ω, which is far less useful. (It should be stressed, however, that these bounds are by no means absolute, but are merely the result of "preferred region" analysis which attempts to ensure that all the resistor properties of interest are acceptable [3.8].)

3.6 The Lange coupler

A number of applications in MMIC engineering require a signal to be split into two signals of equal magnitude. Figure 3.17 is a photograph of a

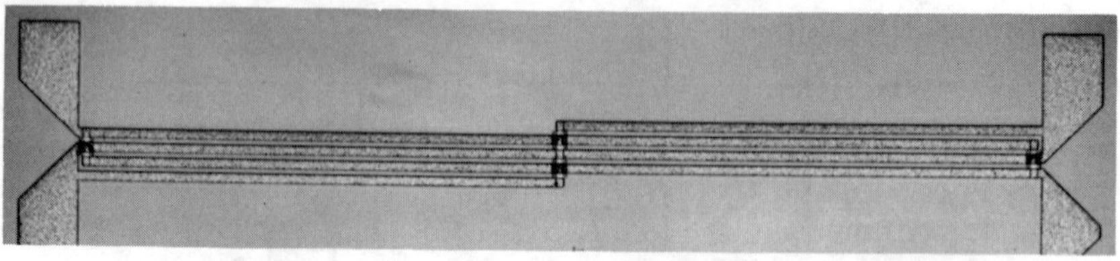

Fig. 3.17 Photograph of a Lange coupler (from [3.8], with permission).

structure designed to do exactly this: a 3 dB coupler named after its inventor, Julius Lange. A brief description of the Lange coupler can be given using Fig. 3.18 to help clarify the construction†.

Lange's intention was to conceive a coupler offering nearly an octave bandwidth with a coupling factor of around -3 dB. The interdigitated coupling section compensates for the even- and odd-mode phase velocity dispersion that would otherwise exist on parallel coupled lines with a non-uniform dielectric filling (air above the strips, dielectric below them). There is a continuous strip connecting the direct port with the input port. Signal transfer from the input port to the direct port and the coupled port, all with the correct phasing, occurs through several coupled paths involving airbridges linking various arms of the structure. The airbridges should as closely as possible resemble electrical short-circuits. Thus the airbridge links force certain points on various fingers of the coupler to be at the same signal amplitude and phase at any instant. The shorter fingers are approximately $\lambda_g/4$ at the highest frequency in the band of interest, whereas the length of the central finger connecting the direct port with the input port is $\lambda_g/4$ at the lowest frequency in the band.

The Lange coupler generally offers a good match, and is useful over a bandwidth of about one octave. Typically, the voltage standing wave ratio (VSWR) may be less than 1.15 over an octave bandwidth, depending upon the frequency of operation, with a phase shift over the band to within $\pm 3°$ of 90° (90° being the nominal phase shift for a quadrature coupler).

3.7 Other components

Referring back to Fig. 3.1, an MMIC generally requires airbridges to form low-capacitance cross-overs, and via-grounds to connect certain points

†For those unfamiliar with the basic functioning of a 3 dB coupler, it may be helpful to refer to Section 10.2.7 on balanced amplifiers, where a summary will be found.

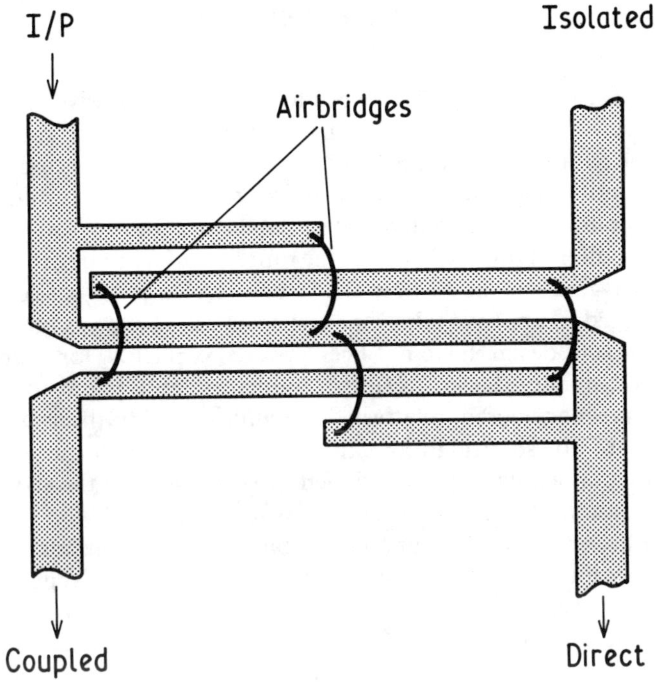

Fig. 3.18 Schematic of a Lange coupler.

in the circuit on the front side of the chip, such as FET sources, to ground, which is the metallized reverse side of the chip.

As an approximation, the inductance, capacitance and resistance of an airbridge can be estimated by regarding it as a short length of microstrip of length and width equal to those of the metal bridge itself, with an air dielectric of thickness equal to the height of the airbridge. Via-grounds, or via-holes, principally have parasitic inductance (see Section 6.11).

Another component not mentioned so far is the spiral transformer. A spiral transformer essentially consists of two interleaved spiral inductors. Airbridge connection may be made to the windings to bring out centre taps if desired. Transformers of the spiral type have been used with outstanding success up to at least 6 GHz by Pacific Monolithics [3.10], resulting in MMIC layouts in which very little of the available GaAs area is unused. Extremely small chips to perform a given microwave function can consequently be realized.

3.8 Final remarks

The purpose of this chapter has been to describe the passive components commonly used in MMICs, and to give some approximate methods for calculating the parasitic reactances and resistances associated with them. Such methods do not supplant the need to fabricate and thoroughly pre-characterize components at microwave frequencies before using them in MMIC design; all they do is help define the range of values to be expected, and thereafter offer guidance as to how the structure of a practical component might be changed to improve it in some respect. Certain aspects important to passive component design have not been covered explicitly, for example the calculation of quality factor (Q) for inductors and capacitors. These calculations are considered relatively straightforward. Others have been omitted that are not so straightforward. The thermal properties of MMIC resistors, for example, have been excluded as being beyond the scope of the present work, and not because the subject is unimportant.

A factor not so far touched upon is coupling between components where no coupling is intended. Probably everyone with an involvement in the MMIC industry has had experience of chip designs failing to meet their specification because of parasitic coupling. They can be the most time-consuming faults to find, and can present a stern challenge to the inventiveness of engineers to "think inside" the chip to discover where the parasitic coupling occurs. Finding the answer may involve innovative experiments in which signals are injected into the MMIC at various points while observing the transmission to various other points. Such probing and dissection of chips also presents a challenge to the practically adventurous. For the most part, however, it is a problem we would prefer to do without.

No useful approximation methods can be offered to assess parasitic coupling. Fortunately, there now exist commercial computer packages which find electromagnetic field solutions in three dimensions, given the multiplicity of boundary conditions imposed by the various components in an MMIC. Having designed an MMIC, therefore, the first-trial layout can be investigated for parasitic coupling effects before committing the design to masks.

In Chapter 4 we shall briefly examine the other non-FET components used in MMICs: Schottky diodes and PIN diodes.

References

3.1 Hoffman, R. K. *Handbook of Microwave Integrated Circuits*, chs 3–8, Artech House, Norwood, 1987.

3.2 Edwards, T. C. *Foundations for Microstrip Circuit Design*, chs 3–4, Wiley, Chichester, 1981.
3.3 Ibid., pp. 42–43.
3.4 Presser, A. RF properties of microstrip lines, *Microwaves*, **7**, 53–55, March 1968.
3.5 Grover, F. W. *Inductance Calculations*, Van Nostrand, Princeton, New Jersey, 1946.
3.6 Pengelly, R. S. GaAs monolithic microwave integrated circuits, in Thomas, H., Morgan, D. V., Thomas, B., Aubrey, J. E. and Morgan, G. B. (eds) *Gallium Arsenide for Devices and Integrated Circuits*, pp. 265–269, Peter Peregrinus, London, 1986.
3.7 Daly, D. A., Knight, S. P., Caulton, M. and Ekholdt, R. Lumped elements in microwave integrated circuits, *IEEE Microwave Theory Tech.*, **MTT-15**(12), 713–721, December 1967.
3.8 Wilson, K. Other circuit elements for MMICs, *GEC Journal of Research (Special Issue on Monolithic Microwave Integrated Circuits)*, **4**(2), 126–133, 1986.
3.9 Williams, R. E. *Gallium Arsenide Processing Techniques*, ch. 14, Artech House, Dedham.
3.10 Podel, A., Moghe, S., Lockie, D. and Ali, F. GaAs real estate—making the most efficient use of semiconductor surface area, *Microwave Journal*, **30**(11), 208–212, November 1987.

Chapter 4

PIN and Schottky Diodes

4.1 Introduction

Diode-based MMICs fulfill very important chip-level functions, notably mixing of two signals, phase shifting of a signal and switching of two signals between alternative transmission paths. Although FET circuits are sometimes implemented to perform these functions, there are two reasons why mixers, phase shifters and switches are generally more effectively realized using diodes. First, the performance of diode circuits is said to be superior [4.1]. Secondly, they are less demanding of fabrication control, thus giving better yields and cost advantages. Wilson [4.1] has argued that diodes should be preferred to FETs for the following reasons:

(i) Improved RF performance due to a several times lower CR product and far less susceptibility to flicker noise.

(ii) The geometry does not use such fine lines, so improving yield at the lithography stage.

(iii) The plan area is less than for a FET.

(iv) In some circuits, no external dc bias is required.

(v) The "no bias" requirement implies fewer on-chip components.

As will shortly become apparent, however, the layer structures needed for Schottky and PIN diodes are quite different from most FET layers. Consequently, the five points above are principally valid when an entire MMIC supports diodes only, and not FETs. When diode-based and FET-based functions must both be supported on the same chip, either a genuine mixed technology must be developed to provide simultaneously the layers required by the individual device types, or it becomes attractive to try and realize the diode-based functions using FETs.

In this short chapter, we shall briefly examine the technological requirements for Schottky and PIN diodes insofar as they are distinct from FETs, and give approximate expressions for the main capacitive and resistive elements associated with their "on" and "off" states.

4.2 Schottky diodes

In its most simple form, the current-voltage characteristic of a Schottky contact considered one-dimensionally is

$$I = I_0 \left\{ \exp\left(\frac{qV}{mkT}\right) - 1 \right\} \tag{4.1}$$

Historically, various theoretical accounts for this dc behavior have been offered, including the thermionic emission theory [4.2], diffusion theory [4.3] and the thermionic-field emission theory [4.4], all of which yield a solution of the type given in eqn (4.1) with the factor m essentially equal to unity. Sooner or later, these theories involve consideration of the so-called quasi-Fermi level [4.5], whereupon a degree of what might be termed "self-inconsistency" appears. The idea of adapting the Fermi level—which is strictly an equilibrium concept—to non-equilibrium situations requires that the gradient in the quasi-Fermi level should be less than kT over one mean free path. The difficulty is that the foregoing theories imply gradients which violate this requirement, especially near the metal–semiconductor interface. For this reason, more advanced descriptions using a momentum and energy conservation (or transport) approach [4.6], or Monte Carlo simulation [4.7], tend to be more satisfying. Even so, dramatically simplifying assumptions are made about the nature of the metal–semiconductor interface, notably that the bulk band structure of the two material components persists right up to the metallurgical interface, and that the interface itself has zero width. More realistic descriptions acknowledge the penetration of metal-like wavefunctions into the semiconductor [4.8], and the existence of Local Densities Of States (LDOS) which are quite different from the bulk band states, and which occupy several monolayers in the vicinity of the interface [4.8, p. 795]. Abandoning the remaining idealized assumptions relating to atom position and bonding that are embodied in these treatments brings the realization that the detailed position of atoms in the interface region (known as interface reconstruction), the atomic species, and the bonding between them, are all technologically dependent and vary with cleaning procedures, metal deposition conditions and so on [4.9]. The upshot of these remarks is that the true complexity of the Schottky diode is at last beginning to be understood in terms which matter to its detailed behavior.

In practice, in the forward direction the factor m may have values up to 2 and, in the reverse direction, the d.c. current does not equal the constant value I_0 that eqn (4.1) would imply for reverse bias greater than about 3 kT/q. The relevance of these non-idealities to microwave circuit applications of the Schottky diode can be illustrated with reference to Fig. 4.1, which shows a typically oversimplified band diagram of the Schottky contact including an interfacial region (which may be oxide or some other contaminant). As bias V is applied, the voltage drop V_i across the interfacial region changes, and so too does the voltage across the resistance of the bulk regions outside the space-charge layer. In other words, these regions absorb part of the externally applied voltage, thereby invalidating the assumption implicit in the simple derivations leading to an equation of the form of eqn (4.1) that all the applied potential V appears across the space-charge layer, i.e. in practice, the actual value of V (say V′) in eqn (4.1) is less than the applied value. Consequently, if eqn (4.1) is fitted to experimental data, m will turn out to have a value larger than unity (typically 1.1 to 1.2, although the value is not constant over the whole range of forward bias). Series voltage drops are thus one of the mechanisms which degrade the m value. Another is "parallel current flow", meaning any mechanism by which current carriers can be transported through the contact other than by emission over the barrier of height ($V_{BO}-V'$). Typical transitions are indicated on Fig. 4.1 as

(i) Capture of electrons by interface states, with subsequent re-emission into the interface and metal.

(ii) Capture by bulk deep levels with subsequent re-emission and transport through the interface.

(iii) Recombination, via bulk deep levels, in the space-charge region.

Considered by itself, for example, current transport entirely by process (iii) leads to an m value of 2. The overall value of m in a practical diode depends not only upon the mechanism involved but also upon the magnitude of the current contributed by that mechanism to the total diode current.

The importance of these questions to the microwave behavior of the device is that the processes which involve charge exchange with interface states and bulk deep levels, thereby contributing to interfacial voltage drop and parallel current non-idealities, operate on a scale of time which is long compared with a microwave signal period. For instance, the characteristic time for emptying of an interface state may be around 1 ms, whereas at 10 GHz the signal period is 100 ps. In contrast, the mechanisms of electron transport over the barrier (be they thermionic emission or diffusion related) are extremely fast, involving timescales of the order of the dielectric relaxation

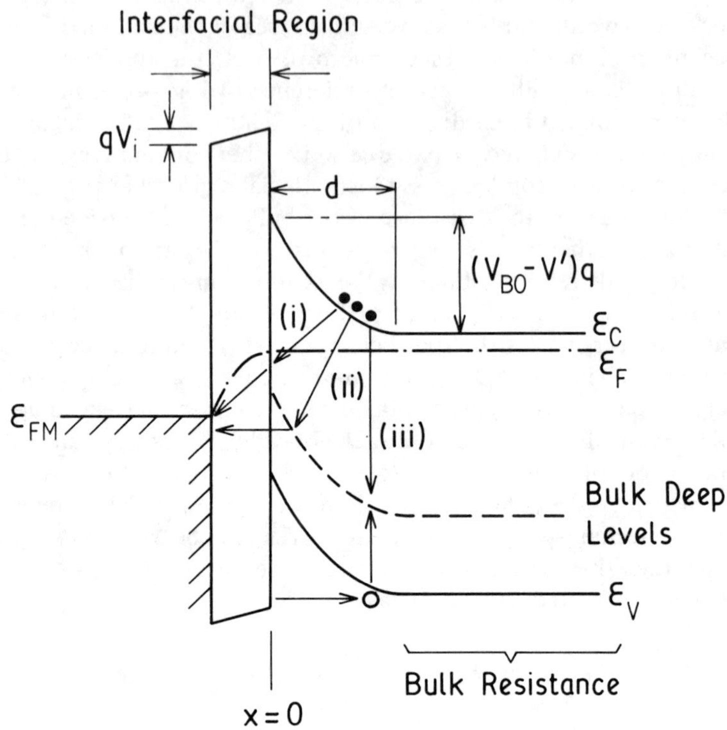

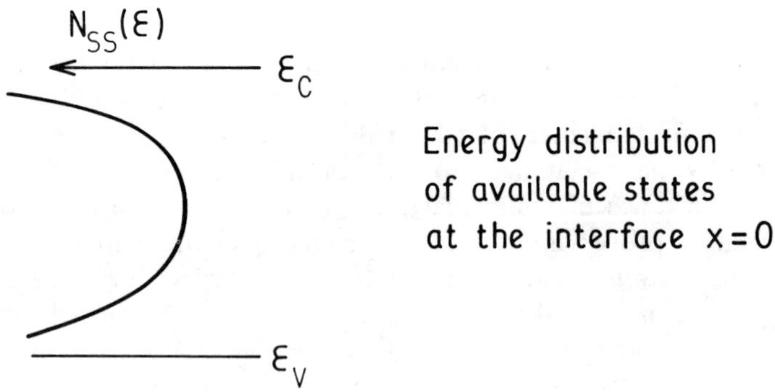

Fig. 4.1 Highly simplified energy-band diagram of a Schottky contact showing possible electron transitions (under forward bias) other than emission over the barrier.

time in the N-type semiconductor. Thus, although eqn (4.1) with m equal to, say, 1.6, may be a reasonable description of the diode's characteristics at dc, the RF behavior may be different, and eqn (4.1) with $m = 1.6$ may not apply at all.

A simple practical expedient to avoid any unpredictability and possibly poor rectifying characteristics under RF conditions, therefore, is to insist upon an m value reasonably close to unity (say $m \leqslant 1.2$). Mixer applications rely on the non-linearity expressed by eqn (4.1), and for this reason alone it is important that the diode characteristics are well defined and non-dispersive. Furthermore, the electrical noise generated by the diode increases if the sources of non-ideality are permitted to operate. Clean technological processing is required for atomically intimate contact between the Schottky metal and N-type GaAs, which should have bulk deep level concentrations at least one and a half orders of magnitude lower than the N-doping concentration. Satisfactory contacts can be produced by sintering the contact metal, e.g. aluminium, to break down the interfacial layer of oxides and other species adsorbed on the GaAs surface prior to metal deposition, or by using a reactive seed layer of metal, e.g. titanium.

Figure 4.2 shows a typical plan-view geometry of an integrated Schottky

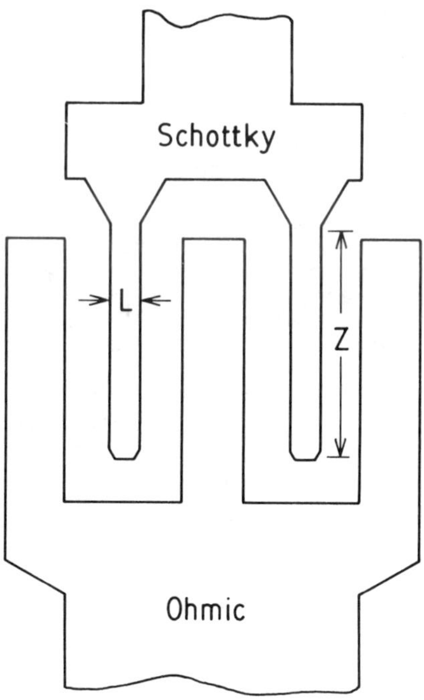

Fig. 4.2 Plan view of typical MMIC Schottky diode.

diode. Figure 4.3 is a cross-section through one of the diode fingers†. Between them, these two figures illustrate the following features:

(i) The narrow Schottky finger contacts, which reduce the series resistance associated with current flow in the layers in the x direction under the fingers, to and from the ohmic contacts.

(ii) The N^+ sub-layer, and the ohmic contacts surrounding the Schottky fingers; both are intended to reduce the series resistance. The N^+ may be 10^{18} cm^{-3} typically.

(iii) The Schottky contact, located in a recess etched in the GaAs N-layer to reduce the influence of surface charge upon the diode capacitance—an effect

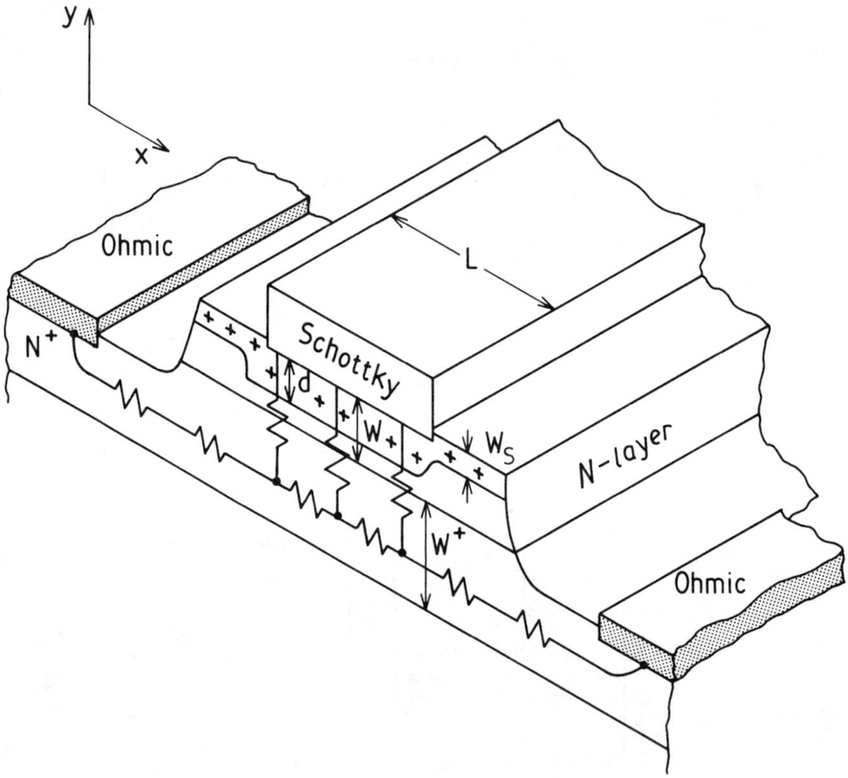

Fig. 4.3 Schematic through one finger of a Schottky diode.

†For consistency throughout the book, we adopt the same nomenclature as for FETs, namely the "long" dimension of a Schottky contact is denoted by the symbol Z, the "short" dimension by L, and the thickness of the N-layer by W (see Chapter 6).

which becomes more prevalent as the finger width L is decreased. Typically, $N \approx 10^{17}\ \mathrm{cm}^{-3}$.

(iv) The separation of the ohmic and Schottky contacts by a distance which is a compromise between minimizing series resistance (requiring narrow separation) and minimizing inter-electrode geometric capacitance (requiring wide separation).

If the finger dimension L is made too small, increases in resistance associated with the metallized finger itself outweigh the reductions in series resistance contributed by the layers ((i) above). Finger dimensions of $L \approx 2\ \mu\mathrm{m}$ and $Z \approx 40\ \mu\mathrm{m}$ represent a good compromise [4.1].

Assuming the depletion depth d remains large enough for the depletion approximation to be useful†, d is given by

$$d = \left[\frac{2\epsilon(V_{\mathrm{BO}} - V)}{qN}\right]^{1/2} \tag{4.2}$$

For an unbiased diode ($V=0$), assuming $N = 10^{17}\ \mathrm{cm}^{-3}$ and $V_{\mathrm{BO}} = 0.8$ V, $d = 0.1\ \mu\mathrm{m}$. Allowing for the contributions of the edge regions of the depletion layer in Fig. 4.3 to the space-charge capacitance, the equations of Chapter 6, eqn (6.13) and Table 6.1, may be used to show that the diode capacitance per finger is given by

$$C \approx \frac{\epsilon ZL}{d}\left(1 + \frac{2d}{L} - \frac{4d}{L+2d}\right) \tag{4.3}$$

Equation (4.3) assumes the Schottky is located in a recess of depth equal to the surface depletion depth W_{S} in Fig. 4.3. (It is worth remarking that the capacitance may be unpredictable if shallower recesses are used in the presence of a high occupancy of surface states.) Taking $L = 2\ \mu\mathrm{m}$, $Z = 40\ \mu\mathrm{m}$, $d = 0.1\ \mu\mathrm{m}$, eqn (4.3) gives

$$C \approx 0.08\ \mathrm{pF\ per\ finger}$$

If $W = 0.2\ \mu\mathrm{m}$, the series resistance contributed by the undepleted part of the N-layer in the shadow of the Schottky contact is, roughly‡,

†No carriers in the "tail" of the S-shaped carrier depletion edge should be in contact with the interface, requiring that d be at least three Debye lengths, L_{D}, where $L_{\mathrm{D}} = (\epsilon kT/2q^2N)^{1/2}$. Assuming $N = 10^{17}\ \mathrm{cm}^{-3}$, d should therefore be $\geqslant 300$ Å at room temperature.

‡The factor of 12 in the denominator of eqn (4.4) is composed of two separate factors, 4 and 3. The factor of 3 derives from the resistance of the layer in question being distributed across the depletion layer capacitance, so that RF current flows into the capacitance at all points distributed across the resistance. The phenomenon is exactly the same as that occurring in FETs (see text just preceding eqn (6.7). The factor of 4 derives from the Schottky finger having an ohmic contact on either side.

$$R_N \approx \frac{(W-d)}{12qN\mu_o LZ} \tag{4.4}$$

where μ_o is the low-field mobility of electrons in GaAs doped to the level N. Taking $\mu_o \approx 3000\ \text{cm}^2\ \text{Vs}^{-1}$,

$$R_N \approx 0.02\ \Omega$$

Depending upon the detailed values of N^+, W^+ and the Schottky-to-ohmic contact separation, current transport through the N^+ layer in the x-direction in Fig. 4.3 may contribute a comparable resistance. Further, if the contact resistance is $\rho_c\ \Omega$ mm, the contact resistance per finger is:

$$R_{\text{contact}} = \frac{1}{2}\frac{\rho_c}{Z} \tag{4.5}$$

(the factor of $\frac{1}{2}$ arises from having two ohmic feeds in parallel). For $\rho_c \approx 0.2\ \Omega$ mm and $Z = 40\ \mu\text{m}$;

$$R_{\text{contact}} \approx 2.5\ \Omega$$

Thus the semiconductor contribution to the series resistance might be around about 3 Ω per finger in total. The CR time constant is then typically $\leqslant 1$ ps. In physical terms, this means that the time taken for the depletion edge to readjust is intrinsically fast compared with a microwave signal period.

When the diode is biased into forward conduction, the capacitance C of the depletion layer is shunted by a resistance whose physical origin is the conduction of electrons through the space-charge region and over the barrier in Fig. 4.1. The incremental (or microwave) resistance may be obtained by differentiating eqn (4.1). When $\exp(qV/mkT) \gg 1$,

$$r \approx \frac{mkT}{qI} \tag{4.6}$$

I is limited to about 10^6 amp cm^{-2} in the metal Schottky finger so as to avoid electromigration [4.1]. Assuming a metal thickness of 0.5 μm, together with $L = 2\ \mu\text{m}$, we find $I \leqslant 10$ mA per finger, implying

$$r \geqslant 2.6\ \Omega \text{ per finger}$$

Thus for a diode consisting of two fingers of 2 μm × 40 μm as in Fig. 4.2, the total "on-state" resistance would be

$$r_{ON} \approx \tfrac{1}{2}(2.6 + 3 + 0.5) \approx 3\ \Omega$$

where we have estimated the microwave resistance of the Schottky metal fingers as 0.5 Ω.

In the "off"-state, a 10^{17} cm^{-3} layer of 0.2 μm thickness would become fully depleted for a reverse bias of approximately 2 V, and the space-charge capacitance would be (from eqn (4.3))

$$C \approx 0.04 \text{ pF per finger}$$

or $C_{OFF} \approx 0.1$ pF for a two-finger diode, including geometric capacitance.

At 10 GHz, the ratio of off-state capacitive reactance to on-state resistance is therefore

$$\frac{X_{OFF}}{r_{ON}} = \frac{1}{\omega C_{OFF} r_{ON}} \approx 50 \text{ at } 10 \text{ GHz}$$

This ratio is of importance in applications involving diodes in MMIC switches, where the concern is to have the smallest insertion loss in the "on"-state, and the greatest isolation in the "off"-state.

4.3 PIN diodes

In Fig. 4.4 is shown a typical PIN diode structure [4.10] as employed in MMICs. Although a lateral- or surface-oriented construction is possible, wherein both the anode and the cathode are located alongside one another in the surface plane, the carrier lifetime attainable in the *I* region is generally shorter than with the vertical structure shown. The disadvantage of the vertical arrangement is that a deep mesa technology has to be developed;

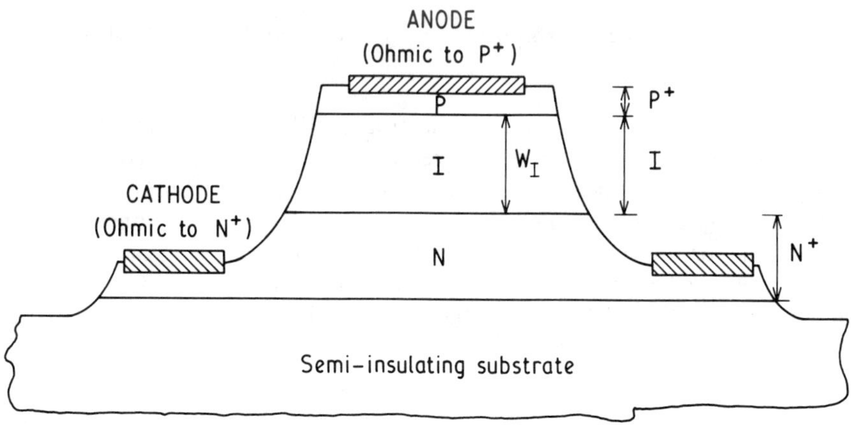

Fig. 4.4 Typical structure of a GaAs PIN diode. $P^+ > 10^{19}$ cm^{-3} 0.2 μm, $I \lesssim 10^{15}$ cm^{-3} N-type 2 μm, $N^+ \geqslant 10^{18}$ cm^{-3} 2 μm.

typically, 5 μm of material must be removed to expose semi-insulating GaAs over which microstriplines can be run. Access to the topmost contact, the anode, then requires a correspondingly deep airbridge technology to be developed (Fig. 4.5). Vapor Phase Epitaxy (VPE) may be used to grow the N^+ layer with sulphur as the dopant. Growth is continued with the dopant turned off to realize the intrinsic (I) layer, which grows unintentionally doped lightly N-type owing to the inability to purge the reactor of sulphur completely. The P^+ top layer may be realized either by ion implantation, or by diffusion of zinc from a compound source in a sealed ampoule. A plan view geometry very similar to that shown in Fig. 4.2 is used, except that the finger width would be perhaps 10 μm or greater.

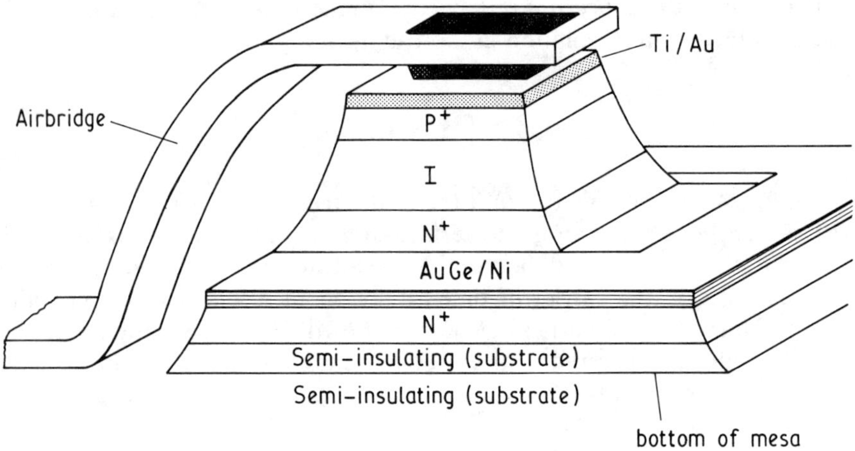

Fig. 4.5 Deep-mesa and airbridge technology required for MMIC PIN diodes.

There are several compromises affecting the PIN element. The purpose of the underlying N^+-layer is in part to reduce the series resistance of the device in the "on" state, but it cannot be so thick that a mesa technology becomes difficult to implement routinely. The doping level is limited by considerations of N-to-I layer interface sharpness and unintentional doping of the I-layer. The I-layer itself must be depleted at zero bias. If we anticipate a barrier voltage between the I-layer and the P^+-region of V_{BO} in equilibrium, the maximum I-layer thickness is governed by:

$$W_I \lesssim \left(\frac{2\epsilon V_{BO}}{qN_I}^{1/2}\right) \tag{4.7}$$

With $V_{BO} \lesssim 1\ V$ and $N_I \approx 10^{14}\ \text{cm}^{-3}$, $W_I \leqslant 4$ μm. W_I needs to be as large as possible so as to be able to sustain a high voltage in the reverse direction before breakdown, and to minimize the "off"-state capacitance. Conversely, because the "on" resistance of the I-region varies as W_I^2, low r_{ON} requires small W_I. Attainment of acceptable carrier lifetime in the I-region requires films of reasonable purity and freedom from defects. Values in the range $0.5 \leqslant \tau \leqslant 1.5$ ns are typical [4.10].

Because recombination is the dominant process, the forward $I(V)$ characteristic of a PIN diode is described by eqn (4.1) with $m=2$. In this "on" state, the resistance to current flow through the I-region is given approximately [4.11] by

$$r_{ON} \approx \frac{W_I^2}{(\mu_n + \mu_p) I \tau} \tag{4.8}$$

where I is the diode current, and τ is the carrier lifetime in the I-region. A simple plausibility derivation of this equation may be given as follows. Generally,

$$\text{Resistance} = \frac{\text{Resistivity} \times \text{Length}}{\text{Area}}$$

Given the PIN geometry,

$$\text{Resistance } r_{ON} = \frac{W_I}{q(n\mu_n + p\mu_p)LZ}$$

Assuming space-charge neutrality, so $p=n$,

$$r_{ON} = \frac{W_I}{q(\mu_n + \mu_p)nLZ}$$

If the carrier lifetime is τ, then $(nW_I LZ/\tau)$ represents the number of carriers which recombine per second in the I-region. Carriers to supply this recombination must flow in via the P^+ and N^+ regions, so the diode current is

$$I = \frac{qnW_I LZ}{\tau}$$

By substitution in the expression for r_{ON},

$$r_{ON} = \frac{W_I^2}{(\mu_n + \mu_p) I \tau}$$

The "off"-state capacitance associated with the I-layer may be estimated by regarding the P^+ and N^+ regions as two highly doped planes separated by W_I, so

$$C_{\mathrm{OFF}} \approx \frac{\epsilon ZL}{W_{\mathrm{I}}} \quad \text{per finger} \tag{4.9}$$

We now obtain, for the ratio of "off"-state reactance to "on"-state resistance,

$$\frac{X_{\mathrm{OFF}}}{r_{\mathrm{ON}}} \approx \frac{(\mu_{\mathrm{n}}+\mu_{\mathrm{p}})J\tau}{\omega\epsilon W_{\mathrm{I}}} \tag{4.10}$$

where J is the current density in the diode. J may be estimated crudely by assuming a near-enough uniform distribution of injected carriers in the intrinsic (I) region†, where all recombination is assumed to take place. Then

$$J \approx \frac{qnW_{\mathrm{I}}}{\tau} \tag{4.11}$$

where n is the density of injected carriers (the I region is assumed to remain space-charge neutral, so $n=p$).

Eliminating J from eqn (4.10) then yields

$$\frac{X_{\mathrm{OFF}}}{r_{\mathrm{ON}}} \approx \frac{q(\mu_{\mathrm{n}}+\mu_{\mathrm{p}})n}{\omega\epsilon} \tag{4.12}$$

This is slightly greater than the ratio of the dielectric relaxation frequency to the operating (signal) frequency; the factor which is favorable is the mobility μ_{n}, which is reduced by carrier–carrier scattering compared with the value in intrinsic material, but not by ionized-impurity and neutral-impurity scattering. If we take

$$\mu_{\mathrm{n}} = 6500\ \mathrm{cm^2\ Vs^{-1}} \qquad \mu_{\mathrm{p}} = 680\ \mathrm{cm^2\ Vs^{-1}}$$

$$n \approx 5 \times 10^{16}\ \mathrm{cm^{-3}}$$

then

$$\frac{X_{\mathrm{OFF}}}{r_{\mathrm{ON}}} \approx 800 \quad \text{at 10 GHz}$$

Substituting eqn (4.11) into eqn (4.8) gives

$$r_{\mathrm{ON}} \approx \frac{W_{\mathrm{I}}}{q(\mu_{\mathrm{n}}+\mu_{\mathrm{p}})nLZ} \tag{4.13}$$

which is the same conventional-looking expression for resistance that we had

†This is generally not a good assumption for GaAs where the ratio of hole-to-electron mobility requires a higher level of hole injection at the P^+ boundary to the I-region than electron injection at the N^+ boundary. The detailed shape of the electron and hole distributions depends upon τ, W_{I} and ω, as well as μ_{n} and μ_{p}.

previously. For $L = 15\ \mu m$ and $Z = 50\ \mu m$, for example, $r_{ON} \approx 0.5\ \Omega$. Resistance associated with the passage of current through the N^+ and P^+ layers, and through the ohmic contacts themselves, may increase the total series resistance in the "on" state to several ohms. For instance, if the series resistances amount to 2.5 Ω all told, then X_{OFF}/r_{ON} is reduced to about 160 at 10 GHz. The ratio is still very favorable, and a factor of several larger than can be achieved with FETs.

4.4 MMIC uses of PIN and Schottky diodes

Schottky diode-based MMICs have been very successfully realized in the form of mixer chips up to over 100 GHz [4.12]. A constant group delay phase-shifting chip has been implemented using an integrated array of Schottky diodes distributed along an artificial transmission line also integrated on the same chip [4.13] (see Section 2.4). Single pole, double throw (SPDT) MMIC switches that will handle 1 W with acceptable insertion loss and isolation over 2–18 GHz have been built [4.10].

An early example of a multifunction chip is the 31 GHz receiver described by Chu, Courtney and Sudbury [4.14]. The receiver chip supports a Schottky mixer and a FET-based IF amplifier. To provide the doping-thickness product required for the MESFET channel layers, the starting material was semi-insulating GaAs with epitaxial layers grown to FET specifications. The buried N^+ and N layers required for the Schottky mixer diodes were selectively grown by epitaxy after etching away the starting material at the diode sites to a depth equal to the combined thickness of the N^+ and N layers. After this second growth, therefore, a near-enough planar surface resulted.

There seems to be little doubt that the evolution of selection area growth and deposition techniques will continue, and that the integration of mixed functions will develop to higher levels of integration and to frequencies well into the millimetre wave bands.

References

4.1 Wilson, K. Other circuit elements for MMICs, *GEC Journal of Research (Special Issue on Monolithic Microwave Integrated Circuits)*, **4**(2), 126–133, 1986.

4.2 Gossick, B. R. *Potential Barriers in Semiconductors*, ch. 3, Academic Press, New York, 1964.

4.3 Ehrenberg, W. *Electric Conduction in Semiconductors and Metals*, pp. 294–304, Oxford University Press, London, 1958.

4.4 Padovani, F. A. and Stratton, R. Field and thermionic-field emission in Schottky barriers, *Solid State Electron.*, **9**, 695–707, July 1966.

4.5 Kittel, C. and Kroemer, H. *Thermal Physics (2nd edn)*, pp. 379–381, W. H. Freeman, San Francisco, 1980.
4.6 Stratton, R. Diffusion of hot and cold electrons in semiconductor barriers, *Phys. Rev.*, **126**(6) 2002–2014, 15 June 1962.
4.7 Moglestue, C. Monte Carlo Particle Modelling of Small Semiconductor Devices, *Computer Methods in Applied Mechanics and Engineering*, **30**(2), 173–208, May 1982.
4.8 Louie, S. G., Chelikowsky, J. R. and Cohen, M. L. Theory of semiconductor surface states and metal–semiconductor interfaces, *J. Vac. Sci. Technol.*, **13**(4), 790–797, July/August 1976.
4.9 Williams, R. H. Physics of heterojunctions, *Advanced Seminar on Phys. Technol. III–V Heterojunction Devices*, Cardiff, 24–25 March 1988 (proceedings unpublished).
4.10 Thomas, L., Hing, A., Hughes, E., Beckerson, J. and Wilson, K. GaAs MMIC broadband SPDT PIN switch, *Electron. Lett.*, **22**(22), 1183–1185, 23 October 1986.
4.11 Curran, J. E. The passage of high frequency currents through a forward-biased PIN diode—(a), recombination in the I-Region, *GEC Res. Lab., Hirst Res Centre, Microwave Division 286 Memorandum*, 1 March 1983.
4.12 Chu, A., Courtney, W. E., Mahoney, L. J. and Clifton, B. J. Millimetre-wave monolithic circuits for receiver and transmitter applications, *RCA Review*, **46**(4), 421–430, December 1985.
4.13 Wilson, K., Nicholas, J. M. C., McDermott, G. and Burns, J. W. A novel MMIC *X*-band phase shifter, *IEEE Trans. Microwave Theory Tech.*, **MTT-33**(12), 1572–1578, December 1985.
4.14 Chu, A., Courtney, W. E. and Sudbury, R. W. A 31-GHz monolithic GaAs mixer/preamplifier circuit for receiver applications, *IEEE Trans. Electron. Devices*, **ED-28**(2), 149–154, February 1981.

Chapter 5

Elementary FET Principles

5.1 Introduction

To those unfamiliar with the GaAs MESFET who are perhaps looking at a cross-sectional diagram of the device for the first time, it might appear that the structure bears a great deal of similarity to the silicon junction FET (Si JFET) treated in so many books at various levels of sophistication. Perhaps the greatest mystery might be what the acronym "MESFET" stands for†. For some reason, the fact that there is an important difference of principle between the way these two devices operate escapes many people for too long. If familiarity came with the Si JFET first, the assumption seems to be that the GaAs MESFET must function in much the same way.

A clear understanding of the principles of the Si JFET is available from such excellent texts as Grove [5.1] and Feldman [5.2]. With this background, a technologist beginning work with GaAs FET circuits might pick up the book by Williams [5.3]. A hybrid microwave integrated circuit designer might consult Hoffmann [5.4], Vendelin [5.5] or Pengelly [5.6]. A device engineer might read Pengelly [5.6], Soares *et al.* [5.7] or Sze [5.8]. All are good examples from the range of texts available. Not unexpectedly, there is a difference in the level of treatment of the GaAs MESFET in these works, and there is a difference also in the clarity of presentation.

There is evidently still room for a clear, stripped down, statement of the basis of the GaAs MESFET as contrasted with the Si JFET, or at least there is room enough to justify the attempt. Our purpose will be to uncover the fundamental issue by retaining only the underlying physical principles that are essential to establishing the central difference in the operation of the two devices. That central difference can be summarized as follows.

†"MESFET" is derived from the structure of the gate electrode, which is a metal-semiconductor Schottky barrier contact. Thus MESFET stands for MEtal-Semiconductor Field-Effect Transistor.

In the long-gate Si JFET, current saturation occurs due to pinch-off of the channel at the drain-end of the gate, whereas in the short-gate GaAs MESFET current saturation occurs due to saturation of the velocity at which electrons drift in the channel, also near the drain-end of the gate.

Rather than deal with the GaAs MESFET alone, we shall deliberately highlight the differences between Si JFET and GaAs MESFET operation to provide continuity for those familiar with the background of the Si JFET. We start with a review of this device.

5.2 Review of Si JFET operation

Figure 5.1 is a schematic diagram of a Si JFET in cross-section. There is an N-layer, typically doped to approximately 10^{16} cm^{-3}, about 1 μm thick. We assume this layer to be uniformly doped, grown by epitaxy rather than being diffused, and we neglect any depletion at the junction of the N-layer with the P-substrate: we take this interface to be a "hard wall". A heavily doped P^+ layer, maybe having a concentration of $\approx 10^{18}$ cm^{-3}, has been diffused to a depth of typically 0.3 μm through a window opened in the silicon dioxide (SiO_2) grown on the N-layer. The depth W of N-layer remaining is then 0.7 μm. When subsequently metallized with aluminium, the P^+ diffused region forms the gate contact of what has now become a P^+-N junction to the N-channel layer. Typically, the dimension L_G, which is the gate length†, is around 10 μm. The source (S) and drain (D) are formed similarly, but this time by diffusing N^+ wells which, when metallized, form ohmic contacts to the N-channel.

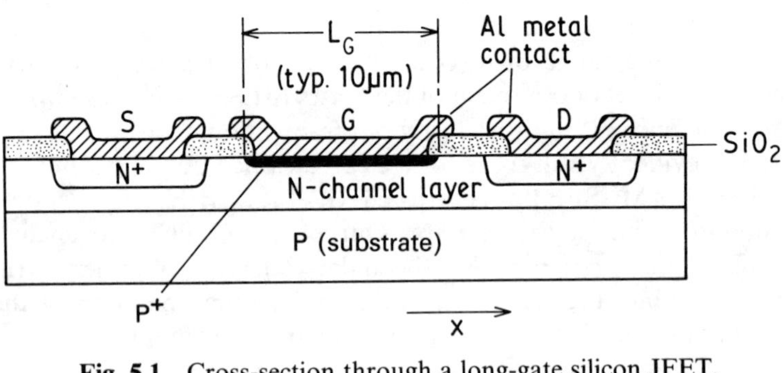

Fig. 5.1 Cross-section through a long-gate silicon JFET.

†We follow the usual convention that the gate length is the dimension of the gate electrode in the direction of electron flow.

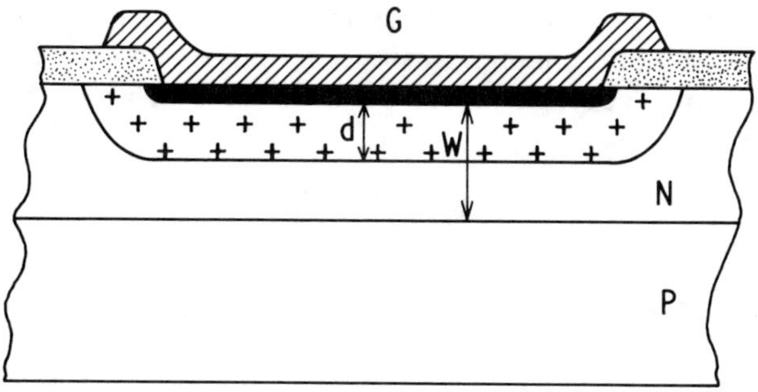

Fig. 5.2 Gate depletion layer in N-channel of silicon JFET with zero drain-source bias.

Because the junction is P^+–N, almost all the junction potential is dropped across a depletion layer which forms in the N-layer (Fig. 5.2). For the doping concentrations indicated, this amounts to approximately 0.8 V [5.1, pp 157–158]. This "built-in" potential we shall call V_{BO}. If the source and drain are connected together and grounded, and a negative voltage is applied to the gate, say $V_{GS} = -1$ V, the depletion layer expands. The total potential change in going from the top of the depletion layer to the bottom, that is from the P^+–N interface to the edge at depth d in Fig. 5.2, is now

$$V_{BO} - V_{GS} = 0.8\text{ V} - (-1\text{ V}) = 1.8\text{ V}$$

Henceforth†, to save carrying unnecessary minus signs, we shall refer to the voltage applied externally between the source and gate by specifying V_{SG}, i.e. the potential of the source relative to the gate. For the example at hand, $V_{SG} = +1$ V. The voltage supported by the depletion layer is now written

$$V_{BO} + V_{SG} = 0.8\text{ V} + 1\text{ V} = 1.8\text{ V}$$

By integrating Poisson's equation twice, the potential across the depletion is given in terms of the material properties and the depletion depth d by the equation

$$V_{\text{across depletion}} = \frac{qNd^2}{2\epsilon} \tag{5.1}$$

where ϵ is the permittivity of Si ($\epsilon = 11.7\,\epsilon_0$). In the present case, $V_{\text{across depletion}} = (V_{BO} + V_{SG})$.

†This convention will be followed throughout the remainder of the book, with reminders from time to time that $V_{SG} = -V_{GS}$.

If a large enough external voltage is applied to deplete the N-channel layer fully, so $d = W$, the channel is said to be "pinched off". The value of V_{SG} required to bring about pinch off is called the pinch-off voltage, or the threshold voltage. Thus

$$V_P = \frac{qNW^2}{2\epsilon} - V_{BO} \tag{5.2}$$

For the device structure under consideration, $N = 10^{16}$ cm^{-3}, $W = 0.7$ µm and so

$$V_P = 3 \text{ V}$$

When the pinch-off voltage is applied between the gate and the source, there is no conducting path any longer between the source and drain, and so we expect no current to flow between S and D in this limit; the whole of the conducting channel has disappeared under the entire length of the gate.

In normal operation the source is grounded and a positive voltage is applied to the drain contact. Except when the channel is pinched off, a current of electrons then passes down the N-layer from the source, S, to the drain, D, in Fig. 5.1. In the device physicist's terms, the electrons drift in the longitudinal (i.e. x-directed) electric field that the drain-source voltage V_{DS} establishes. It is important that we understand what is meant by "drift", and that we fully appreciate the form that the drift velocity versus electric field characteristic takes for Si.

The $v(E)$ curve for Si is shown in Fig. 5.3. When an electric field E is applied to the semiconductor, the electrons experience a force $F = -qE$, and so acquire a component of momentum in the field direction (actually, exactly anti-parallel to the field as expressed by the minus sign). Electrons, however, undergo collisions with the crystal lattice of Si atoms, which are continually in motion due to the thermal energy they possess.

The displacement of lattice atoms about their mean positions occurs in a wave-like manner with quantized states, giving rise to lattice waves referred to as phonons. The result of an electron–phonon collision generally is that the component of momentum in the field direction that the electron has acquired is partly destroyed; the momentum tends to be randomized into arbitrary directions. We thus finish up with the idea that the *total* electron velocity can be, or will be, somewhat larger than its component of velocity along the field direction.

The drift velocity, v, is obtained by finding the average, over all mobile electrons, of the velocity component along the field direction. Referring to Fig. 5.3, at relatively low values of field E, the average component of the electrons' velocity along the field direction increases proportionally to E. The drift mobility, μ_0, is defined as the constant of proportionality, thus

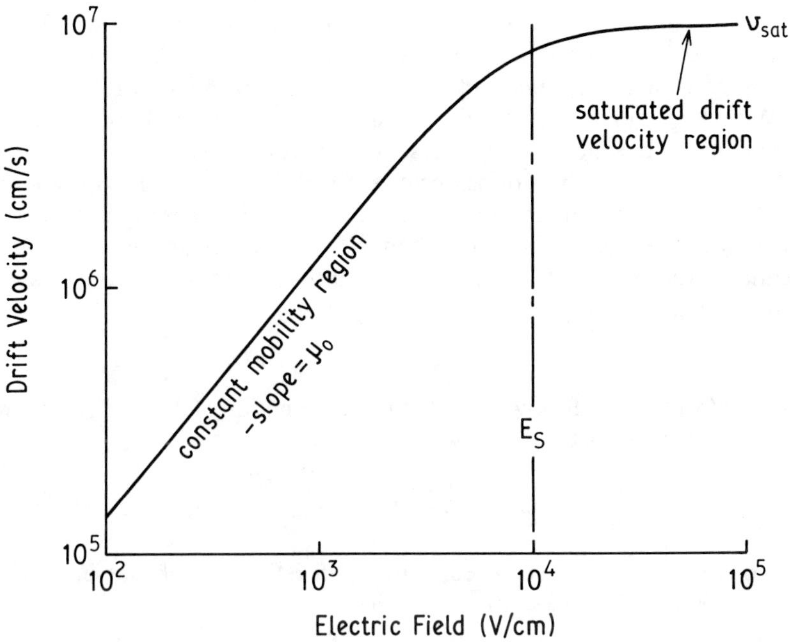

Fig. 5.3 Drift velocity versus electric field for silicon of low N-type doping.

$$v = \mu_0 E \tag{5.3}$$

At high enough field strengths, $E \geqslant 10\ \text{kV cm}^{-1}$ in Si, the drift velocity saturates at a value of about $10^7\ \text{cm s}^{-1}$. We take $E_s \approx 10\ \text{kV cm}^{-1}$ as the boundary point between the constant drift mobility regime and the saturated drift velocity regime.

We now consider the likelihood that the electric field strength in the x direction in the FET channel of Fig. 5.1 could become large enough to cause the electrons moving down the channel to reach v_{sat}. Assuming somewhat simplistically that the field under the gate is uniform along x, the voltage drop in moving from the left end of the gate to the right end, a distance of typically 10 μm, would be

$$V_{DS} = -\int_0^{L_G} E_x\, dx \approx E_x L_G$$

Since E_x must have reached the value E_s for the electrons to travel at the saturated drift velocity, the drain-source voltage required is

$$V_{DS} \geqslant 10\ \text{kV cm} \times 10 \times 10^{-4}\ \text{cm} = 10\ \text{V}$$

Thus, if we only ever operate the transistor at drain voltages up to say, 5 V, the electrons in a Si JFET always move in the constant mobility regime in Fig. 5.3. Given this understanding, we can now proceed with a low-level derivation of the current-voltage (I_D versus V_{DS}) characteristics of the Si JFET.

With V_{DS} applied, there is progressive potential drop along the channel between the source and drain; the closer to the drain an electron gets, the more positive the potential becomes. Because the P^+ region is an equipotential, however, the potential across the depletion layer becomes progressively greater with distance along the channel towards the drain. Calling the potential in the channel $V(x)$, the total potential across the depletion layer at any point x is, from Fig. 5.4,

$$V_{\text{across depletion}} = V_{BO} + V_{SG} + V(x)$$

From eqn (5.1), it then follows that d must be a function of x as well, such that d also increases with distance x towards the drain:

$$V_{BO} + V_{SG} + V(x) = \frac{qNd^2(x)}{2\epsilon} \tag{5.4}$$

The depletion layer is thus tapered, as in Fig. 5.4. (We are, of course, assuming $V_{SG} < V_P$). Figure 5.4 also introduces the channel opening $b(x) = W - d(x)$. Eliminating (q/ϵ), N, etc. from eqn (5.4) using eqn (5.2) then allows us to write for $b(x)$:

$$b(x) = W\left[1 - \left(\frac{V_{BO} + V_{SG} + V(x)}{V_P + V_{BO}}\right)^{1/2}\right] \tag{5.5}$$

Considering an element of the channel of extent Δx (Fig. 5.5), the shaded region has a resistance of

$$\Delta R = \frac{\Delta x}{b(x) \times qN\mu_0 \times Z_G} \tag{5.6}$$

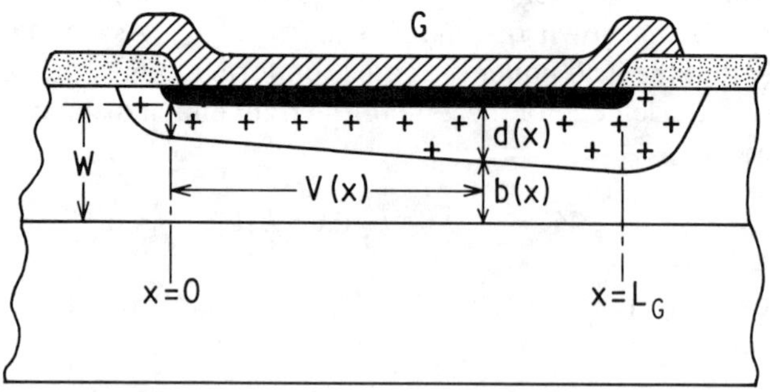

Fig. 5.4 Channel conditions in silicon JFET with positive drain-source bias. IMPORTANT NOTE: Potential across depletion layer at $x = 0$ is $V_{BO} + V_{SG}$.

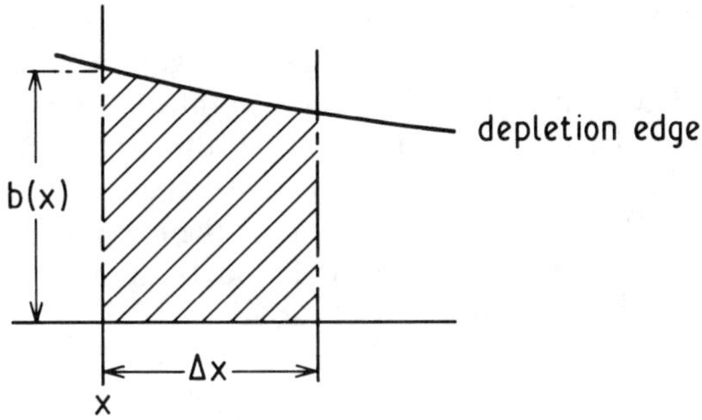

Fig. 5.5 Element of conducting channel for the gradual channel approximation.

where Z_G is the width of the gate, i.e. the dimension into the page in Fig. 5.1.

The current through the shaded element of channel passes into the next element, then the next, and so on until it arrives at the drain. Making the very reasonable assumption that, in the steady state, no electrons are either lost or gained along the way, it follows that the current passing through the element of channel under consideration is exactly equal to the drain current. Thus, if $\Delta V(x)$ is the potential drop in the x direction across the element,

$$I_D = \frac{\Delta V(x)}{\Delta R} = \frac{\Delta V(x) \times b(x) \times qN\mu_0 \times Z_G}{\Delta x} \tag{5.7}$$

Taking the limit as Δ tends to zero, and eliminating $b(x)$ using eqn (5.5),

$$I_D \, dx = qN\mu_0 W Z_G \left[1 - \left(\frac{V_{BO} + V_{SG} + V(x)}{V_P + V_{BO}} \right)^{1/2} \right] dV(x) \tag{5.8}$$

Recognizing that

$$\text{at } x = 0 \qquad V(x) = 0$$

$$\text{at } x = L_G \qquad V(x) = V_{DS}$$

eqn (5.8) can be integrated between limits $x=0$ and $x=L_G$ on the left side of the equation, and between limits $V=0$ and $V=V_{DS}$ on the right side. In performing this integration, we recall also that current continuity requires I_D to be independent of x. After integration, we find

$$I_D = \frac{qN\mu_0 W Z_G}{L_G} \left\{ V_{DS} - \frac{2}{3}(V_P + V_{BO}) \left[\frac{V_{BO} + V_{SG} + V_{DS}}{V_P + V_{BO}} \right]^{3/2} + \frac{2}{3}(V_P + V_{BO}) \left[\frac{V_{BO} + V_{SG}}{V_P + V_{BO}} \right]^{3/2} \right\} \tag{5.9}$$

This is the equation which defines the Si JFET relationship between drain current, I_D, drain-source voltage, V_{DS}, and source-gate voltage V_{SG}. Its limit of validity is when $b(x)$ vanishes at the widest point of the depletion layer, i.e. when the channel becomes pinched off at the drain end of the gate. Equation (5.9) is plotted over its range of validity as the solid part of the curves in Fig. 5.6. Table 5.1 summarizes the data for the example FET.

Because the voltage across the depletion layer at $x=L_G$ is equal to $(V_{SG}+V_{DS})$, the range of validity is

$$V_{SG}+V_{DS}\leqslant V_P$$

or

$$V_{DS}\leqslant V_P-V_{SG} \tag{5.10}$$

The locus of points for which the equality in eqn (5.10) applies is shown by the dotted line in Fig. 5.6. Beyond it, the current remains at a near-enough constant value. Initially, the reason for this behavior need not concern us; the

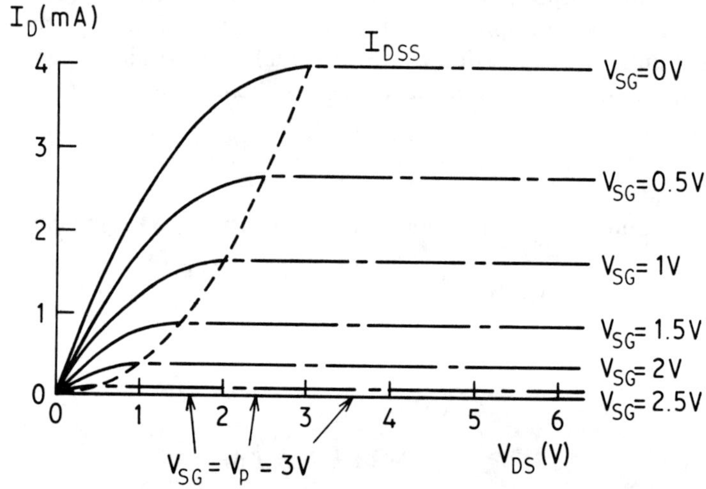

Fig. 5.6 Calculated drain characteristics of long-gate silicon JFET according to the gradual-channel approximation.

Table 5.1 Data for example Si JFET

$N=10^{16}$ cm^{-3}	$L_G=10$ μm
$\mu_0=1000$ cm^2 Vs^{-1}	$Z_G=500$ μm
$W=0.7$ μm	$V_{BO}=0.8$ V
$V_P=3$ V	

important point we have established is that current saturation in the Si JFET occurs because of pinch-off at the drain-end of the gate and that, at least up until this point, the device was operating well down in the constant mobility part of the velocity–field characteristic. Thus velocity saturation played no part in bringing about the onset of current saturation.

As we shall see in the following Section, exactly the opposite is true of the microwave GaAs FET.

5.3 Current saturation in the GaAs MESFET

A schematic diagram of a GaAs MESFET in cross-section is given in Fig. 5.7. As the device is normally used to amplify or switch signals in the microwave frequency range (several GHz upwards), the gate length has to be short: $L_G \leqslant 1$ μm is required. The gate itself is a rectifying metal–semiconductor contact in the form of a Schottky barrier. Ordinarily, the Schottky contact is reverse biased in FET operation, or has zero bias applied in which case the gate is connected to the source via a short-circuit. Consequently, the Schottky gate behaves in a manner entirely similar to the Si JFET gate. Also as in the Si JFET, the source S and drain D are ohmic contacts to the N-channel layer. The N-layer might be doped in the range 10^{17} cm^{-3} to 10^{18} cm^{-3} with a thickness under the gate of less than one-tenth of a micron, to two- or three-tenths of a micron. We shall take typical values of 3×10^{17} cm^{-3} and 0.15 μm.

When a voltage is applied between the source and drain, with the drain positive with respect to the source, an x-directed or longitudinal electric field is

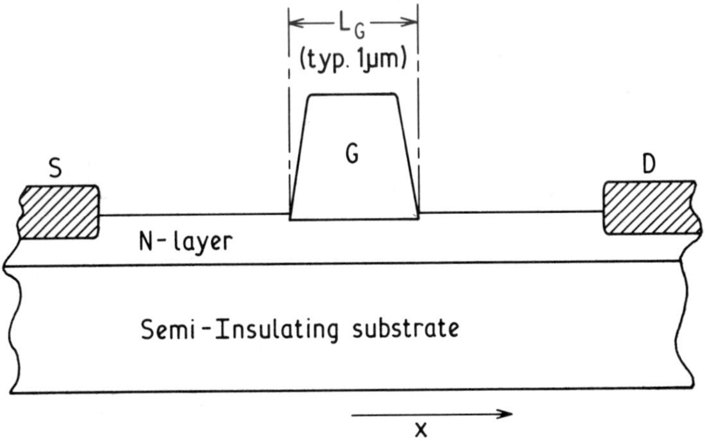

Fig. 5.7 Schematic of GaAs MESFET with short gate for microwave operation.

created in the channel which causes electrons to drift down the channel, from the source to the drain. The velocity-field characteristic for GaAs is sketched in Fig. 5.8. The peak of about 2×10^7 cm s^{-1}, and the region of negative differential conductivity which follows it, apply only in very lightly doped material. At the doping concentrations used in microwave MESFETs, ionized impurity scattering at low fields gives rise to an increase in the rate of electron momentum randomization without, however, affecting the total energy of the electrons. Momentum randomization redirects the component of electron momentum or velocity that was in the field direction, so the x-directed velocity at low fields is reduced by ionized impurity scattering. Put another way, the low-field drift mobility, μ_0, is decreased. The fact that the total electron energy is unchanged by ionized impurity scattering means, however, that electron transfer from the lower (Γ) conduction-band valley to the next higher set of valleys (the L valleys) still occurs at roughly the same field strength as previously, E_S on Fig. 5.8. Because the Γ- and L-valleys are separated in energy, it is merely necessary for the electrons to possess the necessary energy to transfer between valleys, and the direction of their velocity is unimportant to first order. A low-field mobility of μ_0 in the range 2500 cm^2 V s^{-1} to 3000 cm^2 V s^{-1} leads to the constant mobility part of the $v(E)$ curve intersecting the E_S ordinate at approximately 10^7 cm s^{-1}. In other words, the peak drift velocity is roughly v_{sat}, and the velocity-field characteristic in highly doped GaAs looks just like the velocity–field curve for Si with one exception: E_S is now 3.5–4 kV cm^{-1}, not 10 kV cm^{-1}. We require to examine the consequences of this lower value of E_S and the shorter gate length combined.

In Fig. 5.9 we have drawn the gate region showing the depletion layer.

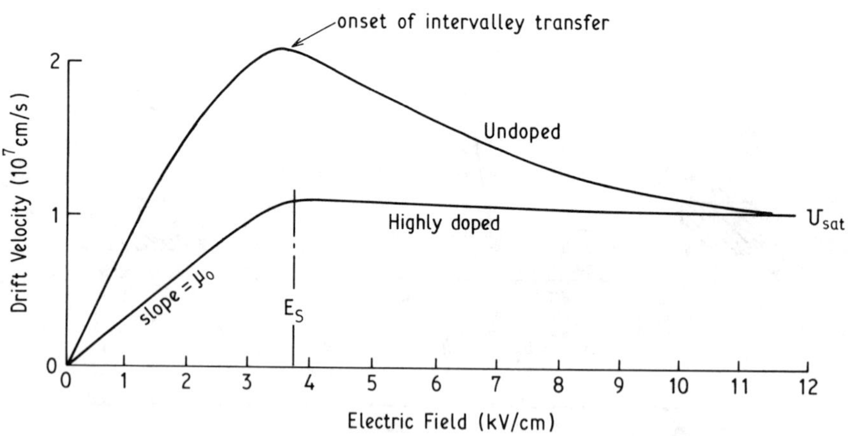

Fig. 5.8 Drift velocity versus electric field for undoped GaAs, contrasted with case for typical highly-doped MESFET layer.

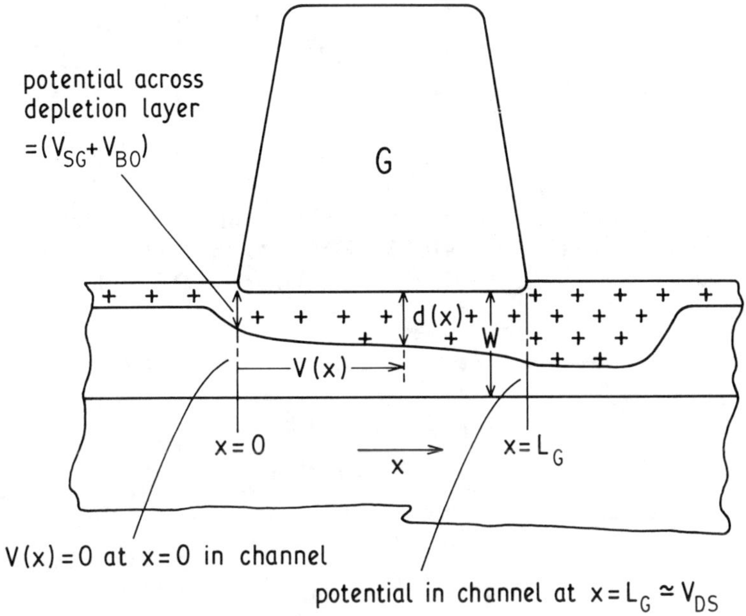

Fig. 5.9 GaAs MESFET in cross-section showing approximate shape of depletion layer with normal operating bias applied.

Again assuming that the x-directed field E_x is initially uniform, and neglecting any potential drop in the regions between the source and the gate, and between the gate and the drain, we once more have.

$$V_{DS} \approx E_x L_G$$

As V_{DS} is gradually increased from 0 V, E_x will increase in value until the field E_S is reached. For fields greater than E_S, the electrons in the channel will move at their saturated drift velocity, and the drain current will saturate as a direct result. The drain-source voltage required for the onset of current saturation is

$$V_{DS} \approx E_S L_G \approx 4 \text{ kV cm}^{-1} \times 1 \text{ μm} = 0.4 \text{ V} \tag{5.11}$$

(This contrasts with the 10 V typically required in the Si JFET.) We now compare this voltage with that required to pinch off the channel, just as we did for the Si JFET.

Equation (5.2) is as valid for calculating the pinch-off voltage of a GaAs MESFET as it was for the Si JFET. For GaAs, $\epsilon \approx 13\,\epsilon_0$ and $V_{BO} = 0.8$ V, where V_{BO} is the height of the Schottky barrier with no external voltage applied. For our typical MESFET,

$$V_P = \frac{1}{2} \times \frac{1.6 \times 10^{-19}\text{C} \times 3 \times 10^{23}\ \text{m}^{-3} \times (0.15 \times 10^{-6})^2\ \text{m}^2}{1.15 \times 10^{-10}\text{CV}^{-1}\ \text{m}^{-1}} - 0.8$$

i.e.

$$V_P = 3.9\ \text{V} \tag{5.12}$$

The earlier result in eqn (5.11) means that the taper in the space charge layer at the onset of velocity saturation is 0.4 V. If at $x = L_G$ the depletion just pinched off the channel, so $d(x = L_G) = W$, the potential across the depletion layer at $x = 0$ would be

$$V_P + V_{BO} - 0.4\ \text{V}$$

Since from Fig. 5.9 we are reminded that, at $x = 0$,

$$V_{\text{across depletion}} = V_{SG} + V_{BO}$$

we have

$$V_{SG} + V_{BO} = V_P + V_{BO} - 0.4\ \text{V}$$

V_{BO} cancels in this equation, so given $V_P = 3.9$ V,

$$V_{SG} = V_P - 0.4\ \text{V} = 3.5\ \text{V}$$

Consequently, providing we apply a source-gate voltage V_{SG} of less than 3.5 V, the channel at the drain-end of the gate will not be pinched off when velocity saturation sets in at a drain-source voltage of $V_D = 0.4$ V. We thus arrive at the following very important conclusion.

> In the GaAs MESFET, over most of the range of gate voltages, current saturation occurs due to saturation of the electron drift velocity at or near the drain-end of the gate. The only exception is as pinch off is approached.

To construct an approximate set of $I_D(V_{DS})$ characteristics for the GaAs MESFET, we calculate the drain current using eqn (5.9) over its range of validity (i.e. below velocity saturation). This range is specified by the conditions

$$V_{SG} \geqslant 3.5\ \text{V} \quad \text{for all } V_{DS}$$

$$V_{DS} \leqslant 0.4\ \text{V} \quad \text{for } V_{SG} < 3.5\ \text{V}$$

For $V_{DS} > 0.4$ V when $V_{SG} < 3.5$ V, the current saturates due to electron drift velocity saturation in the channel. Because the undepleted part of the channel is tapered (Fig. 5.9), being slightly narrower at $x=L_G$, the electric field in the channel, E_x, is slightly larger at this point. Consequently, velocity saturation occurs first at the drain-end of the gate.

The approximate $I_D(V_{DS})$ characteristics for the example GaAs FET data of Table 5.2 are shown in Fig. 5.10. For each V_{SG}, the two regions above and below $V_{DS}=0.4$ V have been connected with a "hand-smoothed" curve to give a reasonable transition†.

Table 5.2 Example data for GaAs MESFET

$N=3\times10^{17}$ cm^{-3}	$L_G=1$ μm
$W=0.15$ μm	$Z_G=300$ μm
$\mu_0=2500$ cm^2 Vs^{-1}	$V_{BO}=0.8$ V
$v_{sat}=10^7$ cm s^{-1}	$V_P=3.9$ V

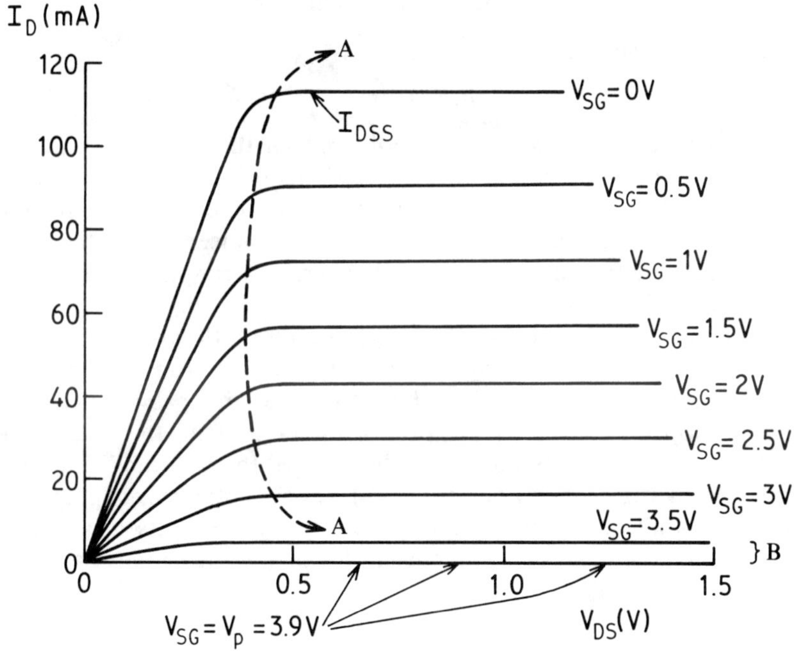

Fig. 5.10 Calculated drain characteristics of short gate GaAs FET; A: region where current saturation is determined by velocity saturation; B: region where current saturation is governed by pinch-off at drain-end of gate.

†In practice, the $v(E)$ curve does not have a sharp knee at $E=E_S$. The mobility gradually decreases as E increases towards E_S in Fig. 5.8 (the highly doped curve) and so, from eqn (5.9), we could expect I_D to make a smooth transition to its saturated value.

Referring to Fig. 5.10, in the whole region of current saturation where $0 \leqslant V_{SG} < 3.5$ V, the drain current is greater than the value at $V_{SG} = 3.5$ V by the amount

$$I_D = qNv_{sat}Z_G[W - d(L_G)] \tag{5.13}$$

The term in square brackets is the channel opening at the drain-end of the channel, $x = L_G$. We can find $d(L_G)$ by rearranging eqn (5.4), and recognizing that $V(L_G) = V_{DS} = 0.4$ V at the onset of current saturation, thus,

$$d(L_G) = \left[\frac{2\epsilon(V_{BO} + V_{SG} + 0.4)}{qN}\right]^{1/2} \tag{5.14}$$

To illustrate the point, consider the case for $V_{SG} = 0$ V. Using the data of Table 5.2, eqn (5.14) gives, for $d(L_G)$,

$$d(L_G) = 0.076\ \mu\text{m}$$

From eqn (5.13) we calculate

$$I_D = 107\ \text{mA}$$

Adding this value to the value of $I_D = 4.7$ mA in the saturated region for $V_{SG} = 3.5$ V gives a total drain current of 112 mA, which is near enough the value evident for I_{DSS} in Fig. 5.10. Similar numerical checks may be made for other values of V_{SG} to illustrate that eqn (5.13) does indeed specify the current increment over the value $I_D(V_{SG} = 3.5\ \text{V})$. There is a simple physical reason why this should be so.

In eqn (5.13), the factor $Z_G[W - d(L_G)]$ represents the area of the channel opening at the drain-end of the gate through which electrons may pass. The factor Nv_{sat} represents the number of electrons per cm^2 passing per second through this opening. Thus $Nv_{sat}Z_G[W - d(L_G)]$ represents the total number of electrons per second passing out of the channel to the drain. Multiplying this number by the charge q per electron gives the total charge passing per second to the drain, i.e. the drain current. Equation (5.13) by itself gives a close enough estimate of the total drain current for most practical purposes.

5.4 Mechanism of current saturation—summary

From Sections 5.2 and 5.3, the following conclusions can be drawn. Given that E_S is the electric field required for saturation of the electron drift velocity, L_G is the FET gate length, and V_P is the pinch-off voltage,

(a) If $E_S L_G > V_P$, current saturation occurs due to pinch-off; the classical Si JFET is an example.
(b) If $E_S L_G < V_P$, current saturation occurs due to velocity saturation; the microwave GaAs FET is an example.

In these criteria, the inequalities are expressed somewhat loosely. They should be taken to mean strong, not marginal, inequalities. E_S is the value of electric field required for velocity saturation before the formation of transverse fields.

A judgment has to be made as to the mechanism of current saturation for each device encountered; all we have done in the foregoing is to draw a distinction with a classical and widely quoted device. Clearly, a Si MESFET could be configured with a sub-micron gate so that current saturation in the device occurred by velocity saturation. Equally, a GaAs FET could be constructed with a gate many tens of microns long so that current saturation in this particular transistor would occur by pinch-off. The fact remains, however, that the microwave GaAs MESFET we are dealing with exhibits velocity saturation in the channel over virtually the entire operational range of interest.

5.5 Essential enhancements

Quite deliberately, the foregoing analyses of the Si JFET and the GaAs MESFET were reduced to the bare physical minimum necessary to carry the argument. Having used those analyses for the purpose for which they were designed, it is important that we at least enumerate the factors that were omitted.

5.5.1 Si JFET

The principal omission was any discussion of the physical mechanism responsible for current saturation when the channel becomes pinched off (the regions indicated by the broken line sections in Fig. 5.6). After all, if the depletion region has reached the bottom of the N-layer, wouldn't it be reasonable to expect the current to fall to zero on the grounds that there are no carriers in the vicinity of $x \approx L_G$ to carry the current? The answer is that it would not, in fact, be reasonable. The idea that the current should fall to zero stems from continuing with the assumption of electron transport by the mechanism of low-field conduction, thereby ignoring other important mechanisms of electron transport, notably the transport of electrons through a high-field region by the process of injection.

Once pinch-off has been attained at the drain-end of the gate (i.e. at $x \approx L_G$), the longitudinal electric field, E_x, becomes highly non-uniform. E_x rises steeply in the channel as x approaches L_G, and falls also quite steeply beyond $x \approx L_G$, thus giving rise to a field dependence upon distance x which is roughly triangular. In this high-field region, the field direction is such as to pull the electrons out of the conducting channel region; for every electron extracted from the right-hand end of the undepleted channel by this mechanism, another carrier flows in at the left-hand end, thereby preserving space-charge neutrality in the conducting channel. Carriers are thus injected into the high-field region, the supply of electrons being limited by the conductance (or resistance) of the undepleted region occupying the region of channel in $0 \leqslant x \leqslant L_G$. We therefore come to the idea that there are indeed electrons passing through the region we had hitherto argued was a region depleted of electrons. There is no paradox; the conclusion that the region should be depleted was based on quasi-equilibrium principles which ignore injection and other non-equilibrium transport mechanisms.

The evolution of the high-field region gives rise to field strengths in the neighborhood of the drain end of the gate well in excess of $E_S \approx 10$ kV cm^{-1}. In other words, the electrons can now reach their saturated drift velocity in the high-field region. The fact that the electrons in the high-field region may travel at v_{sat} is a conclusion which follows only *after* current saturation has set in; drift velocity saturation still plays no part in establishing current saturation in the first place. The changing spatial extent of the high-field region with increasing drain-source bias gives rise to a slight upward slope to the broken curve segments in Fig. 5.6, i.e. it gives rise to a finite value of drain conductance, alternatively termed output conductance. An approximate and very readable treatment is given in [5.2], pp 491–493.

5.5.2 GaAs MESFET

The omissions from the MESFET analysis were rather more serious than the Si JFET. They mainly center on the consequences of electrons becoming hot in the strong two-dimensional fields that occur in GaAs FETs (see Appendix 5A for the background to the term "hot electron").

In the GaAs FET, a high-field region forms in the vicinity of the drain end of the gate when velocity saturation occurs. Looking forward a little, Fig. 6.7 illustrates the high-field region, which is associated with an extension of the depletion layer by a distance X into the gate-drain space. The depletion layer extension supports part of the applied drain-source voltage, so the assumption made in Fig. 5.9 that $V(x = L_G) \approx V_{DS}$ can be a particularly poor assumption for values of V_{DS} well into the current saturated region.

Field arising on the positive ionic charge distributed throughout X gives rise to components of electric field in the channel that are y-directed, i.e. transverse to the main channel axis along x. In the appendix to Chapter 6, we will argue that the presence of transverse electric fields in the undepleted part of the channel degrades the low-field mobility, μ_0, such that E_S in Fig. 5.8 (the highly-doped curve) actually becomes dependent upon the operating point of the FET. E_S initially has a value of 3.5–4 kV cm^{-1}, but effectively increases severalfold once the high-field region, with the attendant saturation of electron drift velocity within it, is well established (see Fig. 6.11). The increase in E_S, or equivalently the decrease in μ_0, gives a more rounded "knee" to the curves in Fig. 5.10 and moves the knee to higher voltage values along the V_{DS} axis.

Another consequence of electrons becoming hot in the high-field region at the drain-end of the gate is that their energy is sufficient for them to escape from the channel into the substrate, following a path through the substrate to the drain roughly in parallel with electrons continuing through the channel. The path followed is sketched in Fig. 6.28, leading to a component of drain current as in Fig. 6.30. There is a "knee" in these curves at about $V_{DS}=2$ V.

For the two foregoing reasons, field-dependent mobility and substrate current flow, we expect the knee voltage in Fig. 5.10 to increase from 0.4 V to approaching 2 V, and we furthermore expect the curves to be more rounded and smoother with finite upward slope. Finally, the potential drops across the source-gate and gate-drain regions have been ignored, to the left and right of the gate electrode respectively. Figure 6.26 will identify the potential drop associated with the first of these two regions. Both affect the drain current relationship $I_D(V_{DS}, V_{SG})$.

5.6 Equivalent circuit of the GaAs MESFET

Figure 5.11 shows a diagram seen extensively in review articles and books dealing with the GaAs MESFET, in which a network of circuit elements is superimposed on a cross-sectional view of the FET. Each element is intended to represent the electrical character of a particular region of the device. For instance, the purpose of C_{gs} is to represent charge storage in the depletion region; R_s represents the bulk resistance of the N-layer in the source-gate region, together with any contact resistance at the source ohmic metallization. Having given some rough justification for the existence of equivalent elements in this way, the next step is to extract the overall equivalent circuit, as in Fig. 5.12. The circuit is required by microwave engineers to describe the electrical behavior of the physical structure, i.e. of the FET, in the microwave frequency bands (say 1 GHz up to several tens of GHz).

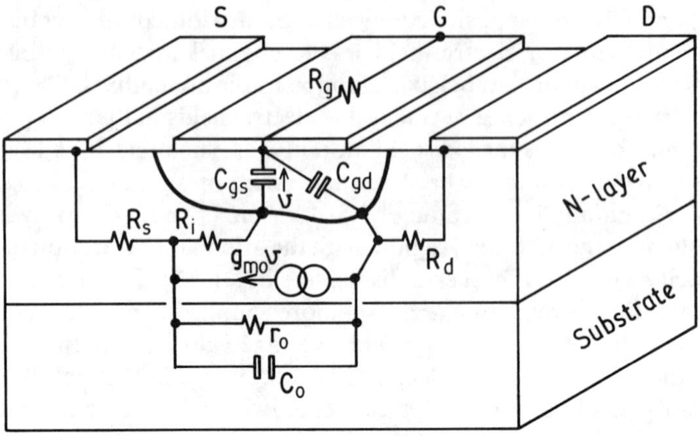

Fig. 5.11 Illustrating qualitative origin of equivalent circuit elements for a GaAs MESFET.

A range of typical values is given in Table 5.3. The range quoted is meant to typify the values we can expect for GaAs FETs having gate lengths of about 0.25 μm to 1 μm, just to give a first idea of the magnitudes involved. In Chapter 6 we shall come to the question of mutually consistent values of the equivalent circuit elements for FETs of a given gate length.

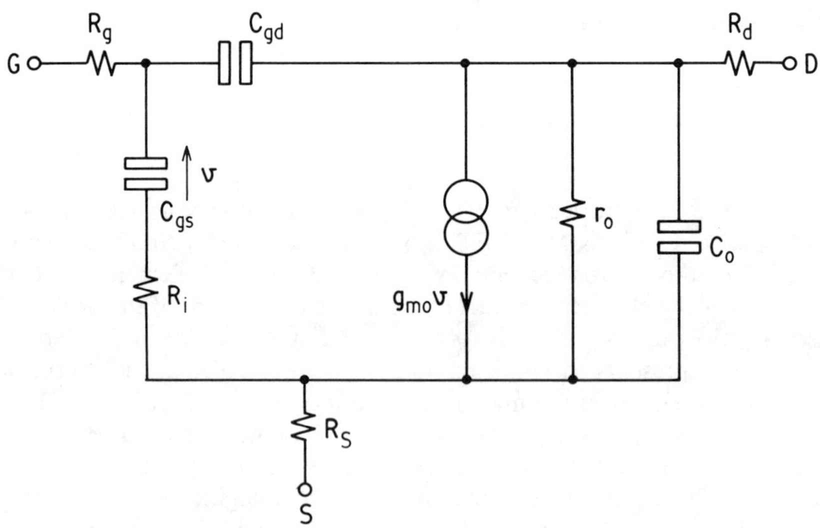

Fig. 5.12 Basic equivalent circuit for GaAs MESFET.

Table 5.3 Range of typical equivalent circuit values for a small-signal GaAs FET

R_g : 0.5–3 Ω	C_{gd} : 0.01–0.03 pF
C_{gs} : 0.15–0.4 pF	g_{m0} : 20–40 mS
R_i : 2–10 Ω	C_o : 0.05–0.1 pF
R_s : 1–5 Ω	r_o : 250–500 Ω
R_d : 1–5 Ω	τ_{gm} : 0–5 ps

The parameter τ_{gm} is the only unobvious parameter. It represents phase delay in the current generator $g_{m0}v$, and is required to achieve consistency between a measurement of the microwave scattering parameters of a GaAs FET and the set of S-parameters calculated from the equivalent circuit. Often, τ_{gm} is argued to equal the transit time of electrons in the channel underneath the depletion region†.

As valuable as this view of the FET may be as a first step, it is seriously lacking for the task of designing and understanding monolithic microwave integrated circuits (MMICs). It is strictly qualitative in outlook, whereas we require a quantitative and well-structured derivation. We must not only justify the circuit topology or interconnection of elements in the equivalent circuit, but we must also deduce expressions for the element values in terms of the physical structure. Chapter 6 addresses this task.

5.7 Concluding remarks

In this chapter we have performed a bare analysis of the dc drain characteristics of the GaAs MESFET and Si JFET and identified the basic mechanism responsible for current saturation in each device, as summarized in Section 5.4. The aim has been to make apparent the fundamental difference in physical operation with an analysis devoid of unnecessary detail that would otherwise obscure the argument.

For many workers the priority has been to develop as accurate a description of the dc characteristics as possible, to which end numerous FET simulators have been written. Less emphasis seems to have been accorded to developing accurate descriptions of the GaAs FET's microwave characteristics, perhaps because it is perceived that the dc behavior is easier to understand, and must be understood as a precursor to trying to understand the microwave properties. The present work takes the opposite view, the view that what matters above all else are the RF characteristics, with relatively little

†As will be shown in Chapter 6, this view is too simple.

emphasis placed upon the dc characteristics. When the initial concern is with small-signal operation of FETs in MMICs, this view is both allowable and profitable, for it avoids the complexities of having to describe the GaAs FET's behavior in the knee region preceding current saturation, which is complicated by the effects described in Section 5.5.2. Thus Chapter 6 seeks to develop a quantitative view of the GaAs MESFET for use in MMIC design.

References

5.1 Grove, A. S. *Physics and Technology of Semiconductor Devices*, ch. 8, Wiley, New York, 1967.
5.2 Feldman, J. M. *The Physics and Circuit Properties of Transistors*, ch. 9, Wiley, New York, 1972.
5.3 Williams, R. E. *Gallium Arsenide Processing Techniques*, Artech House, Dedham, 1984.
5.4 Hoffmann, R. K. *Handbook of Microwave Integrated Circuits*, Artech House, Norwood, 1987.
5.5 Vendelin, G. D. *Design of Amplifiers and Oscillators by the S-Parameter Method*, Wiley, New York, 1982.
5.6 Pengelly, R. S. *Microwave Field-Effect Transistors—Theory, Design and Applications*, Research Studies Press, Chichester, 1982.
5.7 Soares, R., Graffeuil, J. and Obregon, J. *Applications of GaAs MESFETs*, Artech House, Dedham, 1983.
5.8 Sze, S. M. *Physics of Semiconductor Devices (2nd edn)*, Wiley, New York, 1981.

5.A Appendix to Chapter 5

Origin of the term "Hot electron"

If the application of external voltages results in local electric fields within a device of several tens of kV cm^{-1} or higher, the conduction electrons become very energetic. The carriers accelerate, thus acquiring energy from the field. Various scattering mechanisms randomize the electron momentum and, in some electron–phonon collisions, the electrons may lose some of their energy to the lattice so that the lattice energy (or temperature) may increase slightly.

The term "hot electron" arises from expressing the electrons' energy as a temperature, and comparing this "electron temperature" with the lattice temperature. Whereas the lattice temperature may rise by, say, 150 K, the electron temperature may reach several thousand kelvin. It is this imbalance between electron and lattice temperatures which gives rise to the terminology "hot electron". Equivalently, the imbalance can be expressed by saying that the average electron energy is well above the equilibrium energy.

To express the electron energy as a temperature, the rather loose assumption is made that the free electrons obey a Maxwellian velocity distribution:

$$P(v)\,\mathrm{d}v = (4/\sqrt{\pi})\,(m^*/2\,\mathrm{k}T)^{3/2}\,.\,v^2\,.\,\exp(-m^*v^2/2\,\mathrm{k}T)\,.\,\mathrm{d}v \qquad (5.A.1)$$

This expression gives the probability that any electron in an assembly of electrons kept at temperature T has velocity in the range v to $(v+\mathrm{d}v)$ in any direction. The mean square velocity is given by

$$\langle v^2\rangle = \int_0^\infty v^2 P(v)\,\mathrm{d}v \qquad (5.A.2)$$

This involves an integral of the Gaussian Integral family

$$\int_0^\infty v^4 \exp(-\alpha v^2)\,\mathrm{d}v = (3\sqrt{\pi})/(8\alpha^{5/2}) \qquad (5.A.3)$$

We thus find, for $\langle v^2\rangle$,

$$\langle v^2\rangle = (3\,\mathrm{k}T/m^*) \qquad (5.A.4)$$

The mean kinetic energy of the electron is

$$\langle \mathscr{E}\rangle = \langle m^*v^2/2\rangle = \tfrac{1}{2}m^*\langle v^2\rangle = (3\,\mathrm{k}T/2)$$

Often, the subscript "e" is added to T to act as a reminder that we are discussing electron temperature, so,

$$\langle \mathscr{E}\rangle = (3\,\mathrm{k}T_e/2) \qquad (5.A.5)$$

Reserving the symbol T_0 for the equilibrium lattice temperature, the term "hot electron" therefore implies

$$T_e \gg \mathrm{T}_0$$

Examples of hot electron devices are Schottky diodes, Gunn and IMPATT oscillators, FETs and high electron mobility transistors (HEMTs).

Chapter 6

MESFETs

6.1 Introduction

The GaAs MESFET features prominently in a great many MMICs. It performs the basic functions of amplification and switching as required for low-noise amplifiers, broadband amplifiers, power amplifiers and microwave switches. In many of these applications it turns out that the FET must be engineered with a very high degree of precision and accuracy if the chip is to be within specification, which places severe demands upon the process technology. Process tolerances of a few percent must typically be held for an RF yield of a few tens of percent. For this reason, it is important to understand the RF behavior of the GaAs FET through its equivalent circuit, and the way in which the equivalent circuit elements vary with changes to the material and structure of the device.

6.2 Brief outline of structure

In Fig. 6.1 are shown various commonly used plan view designs for small-signal GaAs MESFETs. The layout is important for the following reasons:

(a) For a total gatewidth of Z_G the gate strip must be fed at as many points as possible in the interests of keeping the gate resistance small, and hence the microwave power gain high.

(b) The size and relative placement of the contact pads determines the geometrical components of inter-electrode capacitance.

(c) Some layouts provide better screening than others from the influence of surrounding structures and potentials at which they are held. For instance, in the designs of Figs 6.1(a) and 6.1(b), the free ends of the gate strips are

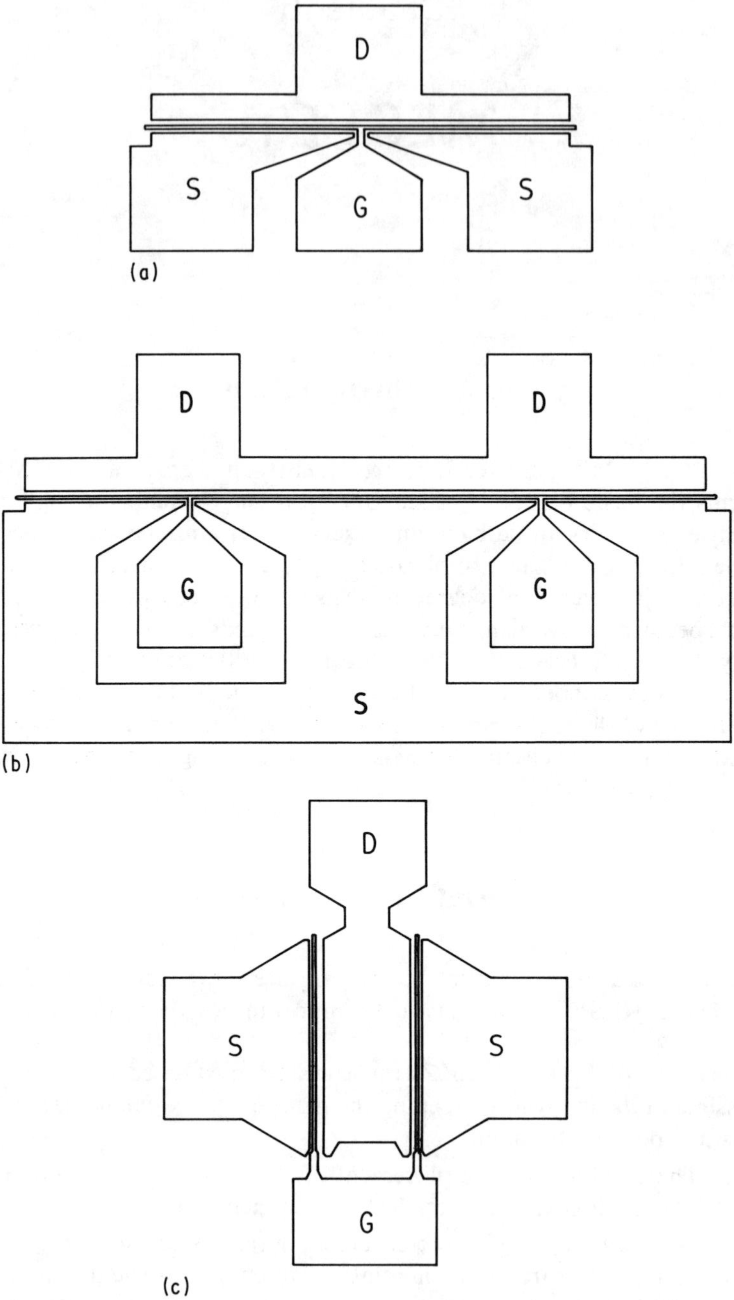

Fig. 6.1 Three plan view structures used for GaAs MESFETs.

exposed to the surrounding GaAs substrate, whereas in Fig. 6.1(c) the ends are screened by the drain pad, which is at a fixed and known potential. In the first two cases, therefore, any electric field fringing between the active region of the FET and the outlying GaAs is not predictable and, consequently, alters the equivalent circuit in an undetermined way.

A cross-section through a gate strip is shown in Fig. 6.2. Under the GaAs free surface in the source-gate and gate-drain spaces is a region where the N-type layer is depleted of mobile electron charge. Electric field arising on the resultant layer of positive, immobile, ionic charge terminates on negative (electron) charge held in surface states, giving rise to a surface potential of typically 0.7 V in equilibrium. These regions are shown hatched in Fig. 6.2. In an operating device, a depletion region forms in the N-layer immediately underneath the gate, extending into the gate-drain space to a degree which is dependent upon, among other things, the surface charge. Since, as we shall see, the shape and extent of the depletion layer determine the gate-channel and gate-drain capacitances in the equivalent circuit, reproducible RF performance requires the surface condition to be very carefully controlled.

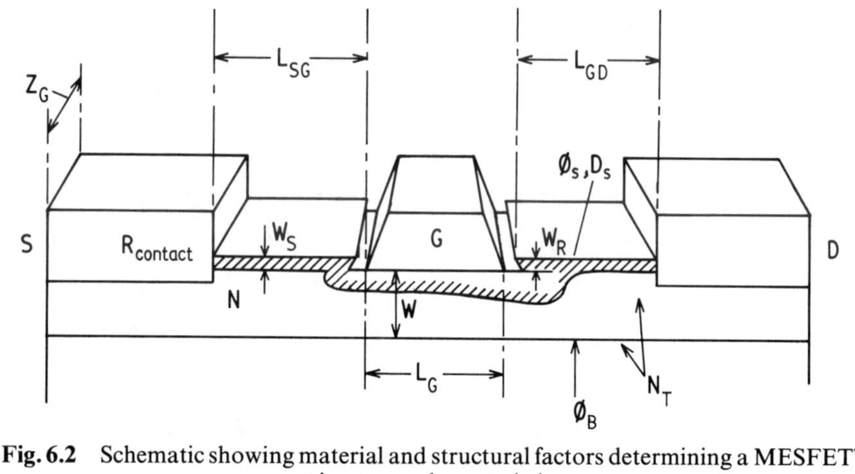

Fig. 6.2 Schematic showing material and structural factors determining a MESFET's microwave characteristics.

The material and device-structural features which determine the microwave behavior of a FET are identified on Fig. 6.2 and are:

N the doping density in the N-channel layer
W the thickness of the N-layer under the gate

W_R the depth of the gate recess
W_S surface depletion depth
ϕ_B the interface potential between the N-layer and the buffer-substrate
L_G the metallurgical gate length
L_{SG} the source-gate separation
L_{GD} the gate-drain separation
ϕ_S the surface potential
$R_{contact}$ the ohmic contact resistance
N_T the density of deep levels ("traps")
Z_G the gatewidth

Generally speaking, the channel doping is a function of depth into the N-layer. In particular, there is a transition region of finite extent between the high doping of the channel and the typically three to four orders lower doping of the buffer-substrate. If this profile tail is gradual, occupying several tenths of a micron, the value of ϕ_B is small (typically 0.1–0.3 V); if, on the other hand, the transition is sharp (i.e. Debye length limited), then ϕ_B is large (approximately 0.7 V). We shall see how ϕ_B governs the magnitude of the current component which flows outside the channel layer proper, the so-called substrate current, which in turn governs the output conductance in the equivalent circuit in Fig. 6.3, g_o, and to some extent affects the transconductance, g_{m0}.

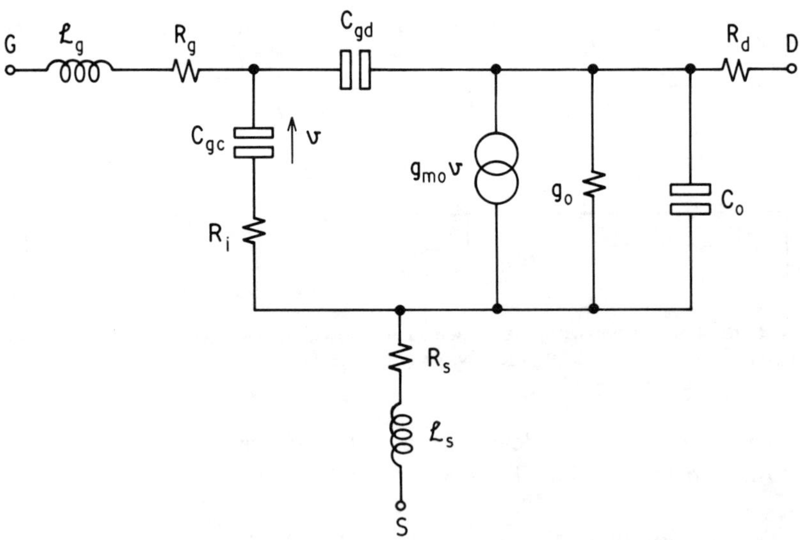

Fig. 6.3 Basic GaAs FET equivalent circuit.

From the foregoing discussion it is clear that ϕ_S and ϕ_B, together with all the more obvious factors (N, W etc.), must be controlled in value if reproducible FETs are to be realized. Both these quantities appear to be substrate dependent, and ϕ_S is obviously dependent upon surface treatment, the presence of dielectric protection layers, and so on. Of all the technological parameters, doping profile tail and surface states are often the least easily controlled.

6.3 Equivalent circuit—physical basis

In MMICs it is primarily the equivalent circuit of the FET which matters; the dc conditions are important only in so far as they establish the element values in the equivalent circuit. Accordingly, we shall deal with the microwave behavior of the FET predominantly.

The equivalent circuit is an abstraction and simplification which yields a representation of the FET that is manageable. Nevertheless, it must represent adequately all the important physical processes which occur in actual devices, including signal delay, charge and energy storage, current modulation and energy (or power) loss. We shall discuss each in turn.

6.3.1 Signal delay

There are three main sources of signal delay in the GaAs FET: propagation delay across the width of the gate; the charging time of that part of the depletion region which operates below velocity saturation; and the charging time of the depletion layer in that part of the device where the electrons are velocity saturated.

In Fig. 6.4 is shown the gate strip of a FET which is assumed to be excited with a signal via a gate pad at the left end. In response to the signal voltage on the metal strip, the charge in the depletion edge changes so that equal and opposite charge increments appear at the metal–semiconductor interface and in the N-layer at a distance d from the interface. Ignoring for the time being the time it takes for the charge in the depletion layer to change, the charge on the metal and its image in the depletion edge form a miniature strip transmission line of width-to-height ratio L_G/d. Because the velocity with which signals propagate along such transmission lines is roughly $c/\epsilon^{1/2}$, the channel is modulated progressively across its width as the signal propagates across the gate [6.1]. Consequently, the drain current, which consists of a summation of all channel current filaments, is delayed with respect to the signal voltage applied to the gate pad.

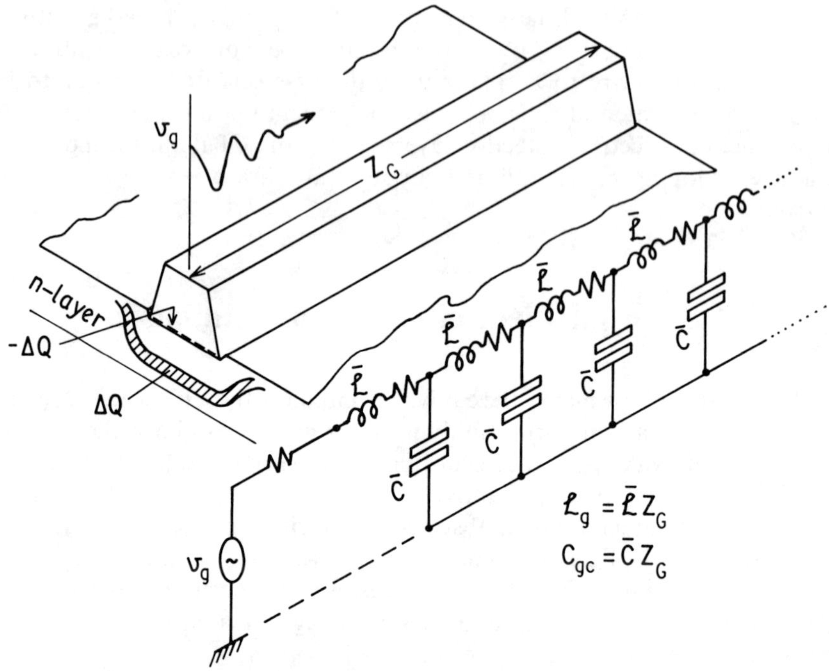

Fig. 6.4 Illustrating wave propagation across a MESFET gate strip. IMPORTANT NOTE: *total* charge changes are shown.

In circuit terms the gate stripline can be represented as in Fig. 6.4 by a distributed inductive-resistive-capacitive network. Assuming the gate to be driven from a source or generator of impedance Z_g equal to the characteristic impedance $Z_0 = (\bar{\mathcal{L}}/\bar{C})^{1/2}$ of the gate strip, it may be proved that the phase delay in i_d relative to v_g is given approximately [6.1] by

$$\theta \approx \beta Z_G \tag{6.1}$$

where $\beta = \omega(\bar{\mathcal{L}}\bar{C})^{1/2}$ is the phase delay per unit distance across the gate strip, and Z_G is the strip width. Equivalently, using the relationship $Z_0 = (\bar{\mathcal{L}}/\bar{C})^{1/2}$, the phase delay may be expressed as

$$\theta \approx \omega(\bar{C}Z_G)Z_0 = \omega C_{gc} Z_0 = \omega(\mathcal{L}_g C_{gc})^{1/2} \tag{6.2}$$

where C_{gc} is the total capacitance of the gate strip to the depletion layer, and $\mathcal{L}_g$ is the total inductance.

If the distributed network of Fig. 6.4 is reduced to a single $\mathcal{L}$–R–C section as in Fig. 6.5, and is also driven from a source impedance equal to the characteristic of the gate strip, then the voltage v, which is the voltage modulating the current generator $g_{m0}v$, has a delay with respect to v_g of

$$\theta \approx \omega C_{gc} Z_g = \omega C_{gc} Z_0 = \omega(\mathcal{L}_g C_{gc})^{1/2}.$$

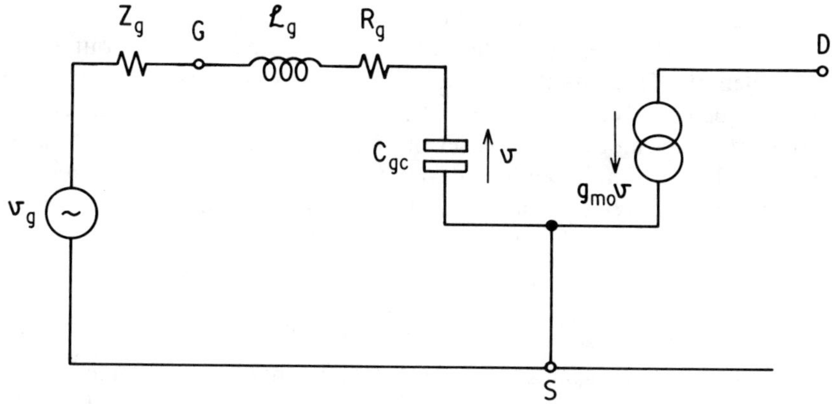

Fig. 6.5 Reduction of distributed transmission line effects in FET gate to a single $\mathscr{L}$–R–C section in the equivalent circuit.

providing $\omega^2 \mathscr{L}_g C_{gc} \ll 1$, and providing θ is small ($\tan\theta \approx \theta$). Hence, given the condition that the source or generator impedance from which the gate strip is driven is equal to the characteristic impedance of the gate strip itself, the lumped equivalent circuit correctly models the phase delay due to wave propagation in the gate.

In order to obtain an impression of how frequently this condition might be satisfied in practice, we can estimate values for $\bar{\mathscr{L}}$ and $\bar{C}$ by regarding the gate as a section of a parallel plate stripline [6.1]. With μ_0 as the permeability of free space,

$$\bar{\mathscr{L}} \approx \mu_0 d/L_G \text{ and } \bar{C} \approx \epsilon L_G/d \tag{6.3}$$

per unit distance in the z direction. Hence,

$$Z_0 = \left(\frac{\bar{\mathscr{L}}}{\bar{C}}\right)^{1/2} = \left(\frac{\mu_0}{\epsilon}\right)^{1/2} \frac{d}{L_G} \tag{6.4}$$

which, for a well designed FET with a gate-length to channel-thickness ratio of 3:1 to 5:1 gives:

$$Z_0 \lesssim \frac{1}{4}\left(\frac{\mu_0}{\epsilon}\right)^{1/2} \approx 25\ \Omega$$

When the source impedance from which the gate is driven is greater then this value, as for example in a 50 Ω system, the lumped equivalent circuit generally models a phase delay $\theta \approx \omega C_{gc} Z_g$ which is too large, with the result that a reduction in the transconductance delay, τ_{gm} (see Section 6.3.4) might be required to compensate.

It turns out, however, that phase delay is not the only consequence of wave propagation in the gate. If the gatewidth is comparable to about a tenth of a wavelength of the signal propagating across the gate strip, the magnitude of g_{m0} can be expected to decrease because the filamentary channel current components entering the drain differ in phase to such an extent that partial phase cancelation occurs. Figure 6.6 summarizes these effects. Compared with this magnitude behavior in the distributed case, the magnitude in the lumped simplification of Fig. 6.5 is given by

$$\left|\frac{g_{m0}\,(\omega)}{g_{m0}\,(0)}\right| = \left|\frac{v}{v_g}\right| = \frac{1}{[(1-\omega^2\mathscr{L}_g C_{gc})^2 + \omega^2 C_{gc}{}^2 Z_g{}^2]^{1/2}}$$

which, given the condition $Z_g = Z_0 = (\mathscr{L}_g/C_{gc})^{1/2}$, may be written as

$$\left|\frac{g_{m0}\,(\omega)}{g_{m0}\,(0)}\right| = \left|\frac{v}{v_g}\right| = \frac{1}{[(1-\omega^2\mathscr{L}_g C_{gc})^2 + \omega^2\mathscr{L}_g C_{gc}]^{1/2}}$$

$$= \frac{1}{[(1-\theta^2)^2 + \theta^2]^{1/2}} \approx 1 + \frac{\theta^2}{2} \quad \text{for } \theta^2 \ll 1$$

For example, for a FET with $\mathscr{L}_g \approx 30$ pH and $C_{gc} \approx 0.2$ pF, $\theta^2 \approx 0.024$ rads^2 at 10 GHz. Although this magnitude behavior is different from that of the real (i.e. distributed) gate strip in Fig. 6.6, it is of no importance provided

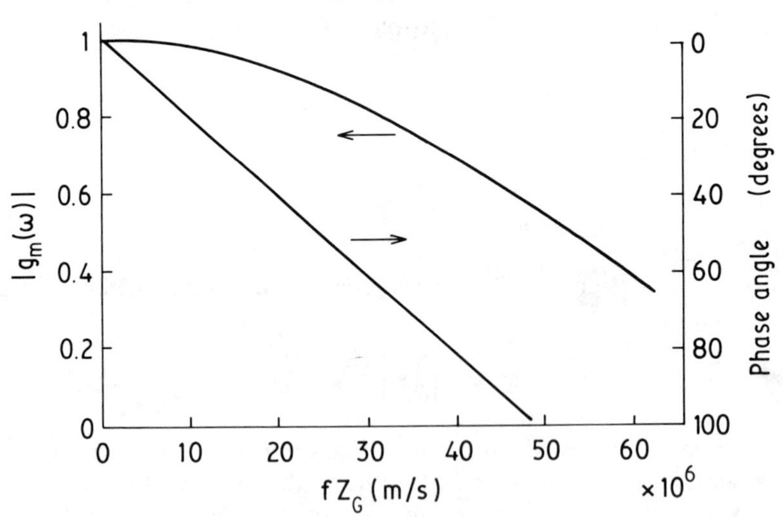

Fig. 6.6 Effect of wave propagation delay in FET gates on magnitude and phase of drain current, here expressed in terms of transconductance magnitude and phase delay.

$$\left|\frac{g_{m0}\ (\omega)}{g_{m0}\ (0)}\right|$$

doesn't deviate more than a few percent from unity. The predominant effect is one of phase delay, which the lumped circuit adequately represents (subject to the condition $Z_g = Z_0$). Since $\mathscr{L}_g$ is required in the equivalent circuit to account for phase delay in the gate in this way, it must be considered an intrinsic property of the FET and not parasitic. Parasitic gate inductance associated with bond wires and gate pads is quite distinct from $\mathscr{L}_g$.

Before leaving the subject of signal delay in the gate, it is important to examine the consequences of having the total gatewidth of Z_G divided into, say, two separate gate strips connected in parallel, as in Fig. 6.1(c) (to which Fig. 6.1(a) is directly equivalent in this regard), or into four equivalent parallel-connected strips, as in Fig. 6.1(b). We first note that the total capacitance is unchanged by this connection, but that the input impedance decreases as

$$Z_{IN} = \frac{1}{m} Z_0 = \frac{1}{m}\left(\frac{\bar{\mathscr{L}}}{\bar{C}}\right)^{1/2} = \frac{1}{m}\left(\frac{\bar{\mathscr{L}} Z_G}{\bar{C} Z_G}\right)^{1/2}$$

where m is the number of parallel strips into which the total gatewidth is divided. Since the total capacitance $\bar{C}Z_G$ is unchanged, the conclusion must be that the effective inductance is reduced to

$$\mathscr{L}_g = \frac{\bar{\mathscr{L}} Z_G}{m^2}$$

One factor of $(1/m)$ derives from the fact that the series reactance of any one strip is now determined by a width Z_G/m instead of Z_G; the other arises from the parallel connection of m of these reduced-width strips in parallel. Thus finally, substituting for $\bar{\mathscr{L}}$, we find the inductance intrinsically associated with the gate strips as approximately

$$\mathscr{L}_g \approx \frac{\mu_0 d Z_g}{m^2 L_G} \tag{6.5}$$

To gain an appreciation of the magnitudes involved here, consider a FET with a gate length of $L_g = 0.5$ μm and single gate strip of width $Z_G = 200$ μm operated such that the depletion depth $d = 600$ Å. Then,

$$\mathscr{L}_g \approx 30 \text{ pH}$$

6.3.2 Charge storage

In Fig. 6.7 is shown the general shape of the space-charge layer in a velocity-saturated MESFET operating at a bias point P. Under the

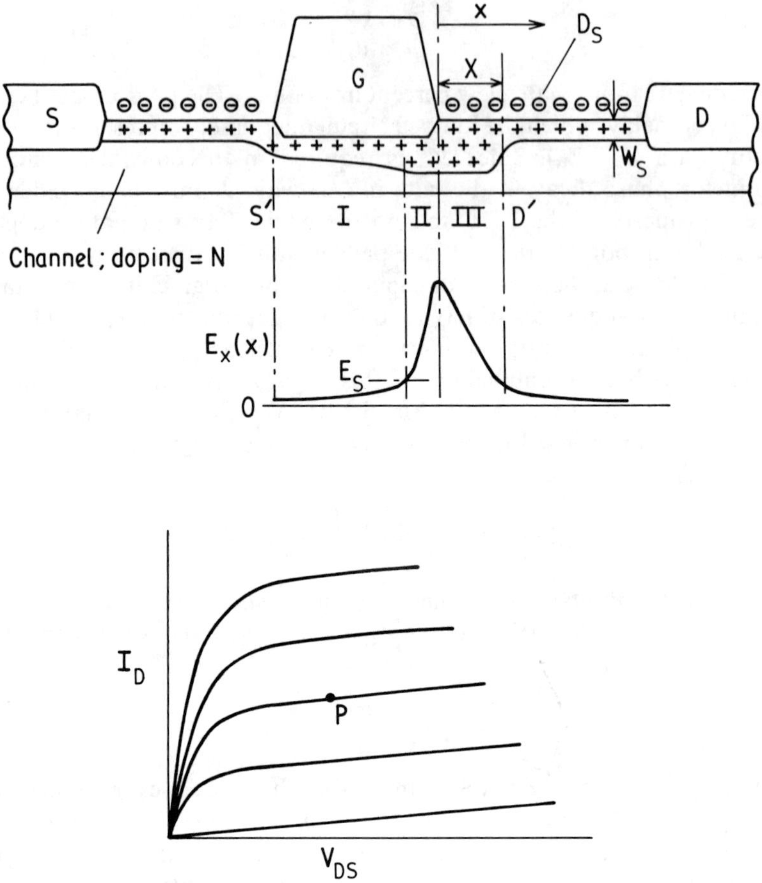

Fig. 6.7 Depletion layer and channel field in an operating GaAs MESFET.

unmetallized source-gate and gate-drain regions there is depletion to a depth W_S such that $NW_S = D_S$, where D_S is the occupied surface-state density. Associated with this depletion is a band-bending giving rise to a surface potential of

$$\frac{\phi_S}{q} = \frac{qNW_S^2}{2\epsilon} = \frac{qD_S^2}{2\epsilon N} \tag{6.6}$$

where, as before, a uniform channel doping N has been assumed. It is important to note that D_S may be spatially non-uniform, with different values in the two regions [6.2]. Under high gate-drain bias conditions, as for example in a power GaAs FET, ϕ_S may rise to well above the equilibrium surface potential of $\phi_{S0} \approx 0.7$–0.8 eV. The active channel region comprises the zones marked I, II and III on Fig. 6.7.

In the forthcoming sections we shall refer to the virtual source and drain electrodes S' and D' as located on Fig. 6.7, and we shall use the symbols

$$V_{S'G} \text{ and } V_{D'S'}$$

to refer to the potentials between S' and the gate G, and between D' and S'. The importance of these virtual electrodes is that their potential differs from the applied potentials due to voltage drops across the source-gate and gate-drain regions where there is bulk resistance R_S and R_D respectively. It is primarily the potentials between these virtual electrodes which govern the equivalent circuit elements of the intrinsic FET, and for that reason it is important that they be distinguished from the applied voltages at the contacts, V_{SG} and V_{DS}.

To grasp the magnitude of this difference, consider a FET operated at $V_{SG}=0.5$ V (i.e. $V_{GS}=-0.5$ V) such that $I_D=30$ mA. If the FET has a source resistance between S and S' of, say, 5 Ω, then,

$$V_{S'G}=V_{SG}+I_D R_S=0.5+(0.03\times 5)=0.65 \text{ V}.$$

The extra 0.15 V will increase the depth of depletion d by an amount which cannot be ignored. Thus the distinction between $V_{S'G}$ and V_{SG}, and between $V_{D'S'}$ and V_{DS}, is important.

In region I, the motion of channel electrons in the longitudinal channel field is governed by the part of the velocity-field characteristic of Fig. 6.A.7, for which $0<E\leqslant E_S$. Over this region, an increase in electric field strength results in an increase in electron drift velocity, i.e. the channel essentially behaves resistively. At the end of region I, the electric field has reached the value E_S at which the electron drift velocity saturates and becomes independent of E. This condition persists until the end of region III, at which point the x-directed channel field has dropped to values below E_S once more. The distinction between regions II and III is that the first is in the shadow of the gate, whereas the second constitutes the space-charge layer extension X into the gate-drain space. Region III is fundamentally necessary to a satisfactory description of FET behavior, for it has a direct bearing on the values of gate-drain capacitance, C_{gd}, output conductance, g_o, and time delay in the transconductance, τ_{gm}. Finally, but importantly, electric field arising from ionic charge in region III gives rise to transverse fields in regions I and II, as illustrated in Fig. 6.8. Transverse fields in region I degrade the longitudinal mobility there, as reference to the appendix to this chapter will show.

If the gate voltage is increased in the negative sense while keeping the gate-drain voltage constant, the depletion layer changes as shown in Fig. 6.9.

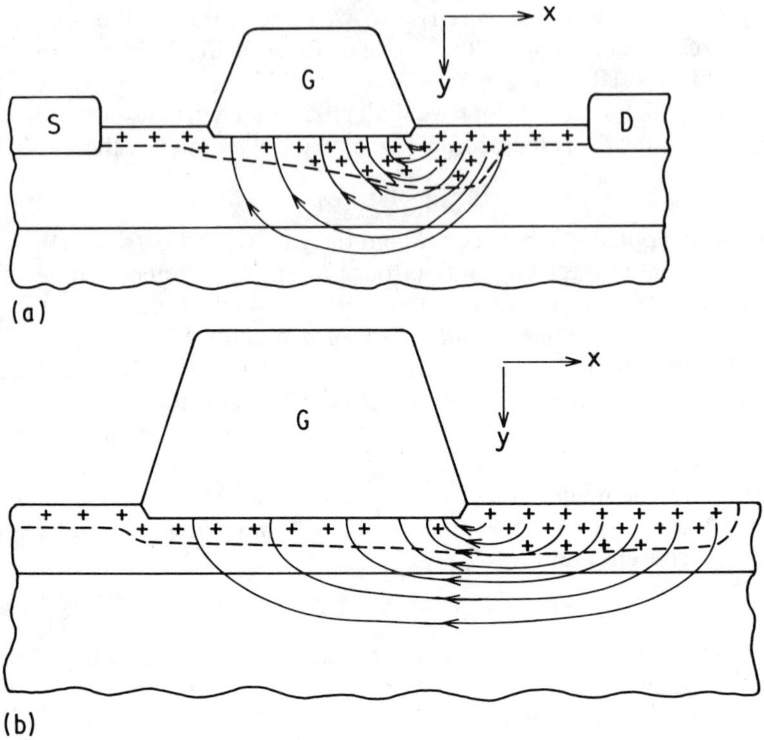

Fig. 6.8 Illustrating y-directed fields in the channel of a MESFET: (a) as frequently drawn in research papers; (b) with more representative aspect ratio.

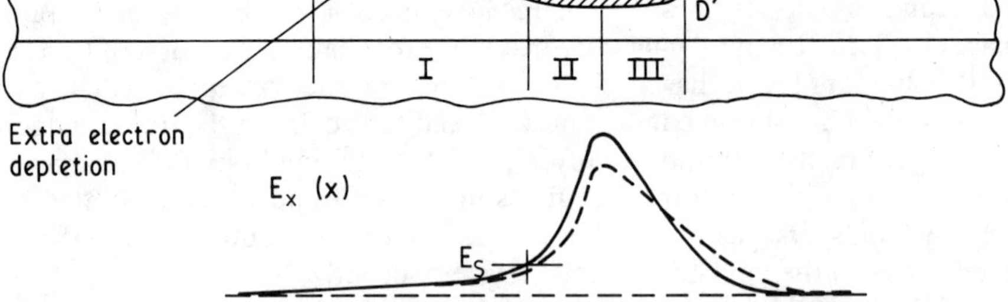

Fig. 6.9 Showing effect on depletion layer of increasing negative bias on the gate while maintaining V_{DS} constant (– – – original depletion layer position; ——— new position).

In region I the depletion layer expands; the electrons which deplete from the shaded region are transported first through the resistive region, region I, then through regions II and III and finally to the drain. Extra positive ionic charge is thus uncovered in the depletion edge in region I in response to the increase in gate voltage, i.e. the depletion layer exhibits charge storage properties which can be described in electrical terms by a capacitance. The combined properties of depletion-layer capacitance and channel resistance in region I can be treated by reference to Fig. 6.10, which recognizes the fact that the capacitance is distributed along the channel resistance.

A reasonable approximation is that region I occupies most of the gate, so the length of region I is less than or equal to L_G. The potential drop in the channel is approximately $E_S L_G$, and the current flowing through the region is just the channel current, I_{CH}, so that the dc resistance of region I is

$$R_{dc} \approx \frac{E_S L_G}{I_{CH}}$$

It is convenient to eliminate E_S from this expression, in favor of more readily determined factors, using the rules-of-thumb from the appendix to this chapter. Neglecting the effects of transverse fields for a moment, a linear extrapolation from low fields with a typical as-doped mobility of

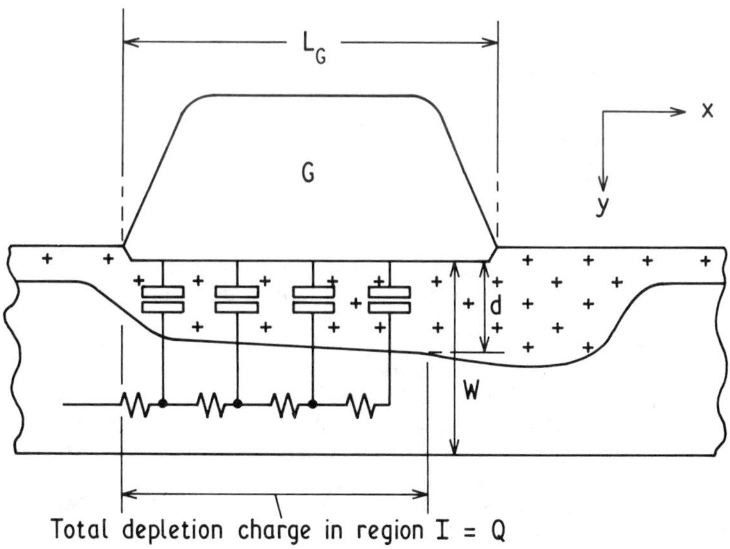

Fig. 6.10 Illustrating distributed C–R network character of depletion layer and channel in region I.

3 000 cm^2 Vs^{-1} intersects the value $v_{sat} = 10^7$ cm s^{-1} at a field of 3.3 kV cm^{-1} (Fig. 6.11). We then have

$$E_S \approx \frac{v_{sat}}{\mu_0}$$

Taking account of mobility degradation due to transverse fields in Fig. 6.11, and applying the rule of thumb from the appendix that μ_0 is reduced by E_y to about one-third of its as-doped value, the previous estimate of E_S is modified to†

$$E_S \approx \frac{3v_{sat}}{\mu_0}$$

so that

$$R_{dc} \approx \frac{3v_{sat}L_G}{\mu_0 I_{CH}}$$

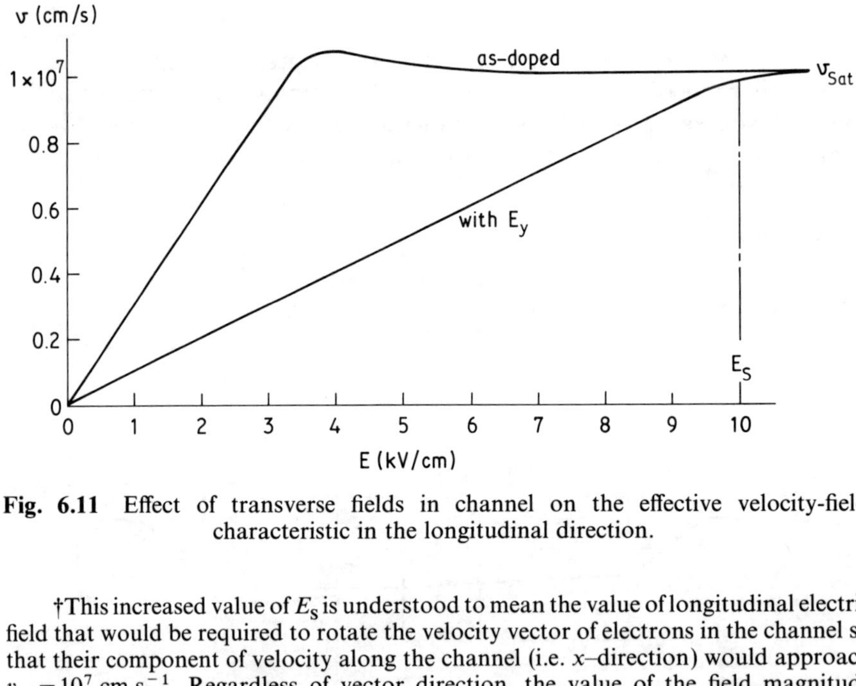

Fig. 6.11 Effect of transverse fields in channel on the effective velocity-field characteristic in the longitudinal direction.

†This increased value of E_S is understood to mean the value of longitudinal electric field that would be required to rotate the velocity vector of electrons in the channel so that their component of velocity along the channel (i.e. x–direction) would approach $v_{sat} = 10^7$ cm s^{-1}. Regardless of vector direction, the value of the field magnitude required to bring about a drift velocity of approximately v_{sat} is, of course, still about 3.3 kV cm^{-1} as given by the as-doped curve in Fig. 6.11. The degenerated curve with transverse fields E_y applies in the low-field region in so far as it describes the resistance to carrier motion in the longitudinal direction in response to applied longitudinal fields. The increased value of E_S is simply a way of describing this increased resistance to carrier motion in the channel direction.

It is a well-known result of distributed network theory that the equivalent RF resistance is one-third of the sum total (or dc) resistance. With reference to Fig. 6.12, this results from the fact that not all current input at the left end of the resistor chain flows through all the resistors; instead, the current flowing through elements of resistance ΔR progressively further down the network diminishes by the currents flowing through the capacitors. Thus we obtain for the RF equivalent value of channel or input resistance, R_i,

$$R_i \approx \frac{v_{sat} L_G}{\mu_0 I_{CH}} \approx \frac{v_{sat} L_G}{\mu_0 I_D} \tag{6.7}$$

The approximation that the channel current I_{CH} equals the drain current I_D assumes that substrate current is negligible. With $v_{sat} = 10^7$ cm s^{-1} and $\mu_0 \approx 3000$ cm^2 Vs^{-1}, a 0.5 μm gate length FET operated at a drain current of 30 mA would have a channel resistance of 5 or 6 Ω.

Continuing with the assumption that the length of region I is approximately equal to L_G, the total charge in the depletion layer in region I is

$$Q \approx qNdL_G Z_G$$

where Z_G is the gate width (perpendicular to the x and y directions). By definition the capacitance is

$$C = \frac{\partial Q}{\partial V_{S'G}} = qNL_G Z_G \frac{\partial d}{\partial V_{S'G}} \triangleq C_I$$

Here, the inter-electrode potential between the gate and source is represented for ease by $V_{S'G}$, which is a positive number, rather than by $V_{GS'}$, which is a negative number. Thus, if the equilibrium contact potential between the gate metal and the N-GaAs layer is V_{BO} (V_{BO} is a positive number),

$$V_{S'G} + V_{BO} \approx \frac{qNd^2}{2\epsilon} \tag{6.8}$$

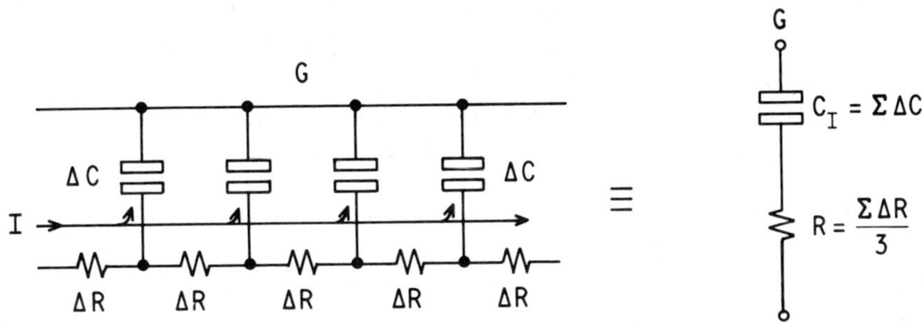

Fig. 6.12 Lumped equivalent of distributed network in Fig. 6.10.

Differentiating and rearranging, we have

$$\frac{\partial d}{\partial V_{S'G}} = \frac{\epsilon}{qNd} \tag{6.9}$$

Eliminating $(\partial d/\partial V_{S'G})$ from the expression for capacitance,

$$C_I = \frac{\epsilon L_G Z_G}{d} \tag{6.10}$$

which is the well-known result that the depletion-layer capacitance below a Schottky contact behaves, under small-signal conditions, as a parallel-plate capacitor. Typically, for a 0.5 μm gate FET, with 200 μm gate width biased to a depletion depth of 600 Å, $C_I \approx 0.19$ pF.

We now have the result that the charging of the space-charge layer in region I can be represented by a series C–R combination, as in Fig. 6.12, with values as given by eqns (6.7) and (6.10). The time constant for this charging is

$$\tau_I \triangleq C_I R_i = \frac{\epsilon v_{sat} L_G^2 Z_G}{\mu_0 d I_{CH}}$$

In a sub-micron MESFET, current saturation in the channel arises from electrons reaching their scatter-limited drift velocity in region II. If the channel opening is the channel layer thickness, W, minus the depletion depth, d,

$$I_{CH} \approx qNv_{sat}(W-d)Z_G \tag{6.11}$$

Then

$$\tau_I = \frac{\epsilon}{qN\mu_0} \times \frac{L_G^2}{d(W-d)} = \tau_d \times \frac{L_G^2}{d(W-d)} \tag{6.12}$$

where τ_d is the dielectric relaxation time for a semiconductor of permittivity ϵ and conductivity $\sigma = qN\mu_0$. Thus, the physical mechanism governing charge exchange with the depletion layer in region I is dielectric relaxation, modified by geometrical factors associated with the channel. For a 0.5 μm FET with a channel layer thickness of 0.1 μm, biased to $d = 600$ Å with a doping of 3.5×10^{17} cm^{-3} and $\mu_0 \approx 2500$ (cm^2 Vs^{-1}), $\tau_I \leqslant 1$ ps.

Turning next to regions II and III, the channel field in these regions is greater than E_S, and the electrons travel at the scatter-limited velocity, v_{sat} (Fig. 6.11). Thus charge exchange with the depletion edge in these regions occurs by electrons moving at v_{sat}. The channel current is governed by $V_{S'G}$ through the depletion depth d, as expressed by eqns (6.8) and (6.11) above, so that regions II and III of the channel behave as a current generator. The capacitance of the depletion layer in regions II and III is distributed across the elements of this current generator, as shown in Fig. 6.13. Physically, when

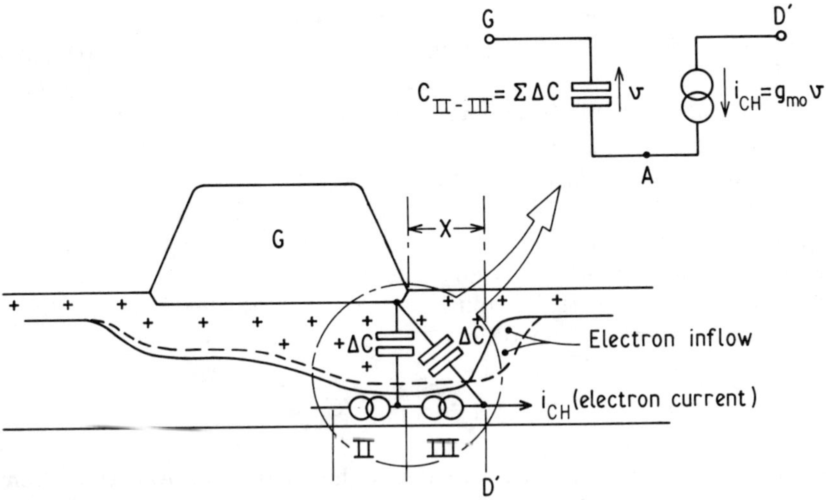

Fig. 6.13 Extraction of lumped equivalent of distributed capacitance and current generator network in the velocity saturated regions of a GaAs FET.

charge exchange with the depletion edge in regions II–III in Fig. 6.13 is taking place, the depletion d is different for each element, i.e., from eqn (6.11), each current generator has a different value by an amount equal to the local charging current. Although a treatment of this distributed capacitance and current generator network is possible [6.3], we shall adopt a simpler approach by reasoning as follows.

(a) Since charge exchange with the depletion edge in regions II–III in Fig. 6.13 involves velocity saturated electrons, we may represent the associated charging delay by summing the elements of capacitance in regions II and III to give a lumped capacitance

$$C_{\mathrm{II-III}} = \sum_{\mathrm{II-III}} \Delta C$$

and assuming this capacitance is charged and discharged by velocity saturated electrons. (This charging delay gives rise to a time delay in the transconductance $g_{\mathrm{m}0}$, as will be dealt with in Section 6.3.4).

(b) A current of velocity saturated electrons flows in regions II and III which is modulated by the depletion layer edge. Precisely this process of modulation, however, requires $C_{\mathrm{II-III}}$ to be charged and discharged, so that charge exchange with the "bottom plate" of $C_{\mathrm{II-III}}$ is effected by the current generator itself.

On the basis of these two physical considerations, a circuit representation of the capacitive and current modulation properties of regions II–III can be made as in Fig. 6.13(b). We shall defer until Section 6.4.1 the question of where the node A should be connected into the remainder of the network of equivalent circuit elements.

An expression for the capacitance $C_{\text{II–III}}$ can be determined by finding the net increase in positive charge in regions II and III in response to an incremental increase in source-gate voltage (the gate being made more negative). From Fig. 6.13, this might first be written as the parallel plate capacitance of regions II–III, minus the capacitance associated with the region where electron inflow occurs, i.e.

$$C_{\text{II–III}} = \frac{\epsilon X Z_G}{d} - C_{gd} \tag{6.13}$$

where C_{gd} is the capacitance associated with the electron inflow at the extreme right edge of the space-charge layer. It turns out that this expression is a useful approximation only for X/L_G values of around 0.2–0.3, with a further restriction on the ratio X/d to around 1.5–2. More extensive considerations show that, for $X/L_G \geqslant 0.5$, a factor of 2 correction is required in the first term, so that a better approximation for higher X/L_G values is

$$C_{\text{II–III}} = \frac{\epsilon X Z_G}{2d} - C_{gd} \tag{6.14}$$

We deal with C_{gd} next.

If the gate-source voltage is held constant and the drain-source (or drain-gate) voltage is increased positively, the depletion extension X increases slightly, as in Fig. 6.14. Electrons deplete from the extreme edge of the space-charge layer to uncover more positive ionic charge, resulting in charge storage which can be represented by a capacitance

$$C_{gd} \triangleq \partial Q / \partial V_{D'G}$$

It may be shown that, for the simple case of a FET in which the gate recess depth, W_R, is equal to the surface depletion depth, W_S,

$$C_{gd} = \frac{2\epsilon Z_G}{1 + 2X/L_G} \tag{6.15}$$

Typically, for a 0.5 μm by 200 μm FET biased with a depletion layer extension of $X = 0.5\ L_G$, $C_{gd} \approx 0.023$ pF.

We may now rewrite the expression for the capacitance associated with regions II and III,

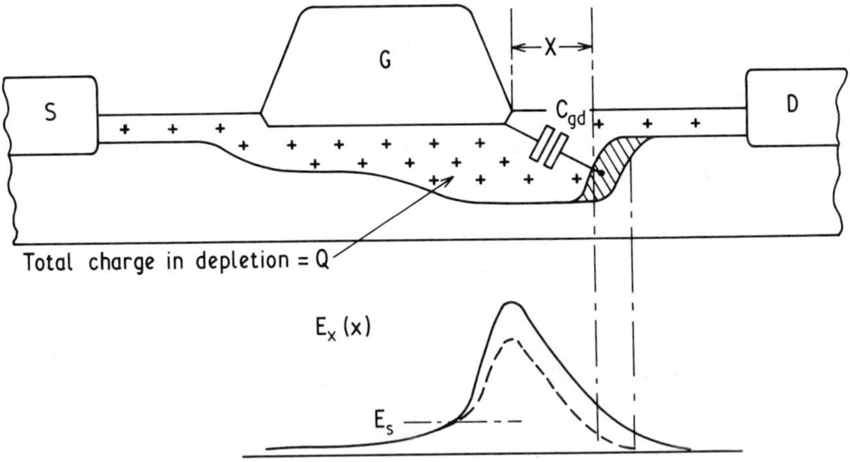

Fig. 6.14 Illustrating increased extension of depletion layer for a positive increase in drain-source bias while maintaining the gate-source bias constant.

$$C_{\text{II–III}} = \frac{\epsilon X Z_G}{2d} - \frac{2\epsilon Z_G}{1+2X/L_G} \tag{6.16}$$

For $X/L_G = 0.5$, $L_G = 0.5$ μm, $Z_G = 200$ μm and $d = 600$ Å,

$$C_{\text{II–III}} = (0.048 - 0.023)\text{pF} = 0.025 \text{ pF}$$

Compared with the capacitance associated with region I, this is smaller by a factor of about 7.5.

6.3.3 Current modulation

Referring once more to Fig. 6.13, any change in the depletion depth d in response to a change in gate voltage changes the cross-section of the channel through which electrons may pass. Recalling that electrons travel in regions II and III at the scatter-limited velocity v_{sat}, the current of electrons is given by multiplying the sheet charge of electrons

$$qN(W-d) \quad \text{electrons cm}^{-2}$$

by the velocity at which this sheet charge is moving, i.e. v_{sat}. Hence,

$$\bar{I}_{CH} = qN(W-d)v_{sat} \quad \text{electrons cm}^{-1}\,\text{s}^{-1}$$

The bar above I_{CH} denotes that this expression gives the dc current per unit gate width. Thus for a FET of gatewidth Z_G, the total current of electrons passing through the channel is

$$I_{CH}=\bar{I}_{CH}Z_G=qN(W-d)v_{sat}Z_G$$

By definition, the intrinsic transconductance is the rate of change of channel current with gate voltage, i.e.

$$g_{m0}\triangleq\frac{\partial I_D}{\partial V_{S'G}}=-qNv_{sat}Z_G\frac{\partial d}{\partial V_{IIG}} \tag{6.17}$$

where V_{IIG} is the voltage in the channel, at the beginning of region II, with respect to the gate. Recognizing that $V_{IIG}=V_{S'G}+E_SL_G$ and that $E_SL_G\approx$ constant, we have $\mathrm{d}V_{IIG}=\mathrm{d}V_{S'G}$. Then, substituting from eqn (6.9),

$$g_{m0}=\frac{\epsilon v_{sat}Z_G}{d} \tag{6.18}$$

To determine a typical value for g_{m0}, we again consider a FET with a gatewidth of 200 μm and a depletion depth of 600 Å: we find $g_{m0}=38$ mS.

To define fully the magnitude of the current generator in the simplified representation of Fig. 6.13, we must specify clearly the voltage in the equivalent circuit which drives the current generator of transconductance g_{m0}. In the above derivation of g_{m0}, the derivative was taken with respect to V_{IIG}, the voltage difference between the depletion edge and the gate metal. This is the voltage across the capacitance associated with region II, C_{II}, and is written as v, for simplicity, in Fig. 6.13.

6.3.4 Transconductance delay

Electron current entering region II from region I either goes into charging the capacitance C_{II-III} associated with the depletion layer in regions II and III, or into establishing drain current. If the gate voltage is changed, a new value of drain current cannot be fully established until C_{II-III} has been charged. Such a delay finds expression in circuit terms as

$$g_m(\omega)=g_{m0}\exp(-j\omega\tau_{gm}) \tag{6.19}$$

An expression for τ_{gm} can be obtained by recognizing from Fig. 6.13(b) that it is the current generator i_{CH} which charges and discharges C_{II-III}, i.e. a charge change of

$$Q=C_{II-III}v$$

results from a current of $g_{m0}v$ flowing for a time τ_{gm}. Thus, from eqns (6.14), (6.15) and (6.18) for $X/L_G\geqslant 0.5$,

$$\tau_{gm} = \frac{1}{v_{sat}}\left(\frac{X}{2} - \frac{2d}{1+2X/L_G}\right) \tag{6.20}$$

For a 0.5 μm gate length FET, biased such that $X/L_G = 0.5$ and $d = 600$ Å, we find $\tau_{gm} \approx 0.65$ psec. Mathematically, eqn (6.20) can give a negative result if $d > (X/4) + X^2/2L_G$; this is merely an indication of the approximation breaking down as negative values of τ_{gm} are of no physical significance.

Several features of the charging delay are worthy of emphasis. First, because τ_{gm} is associated with charge exchange with the depletion layer in the velocity saturated regions, regions II and III, τ_{gm} is zero until velocity saturation in the channel is attained. In other words, there is no delay in g_{m0} until current saturation begins in the $I_D(V_{DS})$ characteristics. Secondly, given that X increases as the drain voltage is increased, τ_{gm} can be expected to increase also. Lastly, because an increase in gate voltage bias (in the negative sense) causes an increase in d but a decrease in X, τ_{gm} decreases for increases in gate bias. These dependences give rise to the general form for τ_{gm} of Fig. 6.15, which shows an experimental set of curves. Finally, τ_{gm} physically represents the effective charging time of the velocity saturated regions of the channel,

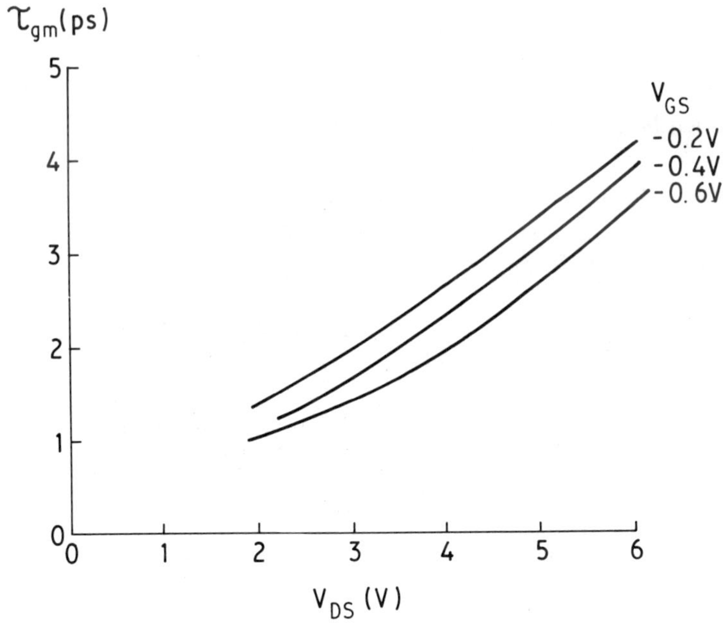

Fig. 6.15 Experimental variation of delay factor in current generator for a 0.5 μm gate MESFET.

regions II and III, which is quite different from the transit time through these regions, X/v_{sat}.

Having derived expressions for the circuit elements associated with the intrinsic part of the FET comprised of regions I, II and III of Fig. 6.7, including inductive effects associated with wave propagation across the width of the gate, we are in a position to abstract an intrinsic equivalent circuit. This discussion will, for simplicity, ignore the output conductance, g_o, which is principally determined by substrate current. That particular topic will be taken up later.

6.4 Intrinsic equivalent circuit

By definition, the intrinsic equivalent circuit neglects any parasitic resistance between the drain-end of region III and the drain metallization itself. Similarly, source resistance is neglected, so that virtual electrodes S' and D' exist as in Fig. 6.16. If we extract from this figure what has already been determined in Figs 6.12–6.14, we arrive at Fig. 6.17. The outstanding question is where should node A be connected for the equivalent network to represent as closely as possible the physical processes occurring within the device itself.

6.4.1 Configuration

Evidently, the options for connecting the node A in Fig. 6.17 are either to the node B between C_I and R_i, or to S'.

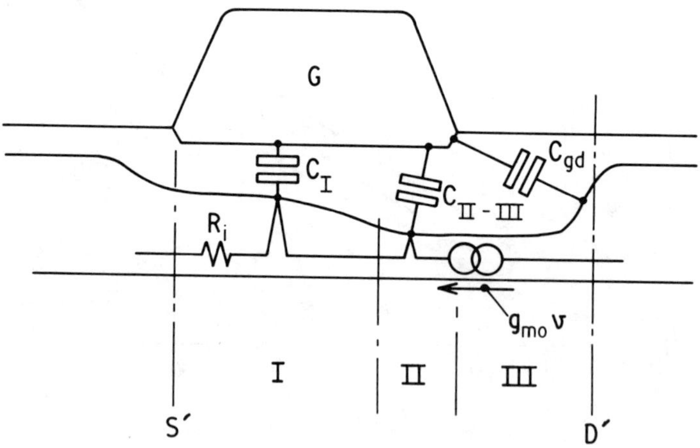

Fig. 6.16 Illustrating virtual electrodes S′ and D′, and the circuit element equivalents of the depletion layer and channel.

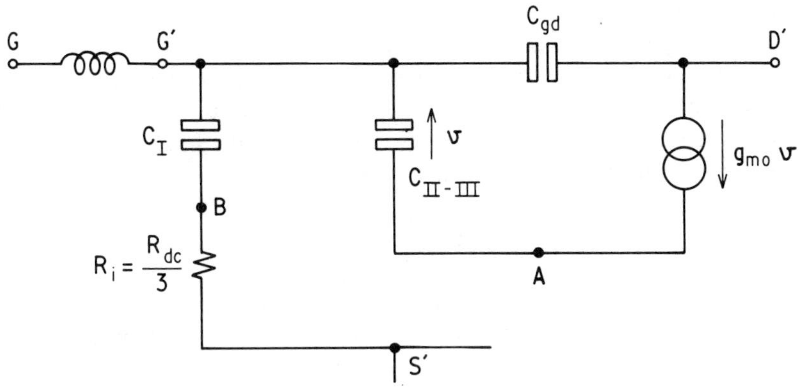

Fig. 6.17 Partial equivalent circuit, as extracted from Fig. 6.16.

At first sight, the logical connection is A–B, as follows from the physical reality that region II adjoins region I. This connection immediately places C_{II-III} in parallel with C_I, which is reasonable. Second thoughts, however, reveal that the connection means that the current $g_{m0}v$ alters the potential across R_i, thereby introducing feedback on the voltage v across the C_I, C_{II-III} combination (which governs the current generator $g_{m0}v$). More correctly, the current through region II (i.e. the generator) is a current of electrons which is extracted from region I, and as such is determined entirely by the conditions existing in region I without any interfering effects of feedback from region II. No such objection arises from the connection of A to S'. This connection, however, implies that the voltage across C_{II-III}, which governs the current generator, is now equal to the voltage across the C_I–R_i combination. Indeed, we have already argued in relation to eqns (6.17) and (6.18) that the voltage which drives the current generator is $V_{IIG} = V_{S'G}$ (the equality coming from the fact that $E_S L_G$ is approximately constant, i.e. the voltage drop across region I is constant, so region I itself behaves as a short circuit at microwave frequencies). Thus, the signal voltage which governs the current generator is *incidentally* the voltage across the C_I–R_i combination; *fundamentally*, it is the voltage across C_{II-III}. A difficulty with this connection is that it makes the signal voltages across C_I and C_{II-III} different, whereas the physical configuration of the depletion layer suggests C_I and C_{II-III} should be immediately in parallel, with the same signal voltage across both. The dilemma arises because region I behaves as a resistance so far as region I itself is concerned, but as a small-signal short circuit so far as region II is concerned.

The correct connection is A to S'. Because C_I and C_{II-III} are not separately known experimentally, the resultant circuit model is not useful. It becomes useful if we transform the network to place C_{II-III} in parallel with C_I to give a single capacitor representing the total gate-channel capacitance. The signal voltage driving the generator must remain that across G'–S', but widespread usage tacitly assumes v remains across $C_{\text{II–III}}$ in its transformed position. To reflect common practice rather than technical accuracy, we also anticipate that v will transform to the position shown in Fig. 6.18.

Considering the part of this network shown in Fig. 6.19,

$$Z_{\text{JN}} = \frac{1 + j\omega C_{\text{I}} R_{\text{i}}}{-\omega^2 C_{\text{I}} C_{\text{II–III}} R_{\text{i}} + j\omega (C_{\text{I}} + C_{\text{II–III}})}$$

From previous sections we have $C_{\text{I}} \approx 0.19$ pF, $C_{\text{II–III}} \approx 0.025$ pF $R_{\text{i}} \approx 5\ \Omega$ as typical values, so at a frequency of 10 GHz the denominator evaluates to

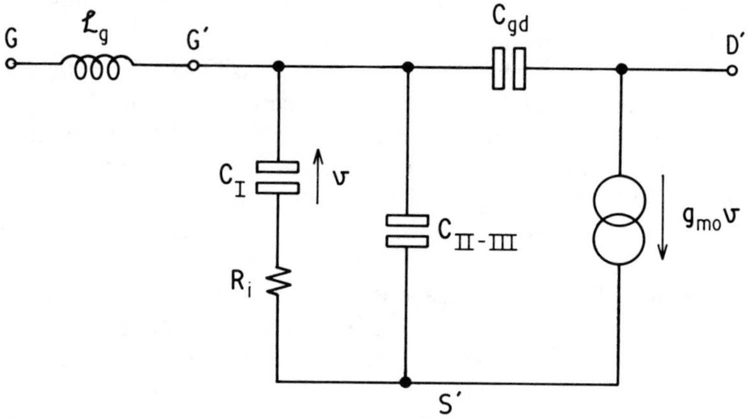

Fig. 6.18 Result of connecting the node A in Fig. 6.17 to the virtual source, S′.

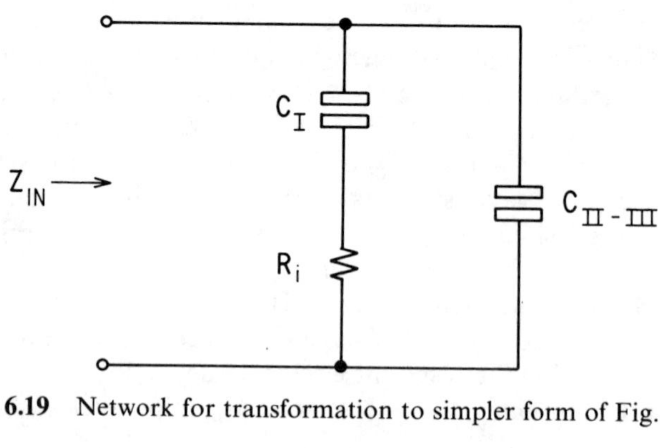

Fig. 6.19 Network for transformation to simpler form of Fig. 6.20.

$-9.4\times10^{-5}+\mathrm{j}1.35\times10^{-2}$. Evidently the leading term can be neglected, and so the expression for Z_{IN} may be simplified to

$$Z_{IN}\approx\frac{C_I R_i}{C_I+C_{II-III}}+\frac{1}{\mathrm{j}\omega(C_I+C_{II-III})}$$

which is the input impedance of a circuit as in Fig. 6.20. We note that $C_I/(C_I+C_{II-III})\approx0.9$ so that, given the uncertainties in the expression for R_i, we may as well simplify the resistive element in Fig. 6.20 to just R_i. We thus finally obtain the equivalent circuit for the intrinsic part of the GaAs MESFET as in Fig. 6.21. Satisfyingly, the capacitance C_{II-III} has ended up in parallel with

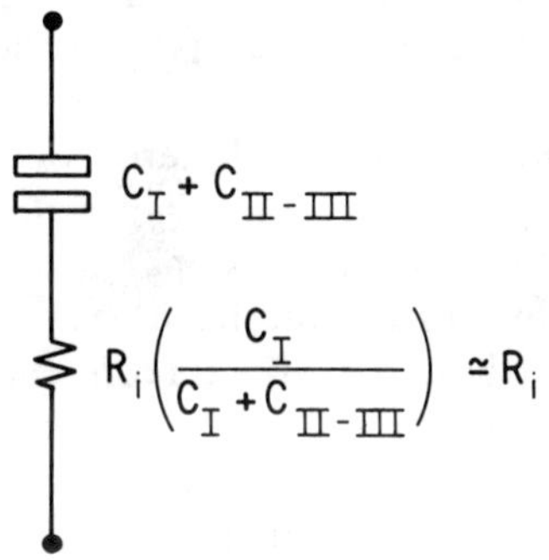

Fig. 6.20 Approximate equivalent of network in Fig. 6.19.

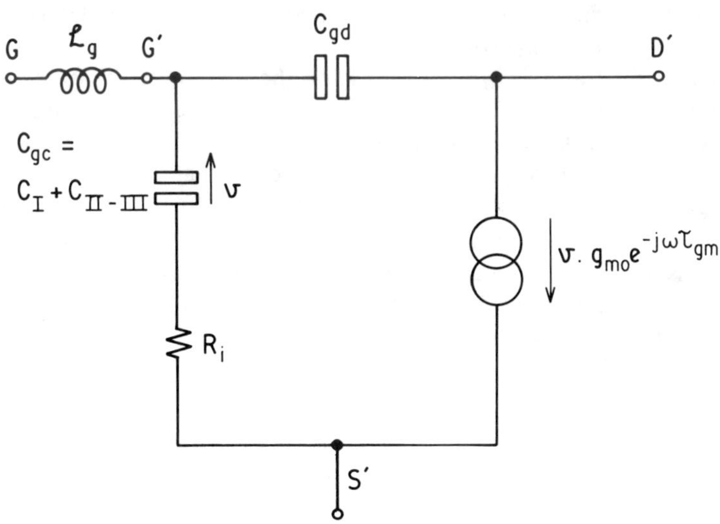

Fig. 6.21 GaAs FET equivalent circuit of Fig. 6.18 with transformed gate-source branch.

C_1, where the connection of node A to node B in Fig. 6.17 would have placed it preferentially anyway. Thus,

$$C_{gc} = \frac{\epsilon Z_G L_G}{d}\left[1 + \frac{X}{2L_G} - \frac{2d}{L_G + 2X}\right] \tag{6.21}$$

6.4.2 Voltage dependence of the space–charge layer extension, X

The expressions so far deduced in many cases involve the extension X of the depletion layer into the gate-drain space. To be of use, these expressions require a knowledge of X by value, particularly as the bias conditions on the FET are changed.

To arrive at an equation defining X in terms of channel doping, N, gate-drain potential $V_{D'G}$, source-gate potential $V_{S'G}$, gate length L_G and so forth, requires a full solution to the two-dimensional field problem which the FET represents. Nearly all solutions to date have been obtained numerically by computer simulation. Furthermore, to be representative of the conditions which actually occur in GaAs MESFETs, such solutions should allow for the occupied surface-state density D_S in Fig. 6.2. Few of them do.

By treating the problem of determining the shape and extent of the space-charge layer as a quasi-electrostatic problem, an approximate two-dimensional Green's function solution has been successfully obtained [6.4], from which guiding principles can be established that are useful in determining an approximate relationship for X.

In Fig. 6.22 are shown solutions for the charge at the gate–metal to N-layer interface, normalized with respect to the depletion charge qNd†. The distribution of electron charge along the gate is found to be non-uniform, rising towards the drain-end of the gate. For higher X/L_G values, $Q_G(x_2)$ generally increases and becomes somewhat more uniform. For the simple case of $W_R = W_S$, the variation of x-directed electric field along the path P in Fig. 6.23 is sub-linear to a degree which depends upon the ratio X/L_G. The area under the field profile defines the total electrostatic potential between the gate and the virtual drain, and includes the equilibrium Schottky contact potential, V_{BO}, thus,

$$V_{D'G} + V_{BO} = \int_P \boldsymbol{E}_x \cdot \mathbf{dx}$$

†These solutions implicitly allow for suppression of the field or charge singularity at the gate corners [6.2].

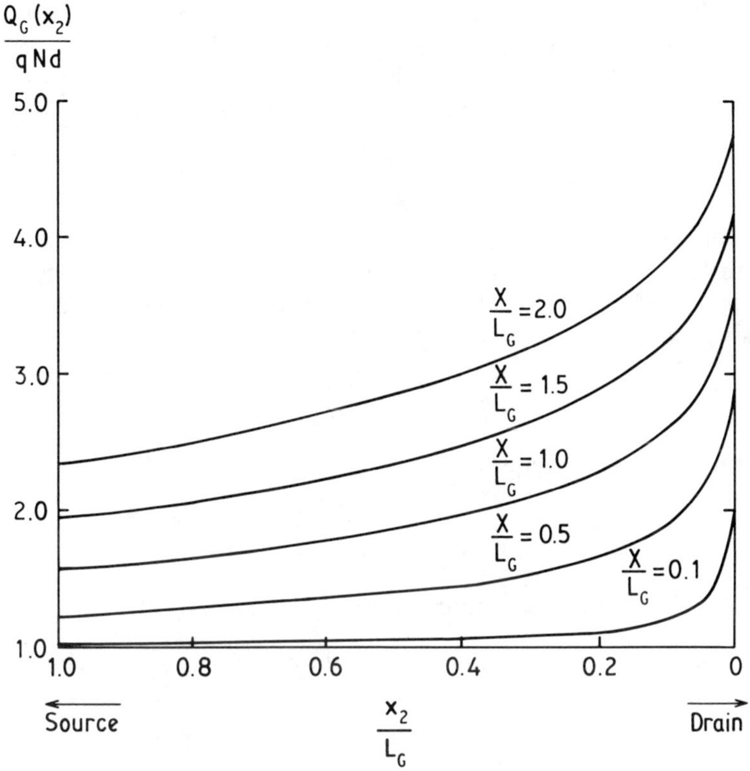

Fig. 6.22 Solutions for electron charge on Schottky gate.

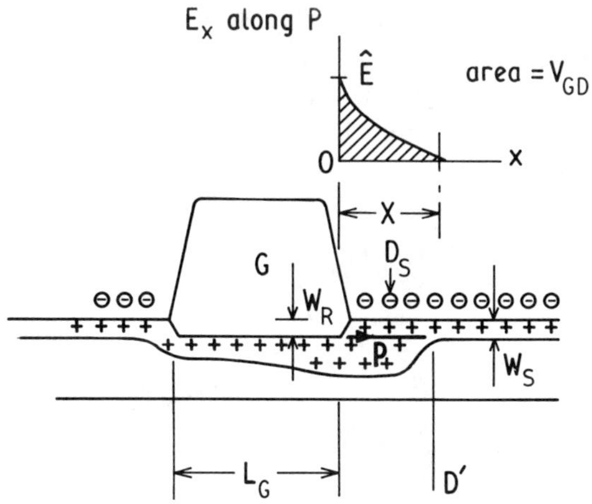

Fig. 6.23 Illustrating form of the x-directed field along a path P.

For values of $X/L_G \approx 0.1$, the relationship between the gate-induced charge Q_G and the x-directed field in the gate-drain space is as shown in Fig. 6.24, in which event a reasonable approximation to the above integral is (taking a triangular approximation for the field profile)

$$V_{D'G} + V_{BO} \approx \frac{qNdX}{2\epsilon} = \frac{qNdL_G}{2\epsilon}\left(\frac{X}{L_G}\right) \tag{6.22}$$

This expression will give an inaccurate estimate for X as X/L_G increases. Nevertheless, it is a useful approximation and allows the generally correct dependences of C_{gc}, C_{gd} and τ_{gm} to be represented for MMIC design and analysis purposes. Thus, as an approximation,

$$\frac{X}{L_G} \lesssim \frac{2\epsilon(V_{D'G} + V_{BO})}{qNdL_G} \tag{6.23}$$

Substituting for d from eqn (6.8),

$$\frac{X}{L_G} \lesssim \left[\frac{2\epsilon}{qN(V_{S'G} + V_{BO})}\right]^{1/2} \times \frac{(V_{D'G} + V_{BO})}{L_G} \tag{6.24}$$

This relationship thus expresses the physical facts that

(a) The depletion layer extension into the gate-drain space, X, increases as $V_{D'G}$ increases.

(b) X decreases as the source-gate voltage increases (i.e. as the gate voltage is increased negatively).

If we note that

$$V_{D'G} = V_{D'S'} + V_{S'G} \tag{6.25}$$

where $V_{S'G}$ is a positive number ($V_{S'G} = -V_{GS'}$), then we are finally in a position to forecast the intrinsic equivalent circuit elements as functions of material and device structure, and bias point. For easy reference, the results are collected together in Table 6.1.

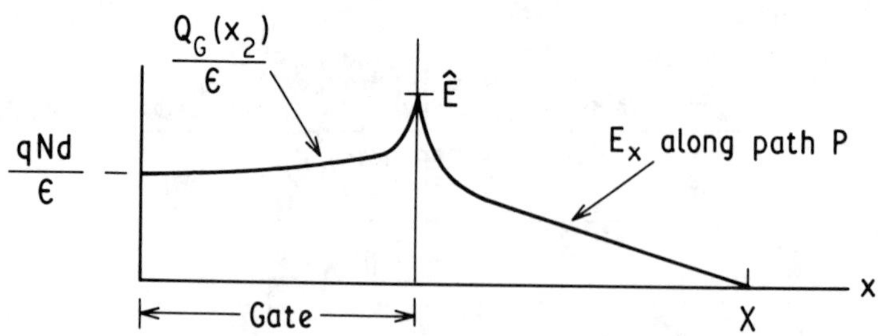

Fig. 6.24 General form of gate charge and electric field solutions.

Table 6.1 Dependences of equivalent circuit elements of the intrinsic FET (Note: These relationships apply only in the current saturation region of operation)

Element	*Equation*	*Eqn no.*
$\mathscr{L}_g$	$\dfrac{\mu_0\, dZ_G}{m^2 L_G}$ (μ_0 = permeability of free space)	(6.5)
C_{gc}	$\dfrac{\epsilon Z_G L_G}{d}\left(\dfrac{1+X}{2L_G}-\dfrac{2d}{L_G+2X}\right)$	(6.21)
C_{gd}	$\dfrac{2\epsilon Z_G}{1+2X/L_G}$	(6.15)
R_i	$\dfrac{v_{sat}L_G}{\mu_0 I_{CH}}$ (μ_0 = low-field mobility at doping N)	(6.7)
g_{m0}	$\dfrac{\epsilon v_{sat} Z_G}{d}$	(6.18)
τ_{gm}	$\dfrac{1}{v_{sat}}\left(\dfrac{X}{2}-\dfrac{2d}{1+2X/L_G}\right)$ ($\tau_{gm}<0$ has no physical significance)	(6.20)
d	$\left[\dfrac{2\epsilon(V_{S'G}+V_{BO})}{qN}\right]^{1/2}$ (pinch-off, V_P, is obtained from this equation as $V_{S'G}$ ($d=W$))	(6.8)
X	$\leqslant\left[\dfrac{2\epsilon}{qN(V_{S'G}+V_{BO})}\right]^{1/2}(V_{D'G}+V_{BO})$	(6.24)
I_{CH}	$qNv_{sat}(W-d)Z_G$	(6.11)
$V_{D'G}$	$V_{D'S'}+V_{S'G}$	(6.25)

In Fig. 6.25 are shown the calculated dependences for the FET considered in previous estimates in this chapter, i.e.

$$L_G = 0.5\ \mu\text{m}$$
$$Z_G = 200\ \mu\text{m}$$
$$W = 0.1\ \mu\text{m}$$
$$N = 3.5\times10^{17}\ \text{cm}^{-3}$$
$$\mu_0 = 2500\ \text{cm}^2\ \text{Vs}^{-1}$$
$$m = 1$$

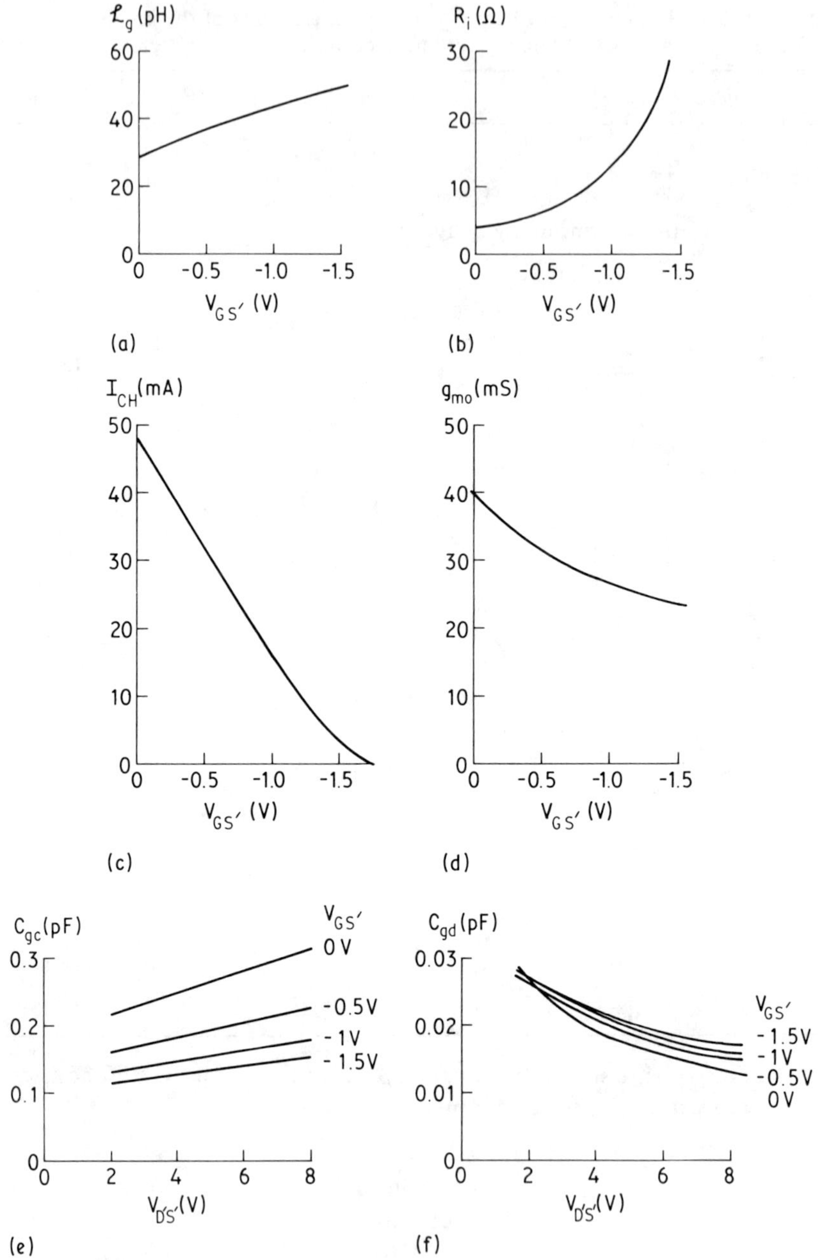

Fig. 6.25 Calculated channel current and equivalent circuit element variations with respect to potentials at the virtual electrodes S′ and D′.

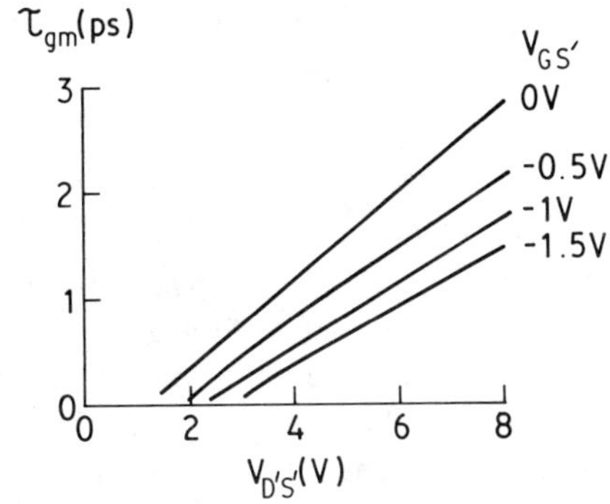

(g)

Fig. 6.25 continued

Figures 25(a)–(d) show those parameters which, at the present level of approximation, exhibit no $V_{D'S'}$ dependence. It is worthwhile to comment upon each briefly in turn.

6.4.3 Gate strip inductance, $\mathscr{L}_g$

Figure 6.25(a) shows there is good reason to expect the series inductance in the gate to vary with bias. Note that the values found here would be divided by four if the FET in question had two gate strips instead of just one. The resultant small variations in inductance may well be masked by uncertainties in measuring *S*-parameters and reducing them to equivalent circuit form.

6.4.4 Channel, or intrinsic, resistance R_i

From Fig. 6.25(b) we see that R_i rises dramatically as pinch-off is approached. In real FETs, R_i is partly determined by substrate current, which is the term generally used to identify current flowing in the semi-insulating material underlying the N-doped channel layer. "Semi-insulating" is a term which applies only to bulk low-field properties. In a GaAs MESFET, the presence of the overlying N-layer, and the existence of high fields (up to several hundred kV cm^{-1}) mean that electrons are injected into the semi-insulating regions, which may be either the substrate itself or an epitaxially-grown buffer layer. Substrate current, and some of its effects, are dealt with in Section 6.6.

6.4.5 Channel current, I_{CH}

Figure 6.25(c) is a typically familiar transfer characteristic showing the FET to have a pinch-off voltage of about 1.7 V. In a practical device, this transfer characteristic is modified by three factors, one of which increases the current reaching the drain for a given gate bias, whereas the others reduce it, as follows.

(a) Substrate current is added to I_{CH} so that $I_D > I_{CH}$; I_{SUBS} imparts a noticeable $V_{D'S'}$ dependence to I_D, as we shall see shortly.

(b) A potential drop $I_D R_S$ occurs across the source-gate "access" region (Fig. 6.26). The channel at the source-end of the gate is thus at a potential $I_D R_S$ which is positive with respect to the source (assumed to be grounded). Hence the potential across the depletion layer at this point is increased by $I_D R_S$, with the result that the depletion depth must be increased. In turn, this implies that the undepleted channel must be reduced in height, thereby reducing the channel (and drain) current.

(c) There is a potential drop $E_S L_G$ across region I (in the x direction) which widens the depletion layer slightly at region II, which is the region where I_D is determined. Hence I_{CH} is slightly reduced from the value (Table 6.1) calculated from the depletion depth at the left end of region I.

6.4.6 Intrinsic transconductance, g_{m0}

Generally speaking, the transconductance of practical devices decreases as the negative bias on the gate is increased. In contrast to the concave-up dependence evident in Fig. 6.25(d), a concave-down shape is frequently seen, at least part of which is due to rapidly falling values of doping in the tail of the profile.

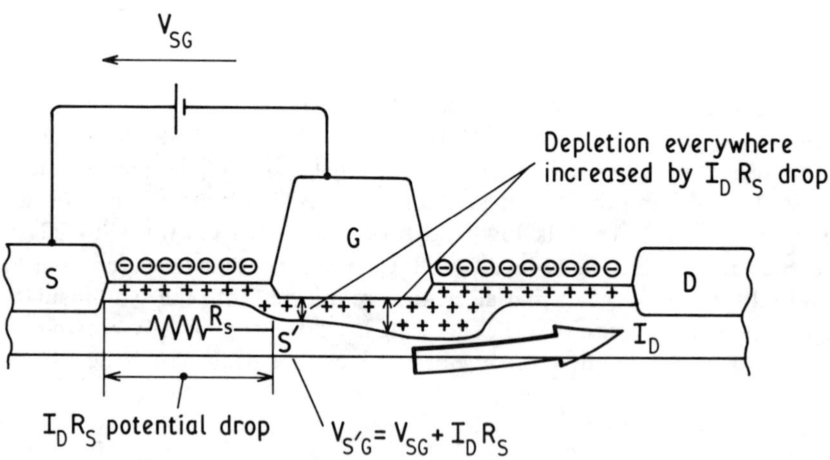

Fig. 6.26 Effect of the $I_D R_S$ drop.

Since g_{m0} by definition is strictly

$$g_{m0} = \partial I_D / \partial V_{S'G}$$

it is clear that substrate current, which adds to the channel current I_{CH} to give the drain current, I_D, will alter the shape of Fig. 6.25(d). Furthermore, because I_{SUBS} is dependent upon $V_{D'S'}$, g_{m0} takes on a dependence upon $V_{D'S'}$ when I_{SUBS} is included, as will become clear in Section 6.6. There are many subtleties involving the drain-source voltage dependence of g_{m0}, particularly the decrease in g_{m0} with increasing $V_{D'S'}$ near open channel which is observed in some FETs, notably those having a shallow gate recess. This decrease would appear to be dependent on changes in interface charge as the FET bias condition is changed, a topic well beyond the scope of this book.

To first order, from eqns (6.18) and (6.8), g_{m0} is a function of channel doping and gate bias only. There is a potential drop $E_S L_G$ along the channel in region I, so the total electrostatic potential across the depletion layer at region II, where current modulation occurs, is not $(V_{S'G} + V_{B0})$ but $(V_{S'G} + V_{B0} + E_S L_G)$. Thus a better approximation for g_{m0}, which expresses this weak dependence upon gate length, is

$$g_{m0} = v_{sat} Z_G \left[\frac{\epsilon q N}{2(V_{S'G} + V_{B0} + E_S L_G)} \right]^{1/2}$$

Note also that in application to practical devices the nominal value of V_{SG} must be corrected for the $I_D R_S$ drop, to obtain $V_{S'G}$.

6.4.7 Gate-channel capacitance, C_{gc}

The increase in capacitance with $V_{D'S'}$, so frequently observed in practical devices, is well represented in Fig. 6.25(e).

6.4.8 Gate-drain capacitance, C_{gd}

Close examination of an experimental plot of C_{gd} versus V_{DS}, with V_{GS} as parameter, reveals a cross-over in the curves at $V_{DS} \approx 2V$. At voltages above the cross-over, the curves are ordered such that C_{gd} is *higher* the more *negative* V_{GS} is made. The fact that this feature is correctly represented in Fig. 6.25(f) means that the space–charge layer in the FET, and the way it responds to the applied gate and drain potentials, is well described by the equations of Table 6.1. A good example of such a plot is to be found in [6.5].

6.4.9 Transconductance delay, τ_{gm}

τ_{gm} can be seen from Fig. 6.25(g) to have the general form anticipated in Fig. 6.15.

6.5 Implications for FET design and usage

Before passing on to matters of energy loss in FETs, parasitic resistance, substrate current effects and parasitic reactances, there is guidance and insight to be derived from the equations of Table 6.1 and the considerations that led to them, especially with regard to the consequences of altering the FET gate length.

It is well known among FET physicists and engineers that the channel doping N is normally higher in FETs of shorter gate length, a practice which stems from the need to maintain a gate-length to channel–thickness ratio of at least 3:1, and preferably 4:1 to 5:1. Such a geometry is required to ensure good current saturation characteristics. However, if a reasonable value of I_{DSS} is to be maintained in the face of reduced values of W, N must be increased (see eqn (6.11)). It is instructive to examine the consequences of smaller L_G and higher N separately.

It should by now be fairly obvious that the really effective parts of the FET are regions II and III. Region II sets the channel opening (W–d) and is the region at which current modulation takes place; region III is the region that supports the gate-drain potential. By comparison, region I performs no useful function. It contributes C_I to C_{gc}, and C_I is the major part of C_{gc}. C_I is, however, unwanted added capacitance which it would be nice to eliminate. Another unwanted property associated with region I is the intrinsic channel resistance, R_i; R_i is proportional to L_G.

Provided the L_G: W ratio remains greater than about 3:1, a reduction in gate length L_G serves the very useful purpose of reducing C_I and R_i without substantially affecting the properties of the useful regions, regions II and III. From Table 6.1 we could anticipate that X will be unaffected, C_{gd} will be slightly reduced, τ_{gm} slightly increased, and g_{m0} unaffected. An almost proportionate decrease in C_{gc} and R_i can be anticipated, however. (The increase in gate inductance $\mathscr{L}_g$ is of no direct importance at frequencies up to J band as $\mathscr{L}_g$ is so small anyway, especially for devices with two or more gate fingers.) In terms that matter to the circuit engineer, it is the reduction in C_I and R_i which are the real benefits to frequency response which follow from using shorter gate length FETs.

When an increase in channel doping accompanies a decrease in L_G, the benefits are partly lost. The depth of depletion d decreases, and so C_{gc}

increases. An increase in transconductance can be expected, however, which is desirable. A decrease in X will also result, so τ_{gm} will decrease, which is helpful. C_{gd} will increase, which is a change we would prefer to do without. Finally, $\mathscr{L}_g$ will decrease, which may be of little significance given its small value anyway.

The foregoing small-signal considerations need to be supplemented by a discussion of the effects of reducing gate length upon output power (in watts per millimetre of gate width). $\bar{P}_O$ is in part determined by the drain-gate breakdown voltage V_{DGB} at which impact ionization first occurs in the gate-drain region. If the only change to the FET structure is a reduction in gate length, both V_{DGB} and $\bar{P}_O$ will decrease in direct proportion. Thus all facets of the microwave characteristics required of, for example, a foundry FET in all its intended applications must be carefully evaluated before finalizing the device material and structural specification.

6.6 Output conductance and other microwave effects of substrate current

Depending upon the layer growth technique used to generate the active channel layer, and the concentration of deep and shallow donor and acceptor levels in the substrate employed, profiles of the general character of Fig. 6.27 can be realized. The steepness of the profile tail depends not only upon the growth method, but also upon the concentration of unoccupied deep acceptors in the substrate which are available to compensate the shallow donor (dopant) in the tail region. If a buffer layer is grown before the active layer, it obviously matters whether the impurities out-diffuse from the substrate and are incorporated in the buffer during growth.

In Fig. 6.28 is shown an operating FET in cross-section. There is a sharp rise in field in the channel across region II such that the peak field near the drain-end of the gate may reach several hundred kV cm^{-1}. (It is a result of solving the two-dimensional Poisson equation in a velocity-saturated FET channel that the sharp rise in field occurs even without electron accumulation in the channel; quite simply, it comes about from field changing vector direction from being largely x-directed to y-directed, as shown in Fig. 6.8.) Electrons are heated by these intense fields, and so the electron occupancy of available electron states in the rising conduction band, towards the buffer or substrate, increases (Fig. 6.29). In other words, the channel of current-carrying electrons tends to widen and spread into the substrate (or buffer), giving rise to a component of current I_{SUBS} in the substrate. The electric field strength along the initial part of the path shown for I_{SUBS} in Fig. 6.28 is easily large enough to keep the electrons moving at the scatter-limited velocity, v_{sat},

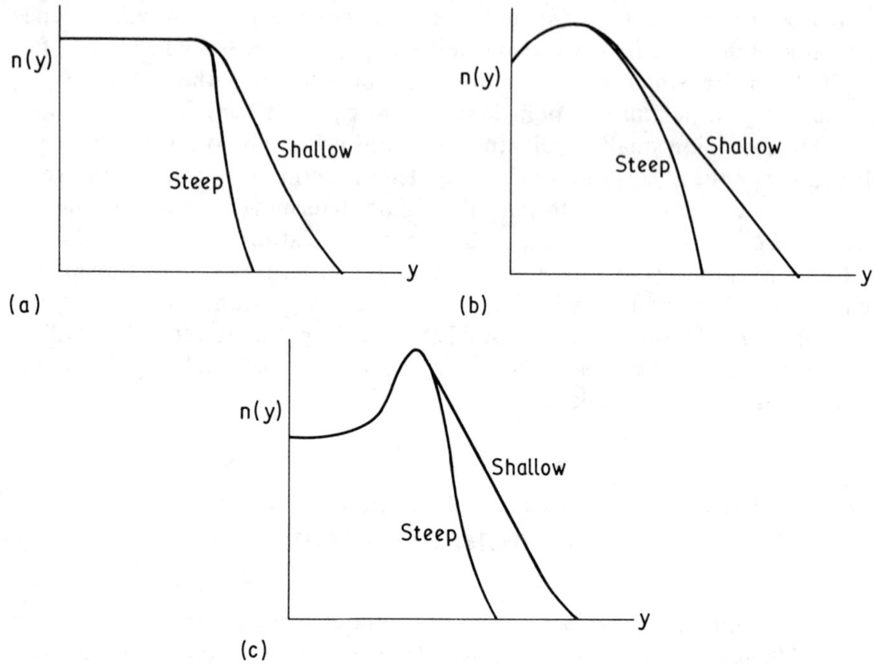

Fig. 6.27 Typical channel doping profiles with alternative steep and shallow tails.

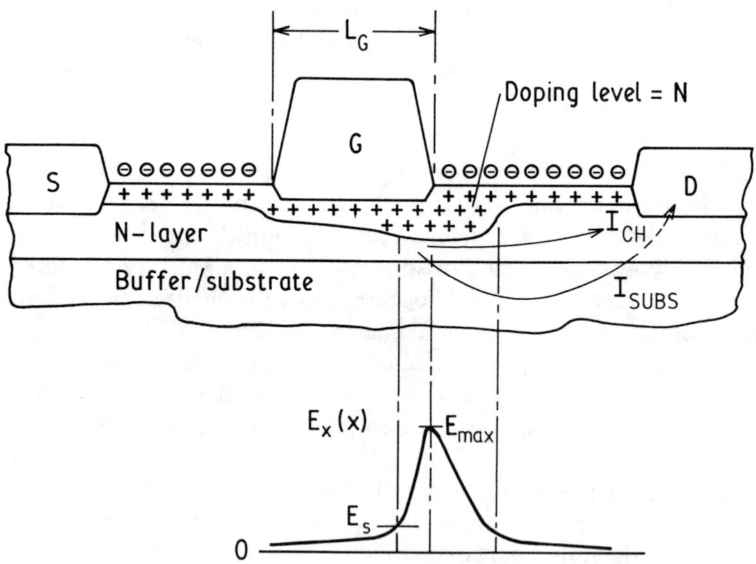

Fig. 6.28 Current paths in layer and through buffer or substrate.

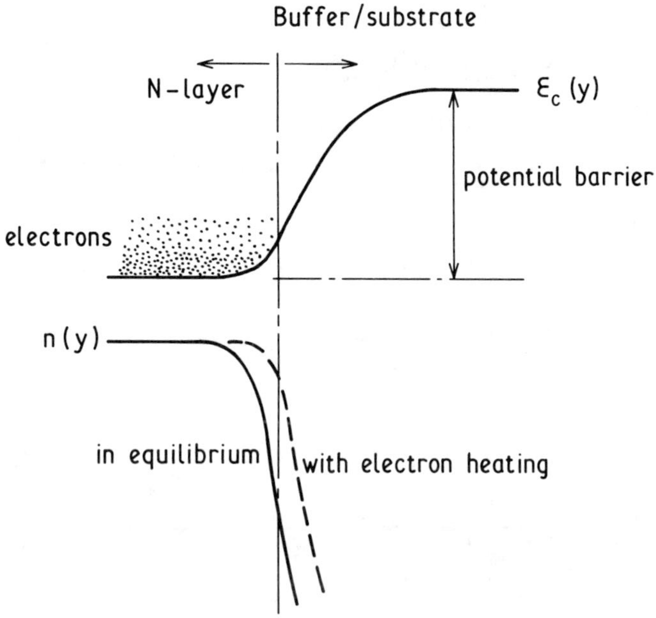

Fig. 6.29 Spreading of carriers in channel towards buffer or substrate due to electron heating in channel.

so that I_{SUBS} may be worked out by calculating the sheet concentration of hot electrons injected into the substrate and multiplying by $qv_{sat}Z_G$. Such electrons are eventually collected by the drain together with the channel electron current I_{CH}. Thus,

$$I_D = I_{CH} + I_{SUBS} = I_{CH}(1 + I_{SUBS}/I_{CH}) \tag{6.26}$$

An approximate analytic theory for the ratio I_{SUBS}/I_{CH} yields the results shown in Fig. 6.30. The interfacial barrier of energy which is effective in confining the electrons to the channel is ϕ_B which, as we shall see shortly, corresponds to the energy rise in the potential barrier of Fig. 6.29 in one mean free path for a hot electron. ϕ_B may have any value up to the full height of the potential barrier itself, which is about 0.7 eV.

Substrate current has a number of very interesting and very important consequences for FET behavior, both at dc and under microwave conditions:

(a) It may constitute several tens of percent of the total drain current.

(b) It is a physical mechanism which contributes to the formation of the knee in the I_D (V_{DS}) characteristics at which current saturation sets in. (dc analyses of the GaAs MESFET based on channel transport alone yield knee

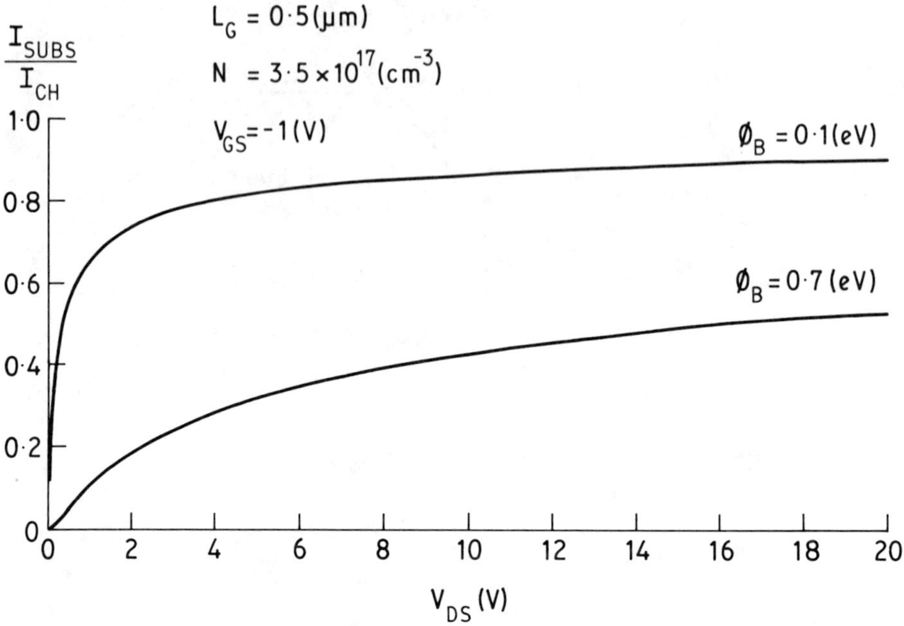

Fig. 6.30 Calculated ratios of substrate current to channel current for 0.5 μm gate FET with shallow profile tail ($\phi_B = 0.1$ eV) and steep profile tail ($\phi_B = 0.7$ eV).

voltages of typically several tenths of a volt, and ascribe the origin of the knee to the onset of velocity saturation in the channel. In sharp contrast, the present theory yields a knee voltage of typically 2 V and ascribes it to saturation of the numbers of electrons that are scattered into the substrate.)

(c) The slope of the total current I_D with respect to V_{DS} defines the output conductance g_o as

$$g_o \triangleq \partial I_D / \partial V_{D'S'}$$

Although analyses neglecting substrate current yield a finite value for g_o due to a shortening of region I as $V_{D'S'}$ is increased (see, for example, reference [6.6]), the magnitudes forecast are often far smaller than those observed in practice, i.e. values of output resistance $r_o = (1/g_o)$ are calculated to be somewhat larger than those typically found in practice. Basing the calculation of g_o upon the substrate component of drain current yields values which are reasonably well in keeping with practical values.

(d) The transconductance is strictly defined as

$$g_{m0} = \frac{\partial I_D}{\partial V_{S'G}} = \frac{\partial}{\partial V_{S'G}} \left[I_{CH} \left(1 + \frac{I_{SUBS}}{I_{CH}} \right) \right] \qquad (6.27)$$

Providing $dI_{SUBS}/dV_{S'G}$ is not zero, g_{m0} will be changed by the presence of substrate current. In the present formalism (Fig. 6.28), the substrate current is taken to consist of hot electrons scattered into the substrate after heating by the high electric field in the channel. The carriers are scattered into the substrate at about the end of region II (i.e. underneath the drain-edge of the gate), in which event they have already undergone modulation by the gate fields. In other words, substrate current is modulated, as well as the channel current, by the applied gate voltage or signal.

Hence substrate current affects g_{m0}. In place of the simple expression in Table 6.1, g_{m0} takes on a dependence upon $V_{D'S'}$. Figure 6.31 shows the calculated form for the example FET operated at a gate bias of 0.5 V, assuming a potential barrier of 0.7 eV and

$$L_G = 0.5\ \mu m$$
$$Z_G = 200\ \mu m$$
$$N = 3.5 \times 10^{17}\ cm^{-3}$$
$$W = 0.1\ \mu m$$
$$\phi_B = 0.7\ eV$$
$$V_{S'G} = 0.5\ V$$

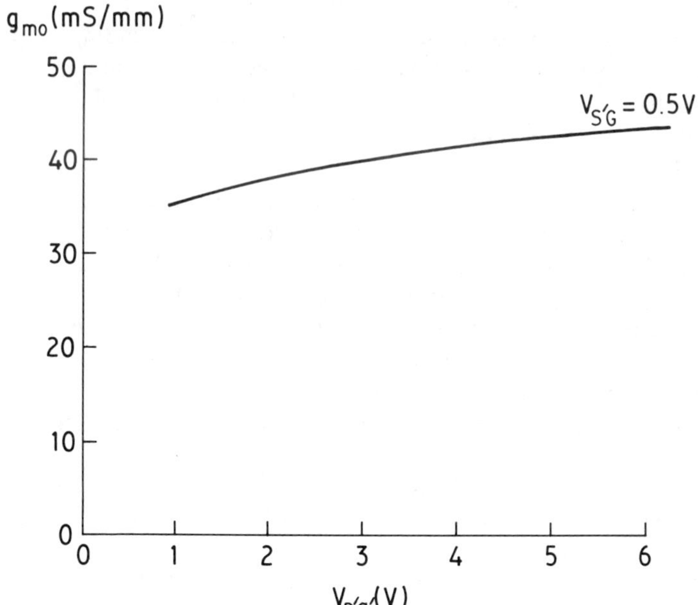

Fig. 6.31 Typical calculated variation of intrinsic transconductance due to substrate current.

As already remarked, g_{m0} can exhibit a decrease with increasing V_{DS}, especially for values of V_{SG} that leave the channel fairly well open, whereas the kind of behavior seen in Fig. 6.31 is more evident at higher gate bias values. It is not possible, however, to account fully for decreasing g_{m0} without taking account of the complex interactions between the shape and extent of the space-charge layer and interface-state charge, and the effect of the $I_D R_S$ potential drop upon the channel opening.

(e) Because the hot electrons which are injected into the substrate occupy regions of semiconductor where there are very few positive ionic charges, they constitute a negative space charge. This space charge alters the shape and extent of the space–charge layer in the channel, because the electrons in the substrate offer a negative space charge upon which electric field arising on positive ionic charge in the depletion layer can terminate. Hence the field and potential distributions are changed. Changes in the shape and extent of the depletion layer show up, in circuit terms, as changes in C_{gc} and C_{gd}. They are not of major importance to the development of yield-driven design principles, and for expediency we shall ignore them.

The value of ϕ_B we use in estimating the magnitude of the injected charge should strictly be a self-consistent value; the injected charge will itself tend to repel electrons from the positions they would otherwise occupy in the tail of the profile under equilibrium conditions, which imparts to the free-electron tail a sharpness more governed by Debye length spreading than by the gradient in dopant ion density. Hence the flow of substrate current tends, if anything, to increase the value of the interfacial potential barrier, ϕ_B. A self-consistent value of ϕ_B cannot be established using the analytic methods employed here.

(f) Finally, it is worth noting that the space charge of electrons in the substrate, near the drain end of the gate, would in many respects have an effect indistinguishable from an accumulation of electrons in the channel itself which has often been discussed in the literature [6.7–6.9]. We simply note that because the buffer or substrate does not constitute a perfectly constraining boundary for electrons in the channel, the boundary conditions which lead to the channel accumulation are to a degree removed anyway, and so we attach no practical significance to the electron accumulation possibility.

The most important of (a)–(f) above from a microwave point of view is (c), the output conductance: (d) and (e), are second order effects, whereas (c) is first order.

Before passing to a calculation of g_o, we will reconsider the barrier energy ϕ_B and the factors which govern it.

Consider, as in Fig. 6.32, an electron at position y' in the channel. When such an electron is scattered in the direction of the substrate, it will travel for

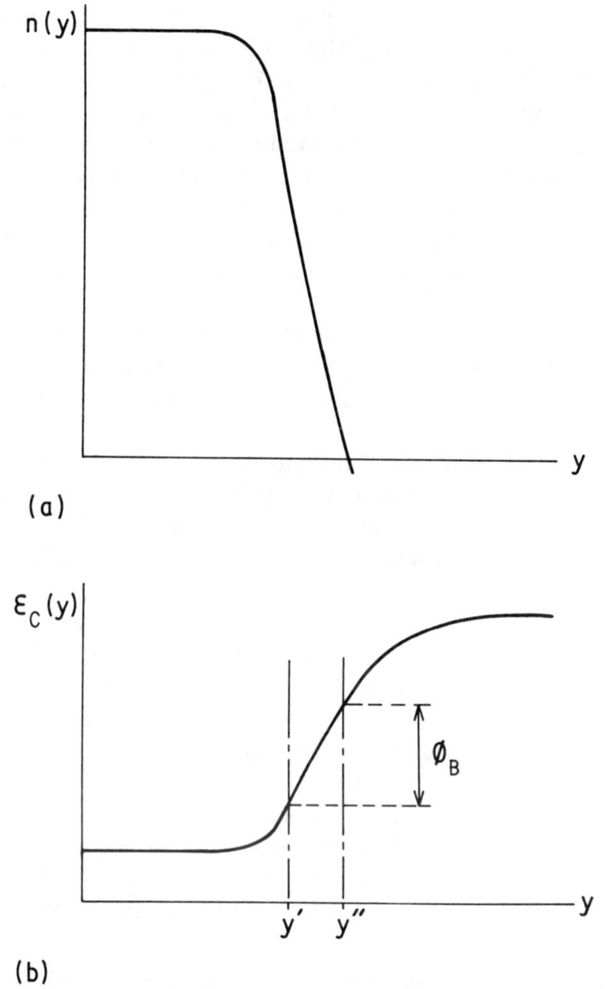

Fig. 6.32 Definition of the N-layer to buffer-substrate interface potential, ϕ_B.

one mean free path into the substrate providing its motion is unimpeded by the barrier of potential represented in Fig. 6.32(b) by the conduction band edge $\mathscr{E}_C(y)$. When this is the case, the electron would again be scattered at y'', a condition which clearly requires that the electron starting from y' should have had energy greater than ϕ_B. If, on the other hand, the kinetic energy of the electron starting at y' is less than the rise in conduction band energy over one mean free path, then the carrier will be reflected by the barrier before being scattered by a phonon. Hence the effective barrier of potential which confines the electrons to the channel is the energy by which the conduction band increases in one mean free path for a hot electron. In turn, the gradient in $\mathscr{E}_C(y)$ is in the first instance reflected in the gradient in the doping profile in the transition from the N-layer to the buffer-substrate. In other words, shallowly graded profiles, as might result from implantation into substrates having a relatively low concentration of compensating acceptors, will give rise to a low value of ϕ_B and high substrate current (Fig. 6.30). Steeply graded profiles, on the other hand, as might result from having a P-layer underlying the N-layer give rise to a high value of ϕ_B and small substrate current.

Taking into account the overall form of the profile, the drain characteristics of FETs fabricated from the profiles of Figs 6.27(b) and (c) could be expected to have the general form shown in Figs 6.33(a) and (b) respectively.

There is a second important mechanism giving rise to substrate current. Referring to Fig. 6.8, there are transverse fields passing across the substrate to N-layer interface whose direction under the gate is such as to lower ϕ_B. The magnitude of this barrier lowering is dependent upon the FET geometry and bias conditions; it gives rise to injection of electrons into the substrate much like injection in P–N junctions as treated in elementary texts on solid-state device theory. Injection by barrier lowering occurs divorced from considerations of electron heating. In the GaAs MESFET, therefore, a mixed cool electron and hot electron substrate current flow occurs.

For the purposes of MMIC analysis and design, we are, above all, concerned with knowing what slope in the $I_D(V_{DS})$ characteristics arises from substrate current, i.e. we need to establish an expression for the microwave drain conductance, g_o†. To calculate the effects of barrier lowering is difficult, whereas a reasonably compact expression can be written for the hot electron component. The first may be allowed for very roughly by halving the output

†It is well known that the output conductance in many GaAs FETs decreases by a factor of typically 3 between dc and microwave frequencies [6.10]. That is to say, at a given bias point, the slope of the $I_D(V_{DS})$ curves might be about 1000 Ω, whereas at 10 GHz the output resistance might be 350 Ω. The break point in this transition is in the range of a few tens of Hz to a few tens of kHz. Charge exchange with deep levels located either at the device surface or in the N-layer to buffer–substrate transition region have been argued to be responsible for this output conductance dispersion [6.11]. Charge exchange with surface states in particular leads to a flattening of the drain characteristics at low frequencies. At frequencies or pulse rates where the states cannot follow the applied voltage, the drain characteristics become steeper, i.e. g_o increases, as may be confirmed by using pulses going down to sub-microsecond durations to measure the $I_D(V_{DS})$ characteristics [6.12, 6.13]. By ignoring charge exchange with surfaces in our treatment, we automatically calculate the microwave value of output conductance, as required.

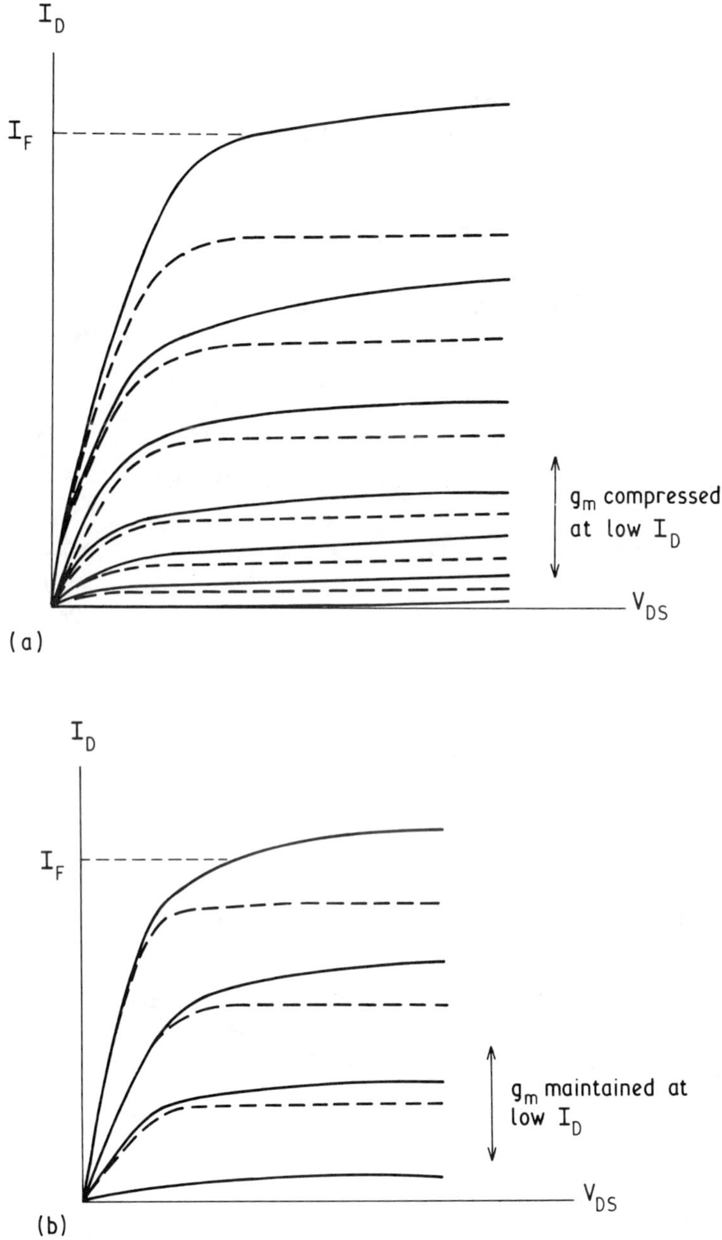

Fig. 6.33 General character of drain characteristics: (a) for profile of Fig. 6.27(b); (b) for profile of Fig. 6.27(c);—shallow profile tail (substrate current large); ----steep profile tail (substrate current small).

resistance arising from the second. Thus we estimate the output resistance, $r_o = (1/g_o)$, as

$$r_o = \frac{1}{2}\frac{V_{D'S'}}{I_{CH}}\left(\frac{kT_e}{\phi_B}\right)\frac{(2\exp(\phi_B/kT_e)-1)^2}{\exp(\phi_B/kT_e)} \tag{6.28}$$

where

$$(3/2)kT_e = f(\tau_{\mathscr{E}}, V_{S'G}, V_{D'S'}) \tag{6.29}$$

is the kinetic energy acquired by an electron in the channel, and $\tau_{\mathscr{E}}$ is an energy relaxation time with a typical value of about 1 ps.

Calculated values of r_o are plotted in Fig. 6.34 for a FET with

$L_G = 0.5$ μm $\qquad \phi_B = 0.7$ eV

$Z_G = 200$ μm $\qquad \tau_{\mathscr{E}} = 1$ ps

$N = 3.5 \times 10^{17}$ cm^{-3} $\qquad W = 0.1$ μm

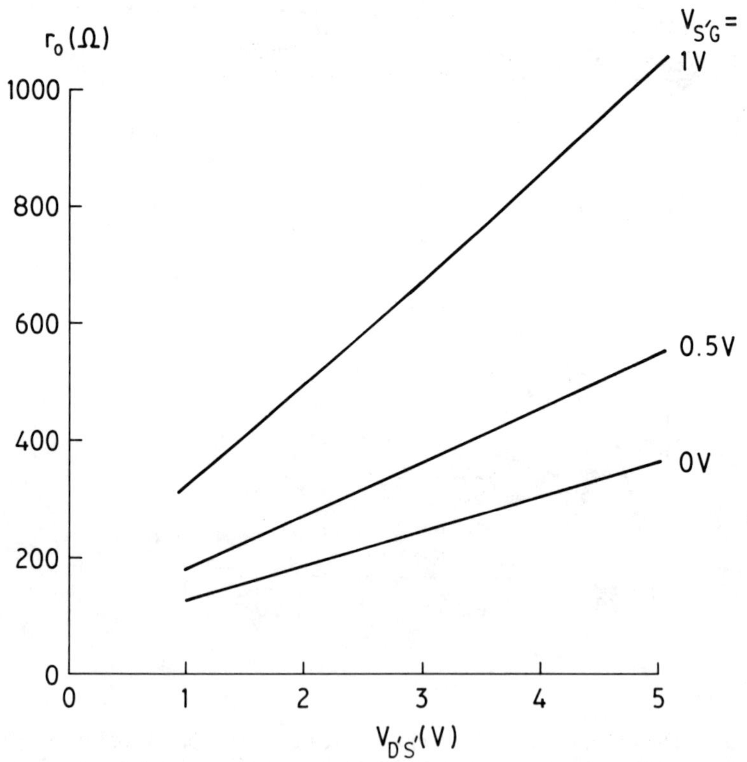

Fig. 6.34 Calculated variation of output resistance for a typical 0.5 μm gate FET.

This is similar to the form measured (see for example [6.5]). As already remarked, however, r_0 is dependent upon charge exchange with deep levels, i.e. it is dependent upon factors which are difficult to control technologically. Even when dispersion in r_0 is circumvented by considering high enough frequencies, uncertainties in surface-state occupancy lead to uncertainties in the depletion-layer size and shape, hence to uncertainties in the fields within the device, and finally to uncertainties in r_0. Until such time as tight control of deep levels is achieved, therefore, it is unrealistic to pretend that close estimates of r_0 can be made. A simpler expression provides values that are often good enough. Careful study of the foregoing equation for r_0 reveals that its form allows the approximation

$$r_0 \geqslant 4V_{D'S'}/I_D \tag{6.30}$$

over much of the range of $V_{D'S'}$. Perhaps the most serious consequence of this approximation is that a plot of r_0 versus $V_{D'S'}$ in the current saturation region of operation passes through the origin, whereas for the full expression it does not.

6.7 Effect of surface charge, non-uniform doping and gate recess depth

All the expressions so far deduced for the equivalent circuit elements assume the channel doping is uniform. At the level of approximation inherent in those equations it may be shown that, for a device with a non-uniform doping profile operated at a given bias under small-signal conditions, there is an equivalent uniform channel doping and layer thickness. Since we are concerned with determining how the small-signal behavior of an MMIC about a given bias point may vary with changes in channel doping, and how MMICs might be designed to be tolerant to such process changes, much useful work can be done with the equations assuming a uniform doping. The principal weakness of the assumption becomes evident when variations in the bias itself are considered; with the equations developed here the doping in the channel at the edge of the new depletion position d will be the same as it was at the original value of d, whereas in the non-uniformly doped channel of a real FET it would be different.

Changes in occupied surface state density have a profound effect upon the microwave behavior of FET-based MMICs through the changes they make to the values in the equivalent circuit. In a manner similar to the dispersion observed in output conductance, the extrinsic transconductance g_m is almost always dispersive in the general manner shown in Fig. 6.35. In the case of g_m, it

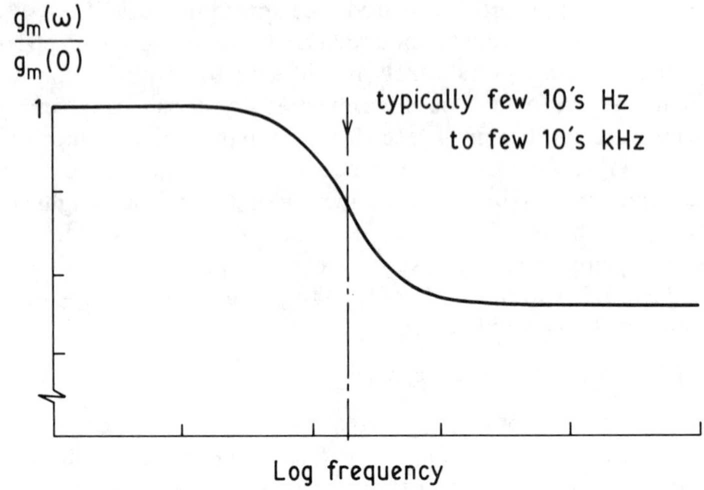

Fig. 6.35 Low-frequency transconductance dispersion as often exhibited by sub-micron gate GaAs FETs.

has been shown that charge exchange with surface states is responsible for the observed dispersion [6.14, 6.15]. The phenomenon is well enough understood to allow the technique to be used as a means of assessing the voltage dependence of occupied surface-state density, D_S, and of assessing various dielectric protection or passivation methods [6.15]. This work shows that D_S may vary with gate voltage in the range

$$dD_S/dV_{SG} \approx 3 \times 10^{11} - 3 \times 10^{12}\ \mathrm{cm}^{-2}\ \mathrm{V}^{-1}$$

Even though such surface states have characteristic frequencies well below the microwave region, they nevertheless are a source of irreproducibility in microwave behavior of a GaAs MESFET or MMIC. The sequence of events is as follows:

(a) Electron charge held in surface states alters the boundary conditions which in part determine the size and shape of the space–charge layer, and hence the field and potential distributions within the device. At microwave frequencies D_S cannot follow the signal but remains constant. Its value will vary with bias, and it may vary from FET to FET, even among FETs on the same MMIC.

(b) The size and shape of the depletion layer, however, determine the equivalent circuit elements through d and X (see Table 6.1), which in turn govern the S-parameters. Hence any uncertainty in D_S is akin to operating the

FET at an uncertain bias point. In other words, the effective bias point varies from FET to FET, even on the same MMIC.

Subject to the reservations made in connection with the way the channel doping, N, is represented in bias variation analyses, tolerance to varying surface conditions can to some extent be assured through designing circuits to be tolerant of changes in bias condition, as follows from item (b) above.

Another omission in our derivation has been the gate recess depth, W_R. Broadly speaking, this has a complementary effect to surface charge. For example, introducing a 200 Å recess in a layer of 3.5×10^{17} cm^{-3} doping would be roughly equivalent to reducing the occupied surface-state density by 7×10^{11} cm^{-2} electrons. A key point to note is that gate recesses do nothing to reduce the uncertainty associated with surface charge; any uncertainty in the gate recess depth actually exacerbates the problem of irreproducibility overall.

As a matter of principle, analytic equations similar to those of Table 6.1 may be written when surface charge, gate recess depth and non-uniform profiles are included in the analysis. Such a treatment is well beyond the aims of the present book. Qualitatively, however, the changes indicated in Table 6.2 can be expected from changing recess depth and surface charge.

Table 6.2 General effects of changes in gate recess depth and occupied surface state density
(Note: The channel layer thickness W under the gate is taken to remain constant as the recess depth changes)

Increases in W_R tend to			
decrease C_{gc}	increase C_{gd}	decrease r_o	decrease V_{DGB} and V_{DSB}
(relatively weak effect)	(relatively weak effect)	(not so weak effect)	(strong effect)
Increases in D_S (or W_S) tend to			
increase C_{gc}	decrease C_{gd}	increase r_o	increase V_{DGB} and V_{DSB}
(relatively weak effect)	(relatively weak effect)	(not so weak effect)	(strong effect)

The last column has been added for completeness, even though gate-drain avalanche has not been treated in this section. The drain-gate voltage at which impact ionization is initiated in the gate-drain space, V_{DGB}, is dependent to first order upon gate recess depth and surface charge. For this reason, it tends to exhibit a wide scatter among nominally identical devices, because surface charge tends to be rather poorly controlled.

For the same reason, sizable deviations from the form and dependences of r_o upon $V_{D'S'}$ and $V_{S'G}$ shown in Fig. 6.34 are often observed. Output resistance depends quite sensitively upon surface charge through the way D_S alters the channel field, and hence the extent to which the carriers become hot (i.e. their random component of kinetic energy, $(3/2)kT_e$ in eqn 6.29). Thus any uncontrolled changes in D_S, which seem to occur particularly with respect to source-gate voltage V_{SG} [6.11], can be expected to give rise to major distortion of the curves in Fig. 6.34.

Leaving aside matters connected with gate recesses, surface charge and non-uniform doping profiles, the set of equations so far developed (Table 6.1, together with eqns (6.28–6.30) offers a sufficient representation of the intrinsic FET, and its microwave behavior, to develop some meaningful routines for MMIC design. The equations are exceedingly compact, thus allowing the development of circuit design and analysis programs that will run on a well-expanded personal computer. They must be augmented, however, by methods to estimate the geometric capacitances associated with the electrode structure, inductance associated with via-holes, and series parasitic resistances in the gate-source and gate-drain regions.

6.8 Series parasitic resistances R_s and R_d and effect on equivalent circuit

In Fig. 6.36 is shown a cross-section of a FET with R_s and R_d identified. Each has two components: the bulk resistance of the access region (of length L_{SG} or L_{GD}), and a contact resistance component. If the specific contact resistance of the ohmic contacts to the N-layer is $\rho_c(\Omega\ \text{mm})$, then

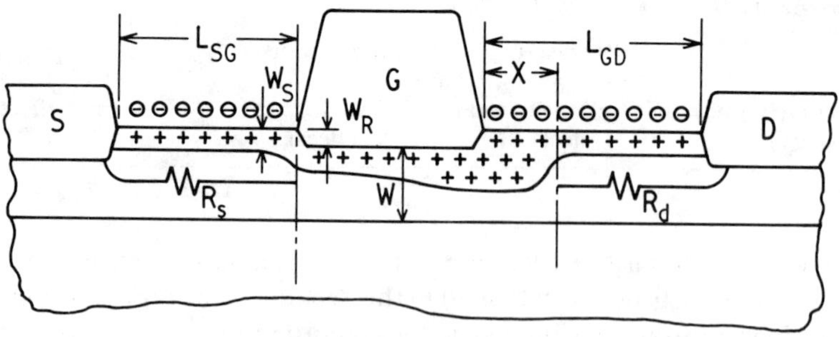

Fig. 6.36 Parasitic resistances R_s and R_d.

$$R_{\text{contact}} = \frac{\rho_c}{Z_G}\ \Omega \tag{6.31}$$

Typically, ρ_c is in the range 0.1–0.5 (Ω mm) for Au–Ge–Ni contacts alloyed to N–GaAs doped to $10^{17}\ \text{cm}^{-3}$, or greater so that for FETs of 200 μm gatewidth,

$$R_{\text{contact}} \approx 0.5\text{–}2.5\ \Omega$$

6.8.1 Source resistance, R_S

Straightforward analysis of the resistance of the source-gate access region yields (assuming $W_R = W_S$ for simplicity)

$$R_{SG} \approx \frac{L_{SG}}{qN\mu_o W Z_G} \tag{6.32}$$

For a FET with

$$L_{SG} = 1\ \mu\text{m}$$

$$N = 3.5\times 10^{17}\ \text{cm}^{-3}$$

$$\mu_o = 2500\ \text{cm}^2\ \text{V s}^{-1}$$

$$W = 0.1\ \mu\text{m}$$

$$Z_G = 200\ \mu\text{m}$$

we find $R_{SG} \approx 3.6\ \Omega$. Thus the total value of R_S might be

$$R_S = R_{\text{contact}} + R_{SG} \approx 4\ \Omega \text{ to } 6\ \Omega \tag{6.33}$$

This resistance must now be included in the equivalent circuit. So far as source resistance is concerned, we do not distinguish between the dc value R_S and the value at microwave frequencies, R_s.

Source resistance modifies the source-gate and drain-gate potentials at which the equivalent circuit elements of Figs 6.25 and 6.34 were evaluated. By simple loop analysis, the following voltage relationships may be written from Fig. 6.37:

$$V_{SG} = V_{S'G} - I_D R_s \tag{6.34}$$

$$V_{D'S} = V_{D'S} + I_D R_S \tag{6.35}$$

$$V_{DS} = V_{D'S} + I_D R_D = V_{D'S'} + I_D(R_S + R_D) \tag{6.36}$$

$$V_{DG} = V_{D'G} + I_D R_D \tag{6.37}$$

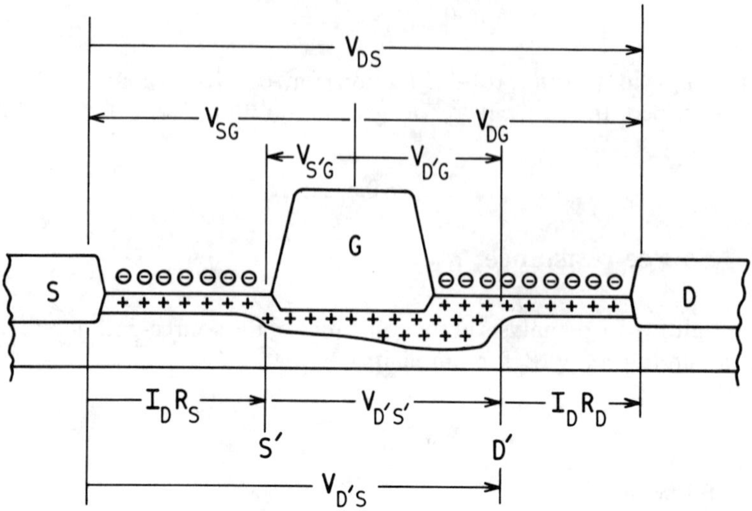

Fig. 6.37 Relationship between GaAs FET voltages.

These equations may be worked in either direction, of course, to give the terminal voltages (undashed set) in terms of the virtual electrode potentials (dashed set), or vice versa. To pursue the example in Fig. 6.25, at $I_{CH} = 30$ mA (and neglecting I_{SUBS}),

$$V_{SG} = 0.55 - (0.03 \times 5) = 0.4 \text{ V}$$

where we have assumed $R_S = 5\ \Omega$. This calculation may be repeated at all I_{CH} values, and Figs 6.25(a)–(d) replotted versus V_{SG} accordingly. Recalling that $V_{GS} = -V_{SG}$, this procedure yields the changes shown in Fig. 6.38. It shows that, for a fixed applied gate-source voltage, increasing R_S causes

(a) an increase in gate inductance, $\mathscr{L}_g$;
(b) an increase in channel resistance, R_i;
(c) a decrease in channel (and drain) current; and
(d) a decrease in transconductance.

The effect of the $I_D R_S$ drop on C_{gc}, C_{gd} and τ_{gm} cannot be determined by simple graphical construction. Since we have

$$V_{S'G} = V_{SG} + I_D R_S$$

or, equivalently,

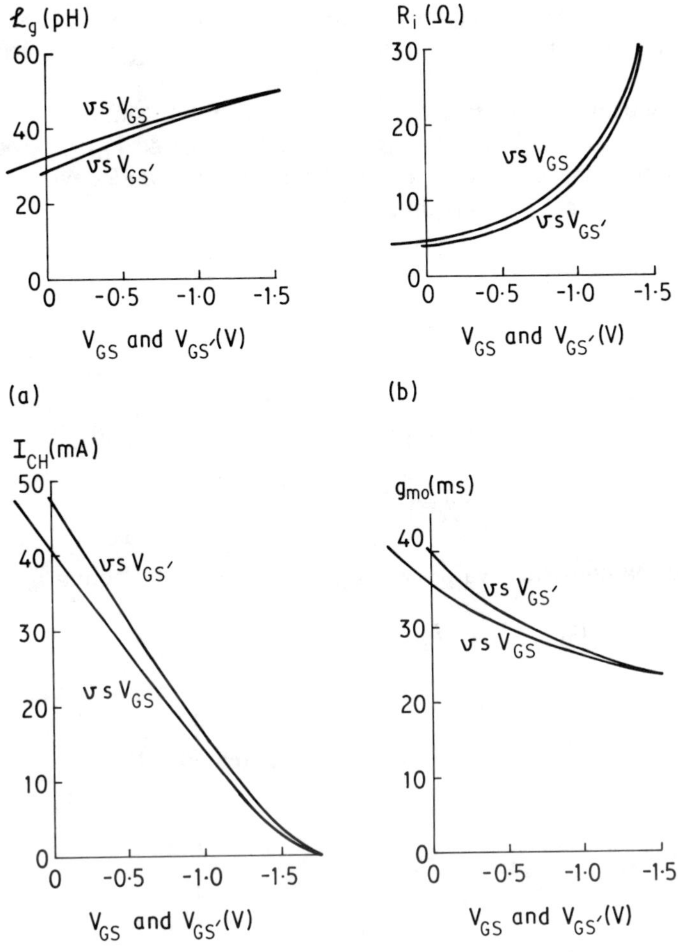

Fig. 6.38 Channel current and equivalent circuit element variations accounting for voltage drops across R_S and R_D.

$$V_{GS'} = -(V_{SG} + I_D R_S) = V_{GS} - I_D R_S$$

increases in R_S have the effect of

(a) Hardly affecting the -1.5 V parametric curves in Figs 6.25(e)–(g) (i.e. the curves near pinch off); and

(b) Moving the parametric curves for $V_{GS'} = 0$, -0.5 V etc. nearer the curves close to pinch off.

Hence the curves tend to be bunched up compared with Figs 6.25(e)–(g). Furthermore, they tend to be stretched out along the V_{DS} axis in accordance with eqn (6.36).

In practical FET structures, therefore, it is not $V_{S'G}$ and $V_{D'S'}$ which are clean, crisp, values such as 0.5 V and 3.0 V; it is V_{SG} and V_{DS}. The procedure for calculating the equivalent circuit from material and device structural specifications in real FETs is thus

(a) from the equations

$$V_{S'G} = V_{SG} + I_D R_S$$

$$I_D \approx I_{CH}$$

$$I_{CH} \approx qNv_{sat}Z_G(W-d)$$

$$d = \left[\frac{2\epsilon(V_{S'G}+V_{B0})}{qN}\right]^{1/2}$$

determine self-consistent values of $V_{S'G}$ and $I_{CH} \approx I_D$ (neglecting I_{SUBS});

(b) determine $V_{D'S'} = V_{DS} - I_D(R_S + R_D)$; (6.38)

(c) evaluate C_{gc}, C_{gd}, g_{m0} etc. at the $V_{S'G}$ and $V_{D'S'}$ values thus determined.

The results are plotted in Fig. 6.39 for the FET treated in Fig. 6.25 for $R_S = 5\ \Omega$ and $R_D = 5\ \Omega$.

6.8.2 Drain resistances, R_D and R_d

We distinguish between the dc resistance, R_D, and the microwave resistance R_d. The two differ because the space–charge layer penetrates into the gate-drain space by a distance X, which itself responds at microwave frequencies. Thus the length of the bulk region $(L_{GD} - X)$, which determines the drain resistance, is itself varying in sympathy with the microwave signal.

Let V be the dc voltage across the region $(L_{GD} - X)$, and let I_{CH} be the dc current passing through it.
Then

$$R_D = \frac{V}{I_{CH}} \qquad \text{(dc resistance)}$$

$$R_d = \partial V/\partial I_{CH} \qquad \text{(microwave resistance)}$$

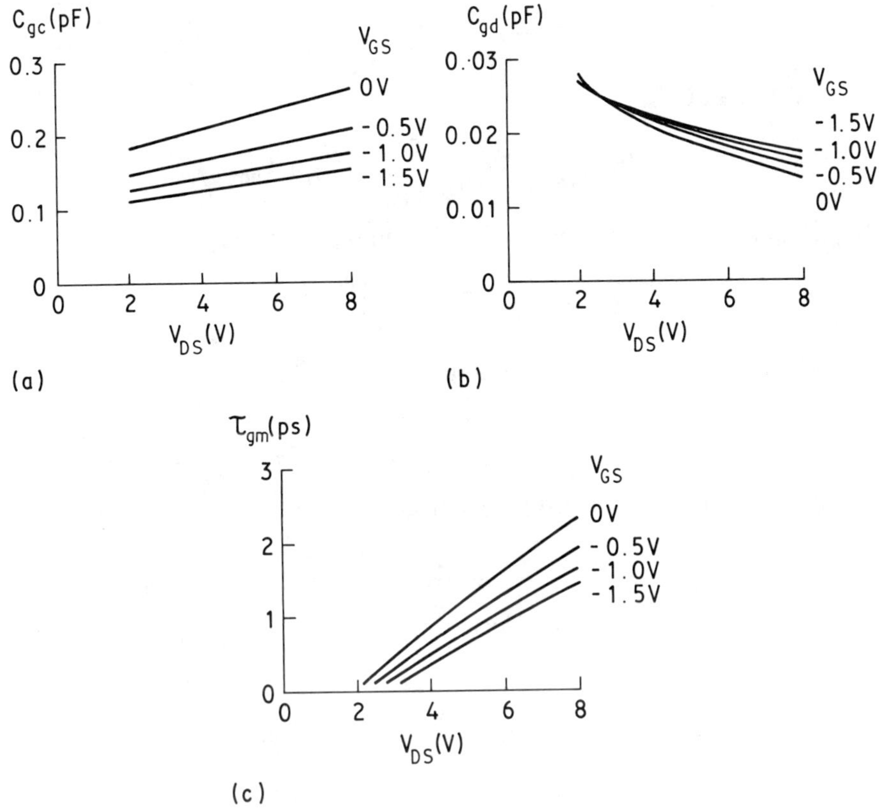

Fig. 6.39 Variations in FET capacitances and transconductance delay with FET terminal voltages V_{SG} and V_{DS}.

We have

$$V = I_{CH}\frac{(L_{GD} - X)}{qN\mu_0 WZ_G}$$

so

$$R_D = \frac{L_{GD} - X}{qN\mu_0 WZ_G}$$

Differentiating the equation for V with respect to itself (i.e. V):

$$1 = \frac{1}{qN\mu_0 WZ_G}\left[\frac{\partial I_{CH}}{\partial V}(L_{GD} - X) - I_{CH}\frac{\partial X}{\partial V}\right]$$

Rearranging, we have

$$R_d = \frac{L_{GD} - X}{qN\mu_0 W Z_G + I_{CH}(\partial X/\partial V)}$$

Table 6.1 gives us

$$X \approx \left[\frac{2\epsilon}{qN(V_{S'G} + V_{BO})}\right]^{1/2} (V_{D'G} + V_{BO})$$

Neglecting the small microwave signal component in $V_{S'G}$, we find

$$\frac{\partial X}{\partial V} \approx \left[\frac{2\epsilon}{qN(V_{S'G} + V_{BO})}\right]^{1/2} \frac{\partial V_{D'G}}{\partial V}$$

Noting that $V_{D'G} = V_{DG} - V$, we obtain

$$\frac{\partial X}{\partial V} = -\left[\frac{2\epsilon}{qN(V_{S'G} + V_{BO})}\right]^{1/2}$$

and so, after substituting also for I_{CH} from Table 6.1,

$$R_d = \frac{L_{GD} - X}{qNZ_G\left\{\mu_0 W - v_{sat}(W-d)\left[\frac{2\epsilon}{qN(V_{S'G} + V_{BO})}\right]^{1/2}\right\}}$$

or

$$R_d = \frac{L_{GD} - X}{qN\mu_o WZ_G\left(1 - 2\frac{v_{sat}}{\mu_0}\frac{W-d}{W}\frac{\epsilon}{qNd}\right)}$$

i.e.

$$R_d = \frac{R_D}{1 - 2\frac{v_{sat}}{\mu_0}\frac{W-d}{W}\frac{\epsilon}{qNd}} \tag{6.39}$$

Inserting typical numbers into the second term in the denominator shows it to be very small compared with unity, so that for present purposes we have the result that

$$R_d \approx R_D$$

The mathematical development to distinguish between R_d and R_D is important. Carrying the self-same reasoning forward to the more complex case when gate recesses and surface charge are accounted for shows that R_d can, as a matter of principle, be negative in some devices (notably those with very shallow recesses operated near open channel). Hence any routine for reducing measured S-parameters to equivalent circuits should allow for the possibility that R_d could be less than zero.

6.9 Gate resistance

Gate strips have bulk resistance determined by the strip's dimension in the direction of gate current flow, which is the width of the transistor, Z_G, and the cross-sectional area of the strip, $(L_G \times h)$.

The height h of the gate strip is normally less than the skin depth

$$\delta = \left(\frac{2}{\omega\mu\sigma}\right)^{1/2} \tag{6.40}$$

where μ and σ are, respectively, the permeability and conductivity of the gate metal. For the usual gate metals, the values as in Table 6.3 apply.

Table 6.3 Permeability, resistivity and skin depth at 10 GHz of the usual gate metals

	μ_0 (H m^{-1})	$\rho = 1/\sigma$ (Ω m)	δ at 10 GHz (μm)
Gold	$4\pi \times 10^{-7}$	2.44×10^{-8}	0.79
Aluminium	$4\pi \times 10^{-7}$	2.83×10^{-8}	0.85

Thus, since $h \leqslant \delta$, the whole height of the gate metal contributes to its conductance. The dc resistance of the strip is

$$R_G = \frac{\rho Z_G}{h L_G} \tag{6.41}$$

Referring back to Fig. 6.4, and neglecting the inductance for a moment (which has already been dealt with in Section 6.3.1), it becomes apparent that not all the current input from the generator and gate pad at the left end of the strip flows all the way across the strip width Z_G. Current is progressively syphoned off into the distributed capacitance represented by the depletion layer, with the result that the voltage drop across the width of the strip is reduced, i.e. under microwave conditions the effective series resistance of the strip R_g is less than R_G, its resistance at zero frequency. It may be shown (see, for example, [6.16]) that

$$R_g = \frac{R_G}{3} = \frac{\rho Z_G}{3hL_G} \tag{6.42}$$

(Note that this argument did not apply to the gate strip inductance $\mathscr{L}_g$ because, as we saw in Section 6.3.1, the principal purpose of including $\mathscr{L}_g$ in the equivalent circuit is to represent the effects of wave propagation delay in the gate; the value of $\mathscr{L}_g$ needed to do this required no factor of 1/3.)

If the gate of width Z_G is divided into m equal strips connected in parallel, then each strip has resistance

$$\frac{\rho(Z_G/m)}{3hL_G}$$

at microwave frequencies so that, upon connecting m of these in parallel,

$$R_g = \frac{\rho Z_G}{3m^2hL_G} \tag{6.43}$$

e.g. for $m=1$, and a gold gate of $h=0.5$ μm, $L_G=0.5$ μm, $Z_G=200$ μm, at 10 GHz,

$$R_g \approx 6.5\ \Omega$$

(For $m=2$, $R_g \approx 1.6\ \Omega$).

6.10 Geometric capacitance

In the physicist's terms, whenever a FET is biased for operation at microwave frequencies, electric field arising on some metal parts of the transistor terminate on other parts. Field arises on positive charge and terminates on negative charge, both on the surface of metals†; the rate of change of this charge with respect to the potential between the two metal parts, which is the line integral of the field along any path between them, gives the capacitance between the two metal parts. In GaAs MESFETs, we refer to such capacitances as geometric in nature, because they are a property of the detailed FET geometry, and to distinguish them from space-charge capacitance.

Geometric capacitance may typically amount to about 10% of both the gate-channel and gate-drain capacitances, as we shall see. It must be properly accounted for if the best work in MMIC analysis and design is to be done. Unfortunately, owing to the diversity of plan view geometries in use, and the variation in pad sizes and separations (see, for instance, Fig. 6.1), only broad

†Actually, the field penetrates into a metal to about three skin depths.

guidelines for methods to estimate the various contributions to geometric capacitance may be given here.

We shall take as an illustrative example the co-planar feed layout shown in Fig. 6.40: such a plan view is used for FETs that can be microwave probed using, for example, the probe equipment marketed by CASCADE Inc. [6.17]. For easy reference the following subscripts will be used, in combination with each other, when discussing geometric capacitances,

a = air path

bp = back plane

dp = drain pad

ds = drain-source

gp = gate pad

gd = gate-drain

gs = gate-source

l = layer (N-layer)

s = source

Starting at the left end in Fig. 6.40, there is capacitance C_{gps} from the gate pad to the adjacent co-planar sources; there is both an air path, and a path through the substrate. There is also capacitance C_{gpbp} from the gate pad to the

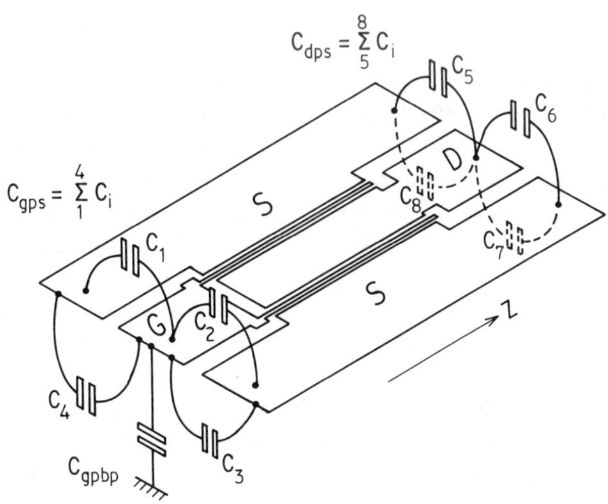

Fig. 6.40 Pad-to-pad, and pad-to-backplane geometric capacitances.

back plane, which is the metallized reverse side of the wafer. It makes a difference whether the back place is grounded or left floating during FET operation. Proceeding to the right end, there are analogous capacitances at the drain pad, C_{dps} and C_{dpbp}, except that the latter now includes capacitance of the drain finger, as will be indicated on Fig. 6.42.

Looking now at the FET in cross-section in the x–y plane (Fig. 6.41), there are capacitances C_{gsa} and C_{gda} between the gate strip and the source and drain contacts respectively. It is most important to note that there is an air path only. Through the semiconductor there is no field path linking the gate to the drain and source metallizations; all field along those paths is screened by charge in the depletion layer.

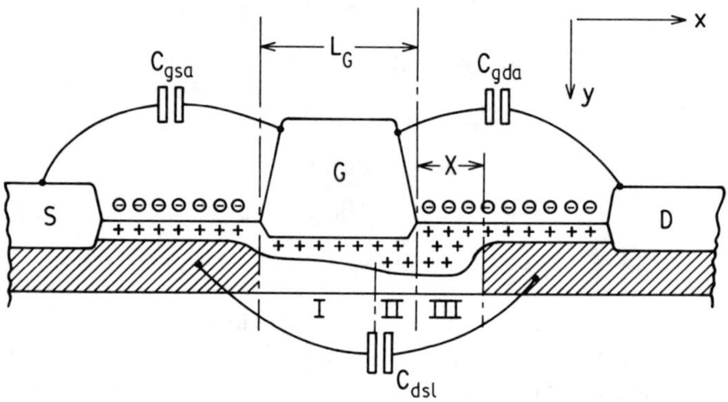

Fig. 6.41 Inter-electrode geometric capacitances.

Either side of the active or intrinsic part of the FET, regions I, II and III, there are bulk regions of the N-layer, as shaded in Fig. 6.41, between which field lines pass. These are the self-same regions that give rise to resistances R_s and R_d. If the FET is drawn more to scale, as in Fig. 6.8(b), it will be seen that the shaded regions constitute two thin, co-planar, sheets facing each other at a distance $(L_G + X)$. We denote the drain-source capacitance associated with these parts of the N-layer as C_{dsl}.

Considering lastly the FET in cross-section in the y–z plane, as in Fig. 6.42, there is capacitance C_{gpdp} between the gate pad and the drain finger-with-pad comprised of paths through both the air and the substrate.

In equivalent circuit terms, the geometric components of capacitance appear as in Fig. 6.43. C_{gsa} and C_{gda} pose a difficulty in that they strictly should be distributed across $\mathscr{L}_g$ and R_g. Their location either at G or G' would serve just as well. For most practical purposes the capacitances can be relocated as shown in Fig. 6.44. Although some of the proofs and transformations are

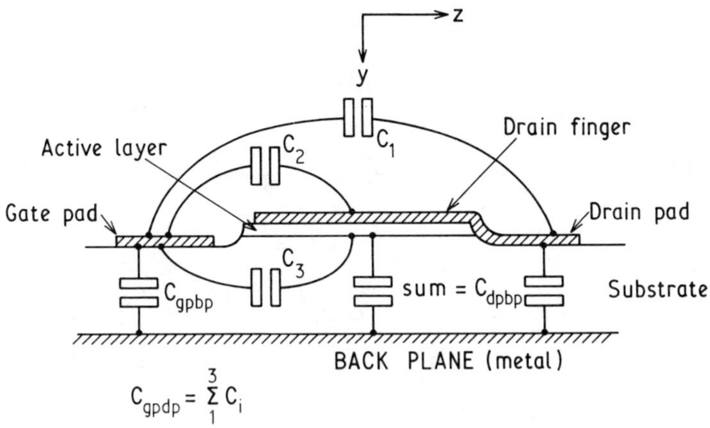

Fig. 6.42 Geometric capacitances in a section across the width of a FET.

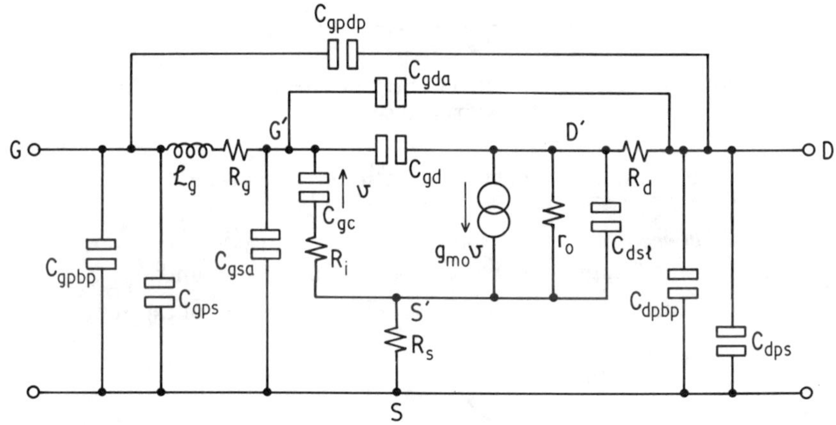

Fig. 6.43 Addition of geometric capacitances to the basic equivalent circuit.

relatively simple (see, for example, Figs 6.19 and 6.20 and the associated text, which can be applied here in connection with the location of C_{gsgeom}), generally it is helpful to study the validity of moving capacitance connections with a circuit simulator such as TOUCHSTONE© or SUPER-COMPACT©. It is also helpful to recognize the likely errors in the estimates of the respective geometric capacitance contributions when making this judgment; for instance, if a shift in connection point requires a 10% change in value, yet the value can only be estimated to 20% in the first place, the connection may as well be made at the most convenient point anyway.

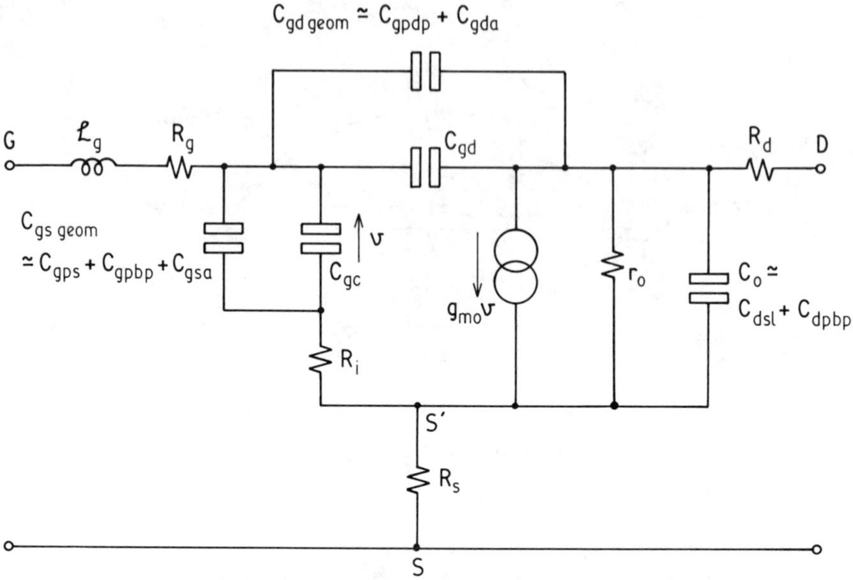

Fig. 6.44 Simplification and reduction of the equivalent circuit including geometric capacitances.

Capacitance data for four basic structures are required.

(a) *Pad between co-planar grounds* (Fig. 6.45), used for estimating C_{gps}, C_{dps} and C_{gpdp}, and may also be used for estimating C_{gsa} and C_{gda}.

(b) *Isolated pad above a ground plane* (Fig. 6.46), used for estimating C_{gpbp}

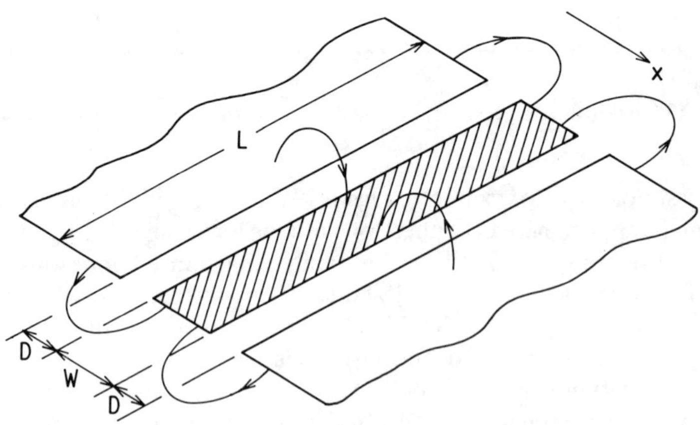

Fig. 6.45 Strip between two semi-infinite planes.

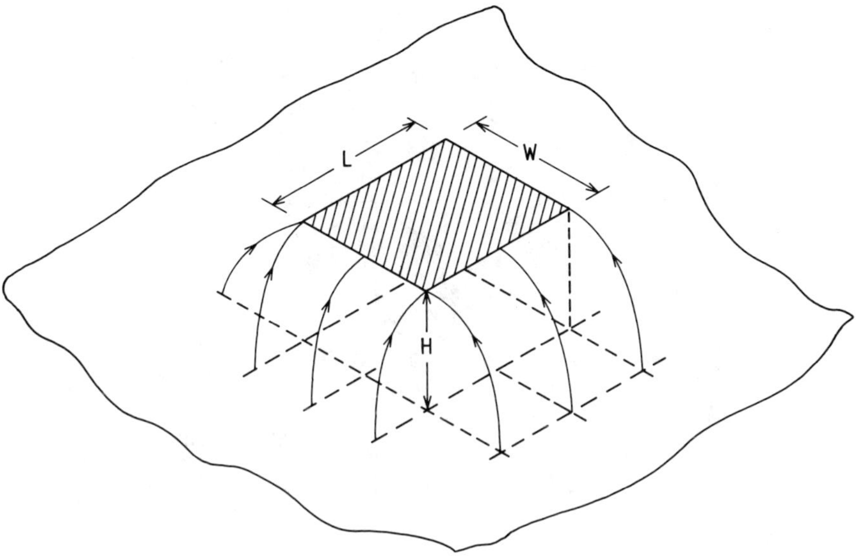

Fig. 6.46 Rectangular plate above a ground plane.

and C_{dpbp} when the gate and drain feeds are not co-planar but are isolated pads.

(c) *Long coupled strips* (Fig. 6.47) (strips of generally unequal width), used for estimating C_{gsa}, C_{gda}, C_{dsl}.

(d) *Parallel plate capacitance in one-dimension* (Fig. 6.48), used for estimating C_{dpbp}, C_{gpbp} when there is a co-planar feed. This simple approach assumes there will be no sidewall contributions to these capacitances because

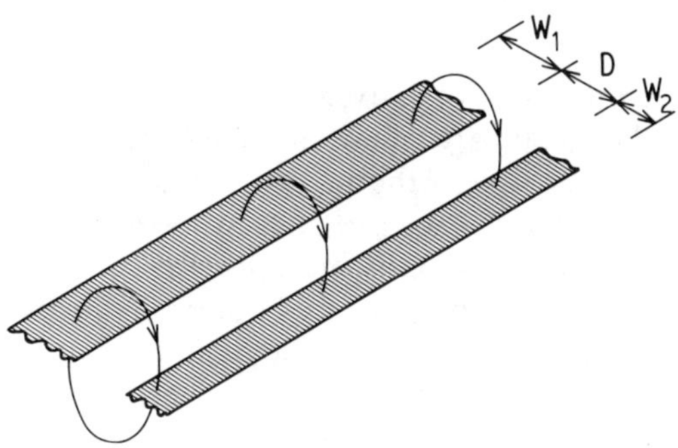

Fig. 6.47 Strips of unequal width.

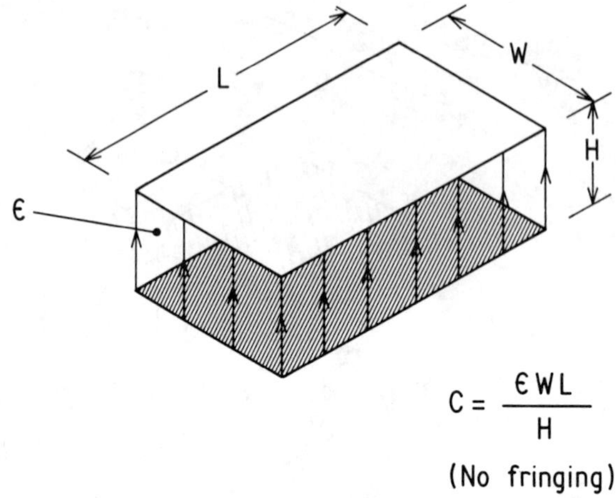

Fig. 6.48 Parallel-plate capacitance.

the fringing field terminates on the co-planar grounds rather than on the back plane.

Estimates of the various capacitances can be made using data for these basic geometries in the following manner.

(a) When L, W and D are expressed in meters, the capacitance of the co-planar pad in Fig. 6.45 may be estimated (in picofarads) from Fig. 6.49, which is a graph given in Barna [6.18]. It assumes the two ground planes extend to infinity to the left and to the right along x. For GaAs FETs where the metallization pattern is on the surface of a GaAs substrate,

$$\epsilon_r = \frac{\epsilon_{r1} + \epsilon_{rGaAs}}{2} = 7$$

i.e. the effective permittivity is the average of the permittivities of the media above and below the metal pattern (air and GaAs respectively). For C_{gsa} and C_{gda} calculations, $\epsilon_r = 1$, and only the air path counts. Note that Fig. 6.49 gives the pad capacitance including two free ends, whereas the actual gate pad, for example as in Fig. 6.40, has only one end free. Consequently, the end capacitance component must be separated out from the remainder of the pad capacitance, consisting of sidewall capacitance and parallel-plate capacitance, and the end contribution halved. The end capacitance can be identified by determining, from Fig. 6.49, how the total capacitance changes as L is increased.

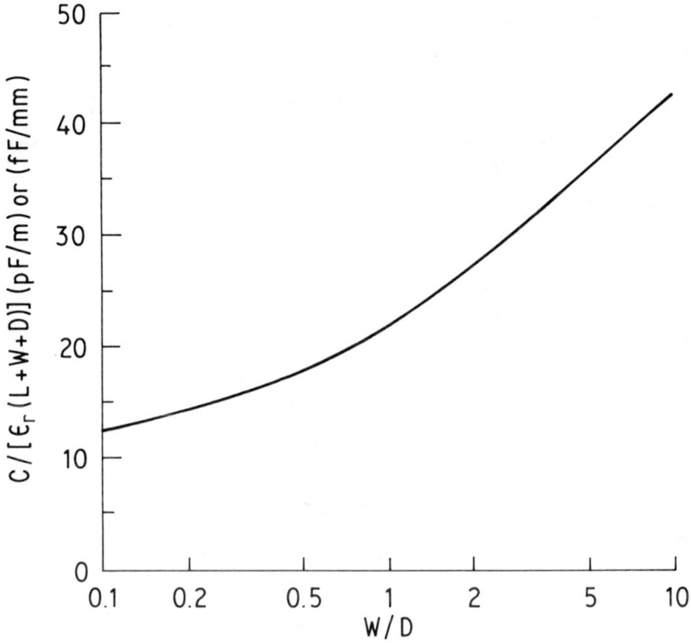

Fig. 6.49 Capacitance data for the geometry of Fig. 6.45 (after [6.18]).

(b) The capacitance of an isolated pad, Fig. 6.46, including just one end contribution as appropriate to a FET, can be estimated from standard microstrip theory in the quasi-TEM approximation (Section 3.2, Fig. 3.4). For a microstrip of strip-width to substrate–height ratio (w/h), Fig. 3.4 allows the capacitance per meter to be found in accordance with the well-known relations

$$\bar{C} = \frac{\epsilon_{\text{eff}}}{cZ_{01}} \qquad \epsilon_{\text{eff}} = 1 + q(\epsilon_r - 1)$$

where c is the velocity of electromagnetic waves in free space, Z_{01} is the characteristic impedance of the microstrip with an air dielectric filling, and q is the microstrip filling factor. By subtracting the parallel plate, one-dimensional, capacitance

$$\frac{\epsilon_{\text{GaAs}} W}{H} \quad \text{or} \quad \frac{\epsilon_{\text{GaAs}} W}{h}$$

the sidewall (or fringing) capacitance contributions to the total can be identified and used to estimate the end capacitance.

(c) For long coupled strips (Fig. 6.47), end capacitance can be ignored, thus simplifying the manipulation required using data from standard curves.

As mentioned previously, the detailed way in which these general procedures are applied is obviously very dependent upon FET geometry. To illustrate the magnitudes to be expected for a general purpose small-signal FET, however, we quote values for a microwave probe-able FET (Fig. 6.40) with the following approximate dimensions.

Source Pads: 300 μm × 60 μm

Gate Pad (Fig. 6.45):

$$L \approx 65\ \mu m$$
$$W \approx 60\ \mu m$$
$$D \approx 40\ \mu m$$

Other Dimensions:

$$L_{SG} \approx 1\ \mu m$$
$$L_G \approx 0.5\ \mu m$$
$$Z_G \approx 250\ \mu m$$
$$L_{GD} \approx 1\ \mu m$$
$$W_{SiN} \approx 1\,000\ \text{Å}$$

The drain pad and drain finger dimensions follow logically as approximately:

Drain Pad (Fig. 6.45):

$$L \approx 65\ \mu m$$
$$W \approx 60\ \mu m$$
$$D \approx 40\ \mu m$$

Drain Finger: 150 μm × 50 μm

The geometric capacitances are then estimated as in Table 6.4.

Table 6.4 Typical geometric capacitances of a small-signal probe-able FET

	UP = Unpassivated, P = Passivated				
	$C_{gs\ geom}$ (pF)		$C_{gd\ geom}$ (pF)		$C_{dpbp\text{-}s}$ (pF) (contribution to C_o)
	UP	P	UP	P	
450 μm substrate Back plane *floating*	0.022	0.023	0.001	0.002	0.008
200 μm substrate Back plane *grounded*	0.023	0.024	0.001	0.002	0.014

Note that C_{dsl} is bias dependent, and so is excluded from the Table. Typically, C_o may amount to 0.04–0.06 pF. $C_{dpbp\text{-}s}$ stands for the drain pad- (and finger-) to-source capacitance, via the back plane. Probing on-wafer is more akin to the 450 μm substrate condition with the back plane floating, whereas a finished MMIC would more likely have the substrate thinned and the back surface metallized and grounded.

The foregoing results include estimates of the effects of silicon nitride dielectric protection, of thickness W_{SiN}, overlying in this case the source, gate and drain in the gate strip area only. Thus C_{gsa} and C_{gda} in Fig. 6.41 become modified because electric flux passing between the electrodes now passes in part through Si_xN_y. Capacitance changes can be estimated by perturbation methods, i.e.

$$\frac{\delta C}{C} = \frac{\int_\tau \delta\epsilon \cdot E^2 \, d\tau}{\int_\tau \epsilon \cdot E^2 \, d\tau}$$

where τ is the volume affected by the nitride and $\delta\epsilon$ is the change in permittivity throughout τ. One difficulty with such methods is that the field distribution $\boldsymbol{E}(x, y)$ throughout τ must be known, or at least approximations to it sufficient for the purpose. A simple alternative is to use Fig. 6.49 to estimate the capacitance elements in Fig. 6.41 with ϵ_r somewhere in value between 1 (for air) and 7 (for Si_3N_4), depending upon the nitride thickness, W_{SiN}. A value of $\epsilon_r \approx 2$ is consistent with the perturbation results given for the example FET considered above.

We note that, in the case of both C_{gc} and C_{gd}, the geometric capacitance has a value very roughly equal to one-tenth of the space-charge capacitance (refer to Fig. 6.39).

6.11 Via-hole inductance

A great many MMICs employ the GaAs MESFET in the grounded-source configuration, thus requiring the source pads marked S in Fig. 6.1 to be connected to signal ground. If the FET can be located at the edge of the chip, source grounding can be effected by multiple wire bonds to the surrounding chip carrier or package. When, as is often the case, the FETs cannot be located near the chip edge, a different chip grounding technique is needed.

A common solution is to etch or mill a hole from the reverse side of the wafer as shown in Fig. 6.50; the reverse side metallization is continued into the hole to make contact with the front-side metallization. Such a hole is known as

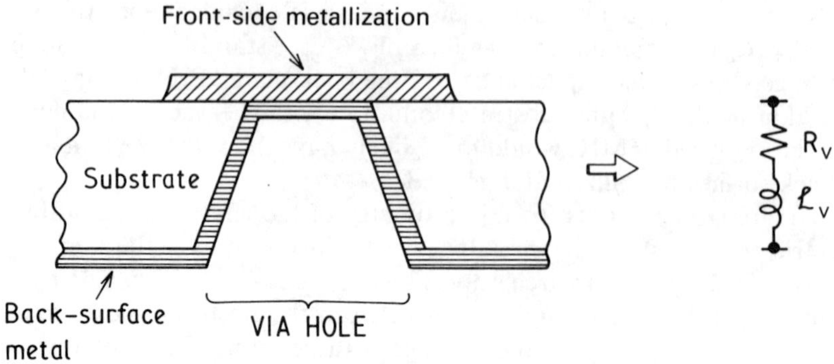

Fig. 6.50 A via ground, with equivalent circuit.

a via hole. Via holes may be metallized on the walls only, as indicated in Fig. 6.50, or, for preference, they may be filled to form a solid metal plug.

In its most simple form, the equivalent circuit of a via hole is an inductance $\mathcal{L}_v$ in series with a resistance R_v. R_v is readily estimated; $\mathcal{L}_v$ is less so. As a coarse guide, via-hole inductance can be estimated using the formula for the internal inductance of a straight length of wire:

$$\bar{\mathcal{L}} = \mu_o / 8\pi \ (\mathrm{H\ m^{-1}})$$

Thus, if the via is a solid plug of metal of length equal to the substrate thickness h,

$$\mathcal{L}_v = \mu_0 h / 8\pi \ (\mathrm{H})$$

e.g. for a substrate of thickness $h = 100$ μm, $\mathcal{L}_v \approx 5$ pH. It should be noted, however, that values of via-hole inductance vary quite widely. We note also that the total source inductance is contributed to by the inductance of the metal source pad between the via itself and the active region of the FET. Source inductances of several tens of picoHenries are typical.

6.12 GaAs FET noise

A prime application of GaAs MESFETs has been in low-noise amplification. A simple analytic expression for calculating the minimum noise figure of a FET is thus extremely valuable†. Before embarking on this

†The word "minimum" refers to a partial minimum, i.e. the minimum noise figure that can be achieved by adjusting the source (or generator) impedance while the FET is operated at an arbitrary bias point. An absolute minimum noise figure can be obtained by adjusting the bias point as well as the generator impedance, as will become apparent later in this section.

development, however, the role of contemporary MMICs in low-noise amplification should be briefly examined.

All FETs are different. Process and material tolerances combine to ensure that no two devices are the same. Consequently, when applications demand the lowest practicable noise figure, the practice of device selection offers an advantage by first identifying the lowest noise transistors. Secondly, because the achievement of minimum noise figure depends upon the generator impedance (or input termination) being adjusted to a fairly critical value (different from the conjugate match), the input circuit needs to be tuned to obtain the best noise performance. As an aside, this principle extends to the same FET used at different temperatures; because the drain current and equivalent circuit elements are functions of temperature, the input circuit will need to be retuned for minimum noise if the device is operated at a different temperature. Thus a FET amplifier cannot be tuned for minimum noise at room temperature and then expected still to give minimum noise upon refrigeration without retuning.

If the implications of these facts for MMICs are examined, we soon conclude that the chances are remote of realizing an integrated FET on an MMIC and, simultaneously, realizing the required driving point impedance to achieve minimum noise from that particular FET. Thus, MMICs are not really the way to approach amplification where the requirement is for the lowest practicable noise figure. Nevertheless, a capability to do the best we can with MMICs is desirable.

To deduce an expression for noise figure, we follow very closely the argument given by Delagebeaudeuf *et al.* [6.19], who in turn based their reasoning on the work of Pucel *et al.* [6.6], Brewitt-Taylor *et al.* [6.20] and Fukui [6.21]. Furthermore, as we have already established expressions for the equivalent circuit elements in terms of the material and structural features of the FET and its bias condition, we can determine $NF_{\min}$ as a function of device structure and operating point if we can secure an equation for $NF_{\min}$ in terms of the device equivalent circuit elements.

Typically, for low-noise operation, a GaAs MESFET is biased at a drain voltage just above the knee voltage, in which case the velocity saturated regions of the channel, regions II and III in Fig. 6.9, are strictly limited in extent, and the greater part of the channel operates in the constant mobility regime. The sources of noise internal to the FET are then principally current fluctuations in region I, which essentially occupies the whole channel, and noise generated by the parasitic source and gate resistances, R_s and R_g. Current fluctuations in region I constitute noise in the drain current, and have charge (or field) fluctuations associated with them which induce charge fluctuations on the gate electrode. (In physicists' terms, the application of Guass' Law, treating the channel, depletion layer and gate metallization as a closed-charge system, requires an equal and opposite charge fluctuation in $Q_G(x_2)$ of Fig. 6.22. In the language of the electrical engineer, the charge fluctuation in the channel is coupled to the gate via the gate-channel capacitance, C_{gc}.) Thus there is an induced noise charge on the gate which gives rise to a noise current in the gate-source circuit that is partly correlated with the drain noise current.

The work of Pucel *et al.* [6.6] showed that under conditions where the gate current noise is small but highly correlated with the drain current noise, the noise figure is given approximately by

$$NF_{\min} \approx 1 + 2\left(\frac{\langle i_{Dn}{}^2 \rangle}{4kT_0 \times \Delta f \times g_{m0}}\right)^{1/2} \times \omega C_{gc}\left(\frac{R_s + R_g}{g_{m0}}\right)^{1/2} \tag{6.44}$$

This is a low-frequency approximation in which $\langle i_{Dn}{}^2 \rangle$ is the mean square drain noise current, and Δf is the noise equivalent bandwidth. Written in this way, the sources of the various contributions to the overall noise are obvious.

The foregoing expression is closely related to a widely used equation due to Fukui. Fukui [6.21] showed empirically that, when the device is biased so as to minimize $NF_{\min}$ (i.e. at the absolute minimum of noise as discussed earlier in the footnote on p. 156),

$$NF_{\min} \approx 1 + K_F \omega C_{gc}\left(\frac{R_s + R_g}{g_{m0}}\right)^{1/2} \tag{6.45}$$

Extensive fitting of this expression to experimental data by workers around the world shows that

$$K_F \approx 2.5\text{–}3 \quad \text{for FETs} \tag{6.46}$$

and

$$K_F \approx 1.5\text{–}2 \quad \text{for HEMTs} \tag{6.47}$$

The factor K_F is evidently a gross simplification of the drain current noise contribution to the overall noise, and as such it is restrictive, especially in light of the fact that an almost equally simple expression (in the algebraic sense) can be deduced which retains far more comprehensive physics, as shown by Delagebeaudeuf *et al.* [6.19]. We follow this route.

Consider a current fluctuation δI in the channel which may be represented, as in Fig. 6.51, by δI being injected at the point $x = x_0$ and removed at $x + \delta x_0$. The drain current response to the current fluctuation $\delta I(x_0)$ is determined by first evaluating the fluctuation in potential drop across region I, δV_I, which results from $\delta I(x_0)$, thus

$$\delta V_I = \delta I(x_0) \times \frac{\delta x_0}{qN\mu_0 Z_G(W-d)}$$

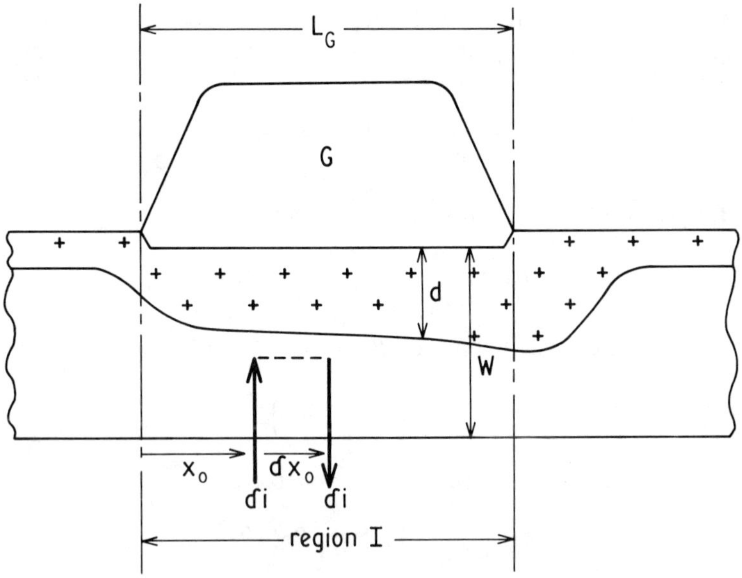

Fig. 6.51 Showing noise current fluctuations in region I of the channel of a GaAs FET.

where the second term is simply the resistance δR of the section of channel of length δx_0. The resulting fluctuation in drain current is found by dividing δV_I by the resistance of region I, which is of length L_G, thus

$$\delta I_\mathrm{D} = \delta I(x_0) \times \frac{\delta x_0}{L_\mathrm{G}} \tag{6.48}$$

Given that the noise power arising from the random (thermal) motion of the electrons is $\mathrm{k}T_0\Delta f$, the mean-squared noise current associated with the section of channel δx_0 having resistance δR is

$$\langle \delta i_\mathrm{n}{}^2 \rangle = \frac{4\mathrm{k}T_0\,\Delta f}{\delta R} = \frac{4\mathrm{k}T_0\,\Delta f q N \mu_0 (W-d) Z_\mathrm{G}}{\delta x_0}$$

$$= \frac{4\mathrm{k}T_0\,\Delta f \mu_0 I_\mathrm{CH}}{\delta x_0 v_\mathrm{sat}} \tag{6.49}$$

where the subscript "n" has been added to remind us that we are dealing with a noise current.

Thus, substituting eqn (6.49) into the mean-squared value of eqn (6.48),

$$\langle \delta i_\mathrm{Dn}{}^2 \rangle = \frac{4\mathrm{k}T_0\,\Delta f \mu_0 I_\mathrm{CH}}{L_\mathrm{G}{}^2 v_\mathrm{sat}} \times \delta x_0$$

Assuming these elemental noise sources to be uncorrelated throughout region I, the total mean-square drain noise current is obtained by integrating over region I:

$$\langle i_{\mathrm{Dn}}{}^2\rangle = \frac{4kT_0\,\Delta f\mu_0 I_{\mathrm{CH}}}{v_{\mathrm{sat}}L_{\mathrm{G}}}$$

so that, upon substitution into equation (6.44),

$$NF_{\min} \approx 1 + 2\left(\frac{\mu_0 I_{\mathrm{CH}}}{g_{\mathrm{m0}}v_{\mathrm{sat}}L_{\mathrm{G}}}\right)^{1/2} \omega C_{\mathrm{gc}} \left(\frac{R_{\mathrm{s}}+R_{\mathrm{g}}}{g_{\mathrm{m0}}}\right)^{1/2} \tag{6.50}$$

Thus, in Fukui's terms, we find for K_{F} in eqn (6.45),

$$K_{\mathrm{F}} = 2\left(\frac{\mu_0 I_{\mathrm{CH}}}{g_{\mathrm{m0}}v_{\mathrm{sat}}L_{\mathrm{G}}}\right)^{1/2} \tag{6.51}$$

Alternatively, we may substitute from Table 6.1 to obtain

$$NF_{\min} \approx 1 + 2\omega\frac{C_{\mathrm{gc}}}{g_{\mathrm{m0}}}\left(\frac{R_{\mathrm{s}}+R_{\mathrm{g}}}{R_{\mathrm{i}}}\right)^{1/2} \tag{6.52}$$

The advantage of eqns (6.50) and (6.52) over Fukui's relationship, which assumes K_{F} to be a constant, is that they allow $NF_{\min}$ to be calculated as a function of the channel current (i.e. the bias point). Delagebeaudeuf *et al.* [6.19] showed the expression applies also to HEMTs. Note that substrate current has been ignored in these equations. Given the view of substrate current as a widening of the channel, a rough allowance for substrate current could have been made in the noise calculation by using an effective channel layer thickness (W) greater than the actual value in eqn (6.49); this would then follow through to give I_{D} in place of I_{CH} under the first square root in eqn (6.50).

In Fig. 6.52 are plotted various noise figure curves for an example FET with

$L_{\mathrm{G}} = 0.5\ \mu\mathrm{m}$ $\quad Z_{\mathrm{G}} = 200\ \mu\mathrm{m}$

$h = 0.5\ \mu\mathrm{m}$ $\quad L_{\mathrm{SG}} = 1\ \mu\mathrm{m}$

$\rho_{\mathrm{c}} = 0.2\ \Omega\ \mathrm{mm}$ $\quad W = 0.1\ \mu\mathrm{m}$

$N = 3.5\times10^{17}\ \mathrm{cm}^{-3}$ $\quad \mu_0 = 2500\ \mathrm{cm}^2\ \mathrm{Vs}^{-1}$

$V_{\mathrm{DS}} = 2$ or 3 V (as noted on Fig. 6.52)

$m = 1$ or 2 (as noted on Fig. 6.52)

There is a very important distinction between the form deduced here (solid lines, continued as dashed curves passing through the origin) and

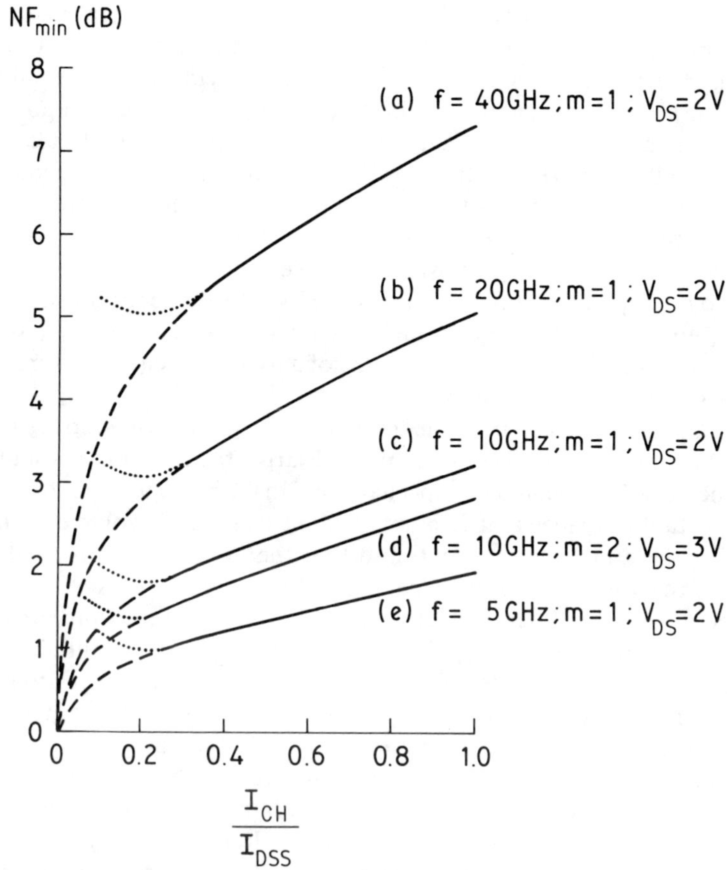

Fig. 6.52 Calculated dependence of minimum noise figure upon channel current, drain bias, and gate finger width. Broken line: continuation of theoretical noise figure for FET with flat doping and infinitely steep profile tail. Dotted line: typical result for practical FETs.

typical practical plots, typified by the dotted curves. Referring back to the shapes of profiles which can be realized in practical FETs (Fig. 6.27), at low values of I_{CH}/I_{DSS} (or I_D/I_{DSS}) the depletion edge enters the profile tail and g_{m0} drops rapidly. From eqn (6.52) we could then expect NF_{min} to increase sharply. This phenomenon does indeed occur in practical FETs, typically at channel currents in the range.

$$0.1 < \frac{I_{CH}}{I_{DSS}} \lesssim 0.2$$

The steeper is the profile tail, the smaller are the currents at which g_{m0} can be sustained and the lower is the absolute minimum noise figure. For good low-noise performance, therefore, a sharp profile is required with the shape of Fig. 6.27(a), perhaps tending slightly towards Fig. 6.27(c). Vapor phase epitaxy, metal–organic chemical vapor deposition and molecular beam epitaxy are all appropriate to the growth of such profiles with sharp interfaces. It should be emphasized that the detailed point at which the upturn in NF_{min} occurs is thus circumstantially dependent upon the substrate or buffer properties, and the spatial disposition of defects and impurities incorporated in the growing epitaxial film at the transition from the buffer or substrate to the N-layer. In other words, good noise behavior (which is often judged in differences as small as one- or two-tenths of a dB) is difficult to assure, so the practice of device selection persists.

The appearance of R_i in the denominator of eqn (6.52) is contrary to what one might at first sight expect. Evidently, the larger the value of a resistance the larger the noise figure could be expected to be. In the present context, however, it is important to realize that R_i is simply a measure of the volume of channel which is generating noise current, and that the noise current is smaller the smaller is the volume of undepleted channel (i.e. the higher is R_i).

There are many other interesting deductions to be made from eqns (6.50) and (6.52). One is that NF_{min} increases with frequency, as illustrated on Fig. 6.52. Another is that benefits accrue from structural changes to source-gate spacing, L_{SG}, gate metallization thickness, h, channel parameters N and W, and so on. The effect of using two gate strips instead of one, while keeping the total gate width Z_G, constant, is also illustrated in the example of Fig. 6.52. One of the more revealing findings is that a reduction in gate length does not, of itself, necessarily lead to a reduction in NF_{min}. Referring to eqn (6.50), NF_{min} increases on account of R_g and the factor $1/(L_G)^{\frac{1}{2}}$, but it decreases on account of C_{gc}. Whether there is a net increase or a net decrease depends, therefore, on a balance of these factors. Finally, of course, measured values can instead be substituted for C_{gc}, g_{m0}, R_s, etc. in the equations for NF_{min} to predict (or check) the noise figure for specific practical devices. For instance, for a general purpose FET, on-wafer probing gave

$$R_g = 5\ \Omega \qquad R_s = 3\ \Omega$$

$$R_i = 7.6\ \Omega \qquad C_{gc} = 0.216\ \text{pF}$$

$$g_{m0} = 37\ \text{mS}$$

Correcting the measured value of capacitance for the geometric component (0.023 pF) which strictly makes no contribution to the noise, we have $C_{gc} = 0.193$ pF and

$$NF_{min} = 2.23 \text{ dB at } 10 \text{ GHz}$$

A device designed specifically for low-noise operation, such as a 0.5 μm gate HEMT, can give noise figures as low as 0.7 dB at 10 GHz with 10 dB associated gain (at room temperature).

The attainment of the minimum noise figure, NF_{min}, at any given bias point requires the generator impedance to be adjusted to a specific value. Given the same assumptions as those under which eqns (6.50) and (6.52) were derived, principally that the induced gate noise is small, it may be shown that the required input termination impedance, or generator impedance, is

$$Z_g|_{OPT} = \left[(R_s + R_g + R_i)^2 + \frac{g_{m0}^2 (R_s + R_g) R_i}{\omega^2 C_{gc}^2} \right]^{1/2} + \frac{j}{\omega C_{gc}} \qquad (6.53)$$

To put example numbers to this, for curve d of Fig. 6.52 at $I_{CH}/I_{DSS} = 0.22$ (and $V_{DS} = 3$ V),

$$C_{gc} = 0.132 \text{ pF} \qquad R_i = 19.1\ \Omega \qquad g_{m0} = 25.4 \text{ mS}$$

$$\text{and } R_s = 4.57\ \Omega \qquad R_g = 1.63\ \Omega$$

And so, at 10 GHz,

$$Z_g|_{OPT} \approx 42 + j121\ \Omega$$

No general deduction as to the magnitude of the impedance required for minimum noise can be made from these particular numbers, and each individual case must be treated on its merits. For instance, for the practical general purpose FET data considered above,

$$Z_g|_{OPT} \approx 26 + j74\ \Omega \text{ at } 10 \text{ GHz.}$$

As mentioned previously, it is rather difficult to present an integrated device on an MMIC with $Z_g|_{OPT}$ in the fixed format of integrated circuits. The required impedance depends upon five of the FET equivalent circuit elements (R_s, R_i, R_g, g_{m0} and C_{gc}), all of which can be expected to vary with process tolerances.

6.13 Power MMICs

Power MMICs make use of integrated MESFETs in the output stage which are configured to handle large voltage and current swings. The gate length and layer parameters needed to achieve such a large-signal capability are normally very different from those required for low noise and high gain, so power MMICs usually require a driver chip whose main purpose is to provide gain with a moderate signal level. Suppose, for instance, a chip set is required to provide 500 mW with 20 dB overall gain. The power MMIC chip itself may be capable of giving about 6 dB gain (i.e. a gain of 4), so that the driver chip would need to provide 14 dB gain with a signal level of 125 mW, or 21 dBm.

Unfortunately, when it comes to the design and analysis of power FETs, the truth of the difficulty has rarely been told. The expectation has been driven by numbers reported in the literature, which for perhaps the last 10 years have been one-off results from individual device chips. It is no accident that the one or two leading merchant suppliers of power MESFETs, who have a high reputation for consistency and reliability, claim conservative output powers, (or, equivalently, output power per millimetre of gatewidth, $\bar{P}_O$). The reason is that output power density depends to first order on two factors which are barely even recognized as important by some organizations and which, judging by the state of captive and merchant supply of GaAs power FETs and MMICs, exceedingly few organizations have control over. These two factors are surface charge and the N-layer to buffer–substrate interface (as characterized in this book by the interface energy, ϕ_B). It is likely that both depend upon the detailed structure of the starting GaAs semi-insulating substrate [6.22], and the first is process dependent, particularly on the composition and deposition conditions of dielectric protecting layers (SiO_x, Si_xN_y, etc.).

The work summarized here is the result of well over 30 man-years of careful physics research at the experimental, analytical and numerical simulation levels. Although much remains to be understood, particularly so far as the formation of available electronic states at surfaces and interfaces is concerned, and the detailed manner in which electrons are transported to and from such states, enough is known at the device and circuit level to construct physical and mathematical descriptions which are meaningful. Among other things it has been shown that the occupied surface-state density in a power FET is likely to be neither static nor spatially uniform during device operation [6.2].

To illustrate the problem, we show in Fig. 6.53 the dependence of output power density, $\bar{P}_O$ watts per millimetre of gatewidth, versus occupied surface-state density, D_S. Bearing in mind that the available state density at surfaces and interfaces can be as high as 10^{14} cm^{-2} eV^{-1} [6.23], we are concerned with occupancies of only a few per cent. What happens in very many power FETs is that energetic electrons generated under the high voltage, high field, conditions of operation fill the surface states to well beyond the maximum power point in Fig. 6.53, so $\bar{P}_O$ collapses or, rather, is never realized. Only one or two organizations appear to have found an answer to this problem.

The output power density is governed in the first place by the limits of

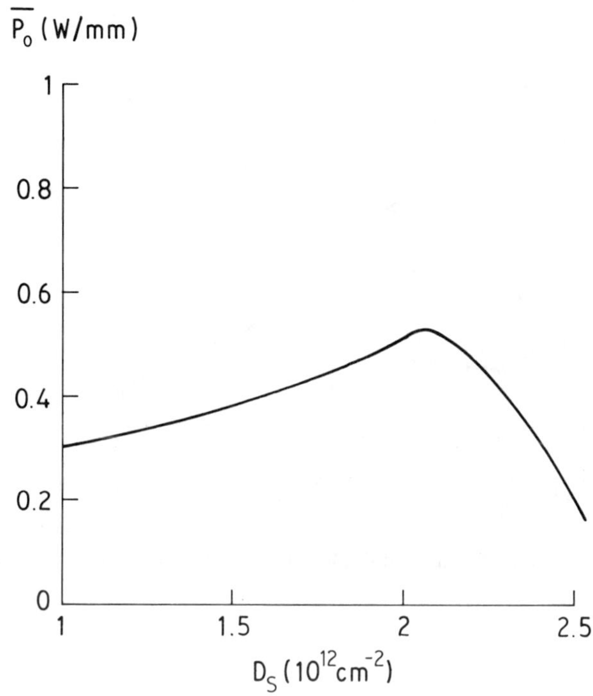

Fig. 6.53 Typical dependence of 0.5 μm GaAs FET output power density per millimetre of gatewidth upon occupied surface-state density.

V_{DSB} and I_F (or $\bar{I}_F$) marked on Fig. 6.54. The bar over I_F denotes the current per millimetre of gate width. The initial drain-source breakdown voltage V_{DSB} is due to the onset of impact ionization in the gate-drain space. (The remark applies only to well designed and fabricated FETs where channel recesses and perhaps N^+ contact layers overcome the problem of avalanche breakdown due to current crowding at the drain (or anode) contact. Such a double recess structure is shown in Fig. 6.55.) Generally speaking, the impact ionization occurring under pinched channel conditions is self-regulating, leading to a stable carrier generation rate rather than to catastrophic and destructive avalanche. Thus the FET can be driven into this region under large-signal microwave conditions without apparent harm in the short term. The form of the I_D (V_{DS}) relationship in the self-regulating avalanche region can be calculated analytically. An example is shown in Fig. 6.56 and has a form very reminiscent of practical curves (Fig. 6.57).

Another apparent anomaly is that V_{DSB} depends upon the potential of a sidegate in some devices (Fig. 6.58) [6.15]. The transistor can be designed in plan view to minimize effects such as these to some extent.

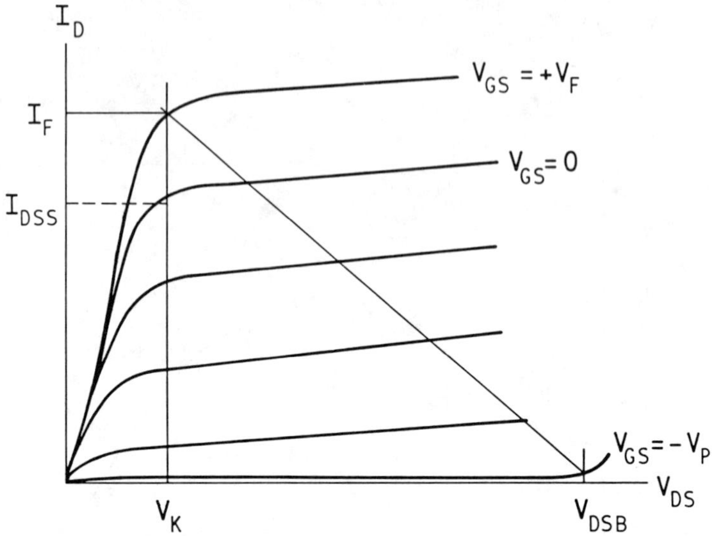

Fig. 6.54 Drain characteristics of GaAs FET identifying current and voltage limits.

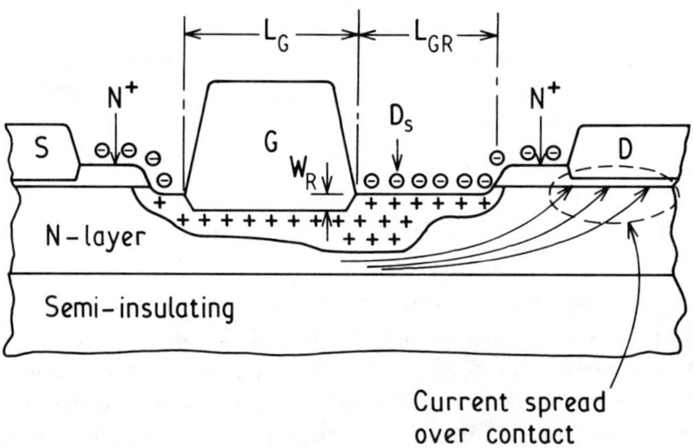

Fig. 6.55 Double recess structure power FET.

The purpose of the foregoing discussion is to give credence to the idea that power FETs demand the utmost sophistication in technology for them to be reproducible. Such are the demands of MMIC engineering. When other than the best and most extensively (and expensively) researched technology is available, widely varying or simply poor performance can be expected as a

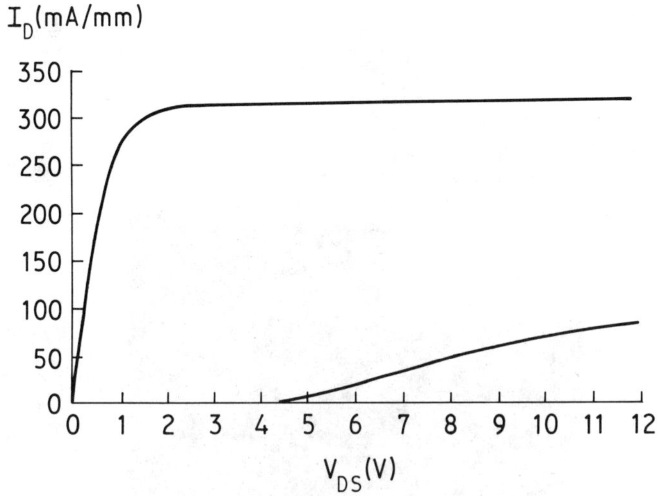

Fig. 6.56 Form of calculated self-regulating impact ionization current.

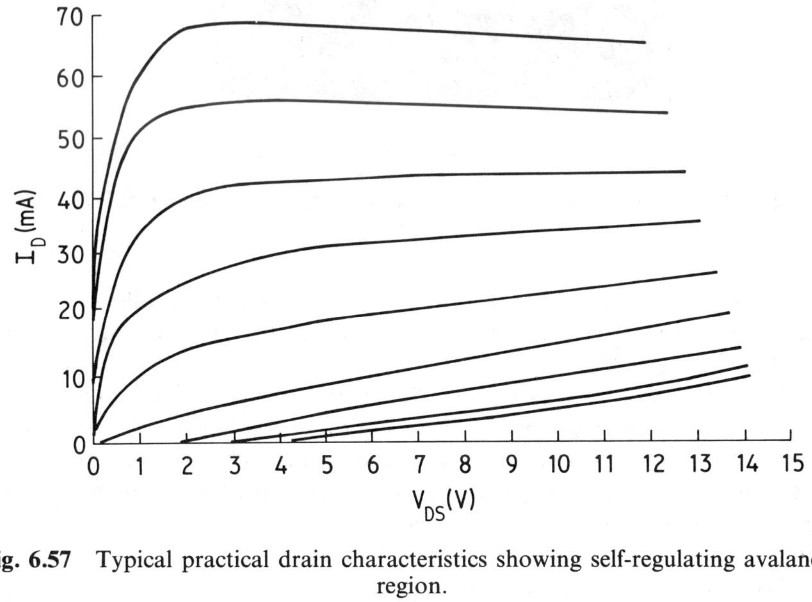

Fig. 6.57 Typical practical drain characteristics showing self-regulating avalanche region.

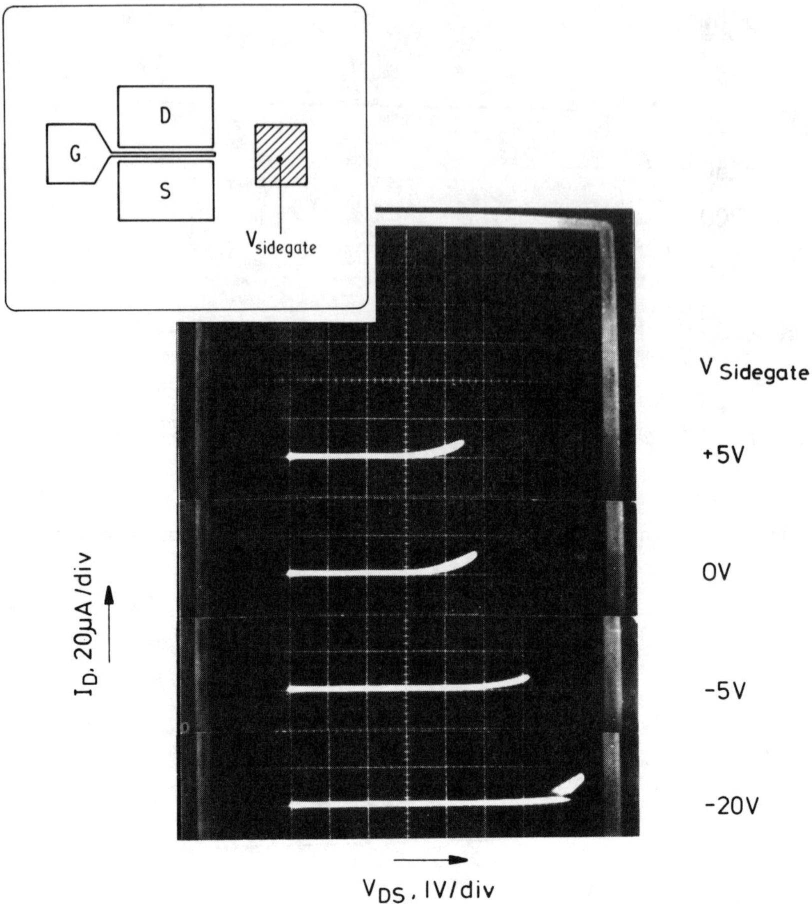

Fig. 6.58 Illustrating dependence of gate-drain breakdown voltage of 0.5 µm GaAs FET on sidegate bias for a sidegate located 20 µm from the free end of the gate.

perfectly reasonable outcome. It follows that any attempt to offer theoretical design guidelines in a work such as this should be approached cautiously and with due reservation.

With appropriate reservation, therefore, we now offer several design equations with a clear statement of the conditions under which they may be expected to apply. The best that can be said is that such equations may offer a rough guide to the numbers which can be expected in some practical cases, recognizing that rarely will the experimental conditions conform to those assumed.

We assume (with reference to Fig. 6.55)

(a) D_S = constant (both spatially and temporally);

(b) a channel layer, of thickness W under the gate, uniformly doped with N donors per cm^3;

(c) a gate recess of depth W_R and length L_G (so the recess is filled by the gate metal);

(d) an outer recess extension on the drain side, L_{GR}, which is typically $2L_G$ or greater and

(e) the gate recess depth $W_R = (D_S/N)$.

With these assumptions, it may be shown that a good approximation for the drain-source breakdown voltage is

$$V_{DSB} = 0.5\left[\frac{E_B L_G}{qNW}(\epsilon E_B - qNW) - \frac{qNW^2}{\epsilon}\right] \tag{6.54}$$

where E_B is a two-dimensional breakdown field obtained by evaluating the ionization integral along the path of maximum ionization rate. $E_B \approx$ 640 kV cm^{-1}. Note that V_{DSB} is approximately proportional to gate length, L_G, all other factors unchanged.

Assuming substrate current can be neglected, and assuming also that the gate can be forward driven by V_F without substantial flow of (forward) gate current, we may write from the equations of Table 6.1,

$$\bar{I}_F = qNv_{sat}(W - d_{min}) \tag{6.55}$$

with

$$d_{min} = \left[\frac{2\epsilon(V_{BO} - V_F)}{qN}\right]^{1/2}$$

To find an expression for the ouptut power (or power density, $\bar{P}_O$), we assume:

(a) The self-regulating avalanche region is to be avoided so as not to jeopardize the device mean-time-to-failure. The peak-to-peak voltage swing is then limited to $(V_{DSB} - V_k)$, so the root-mean-squared signal voltage is

$$V_{rms} = \frac{V_{DSB} - V_k}{2\sqrt{2}} \lesssim \frac{V_{DSB}}{2\sqrt{2}}$$

(b) The rms current is

$$\bar{I}_{rms} = \frac{\bar{I}_F}{2\sqrt{2}}$$

(c) a quasi-static class A estimate of output power may be used to gauge what the FET will be capable of up to its cut-off frequency f_T [6.24].

Thus the power density comes to:

$$\bar{P}_O = V_{rms}\bar{I}_{rms} \leqslant (V_{DSB}\bar{I}_F/8)\,(\mathrm{W\ mm^{-1}}) \tag{6.56}$$

In Fig. 6.59, $\bar{P}_O$, V_{DSB} and $\bar{I}_F$ have been plotted for

$$L_G = 0.7\ \mu\mathrm{m} \qquad N = 1.5 \times 10^{17}\ \mathrm{cm^{-3}} \qquad V_F = +0.3\ \mathrm{V}$$

It must be emphasized that, were this a real FET,

(a) the value of $\bar{I}_F$ at any given value of W would be increased by substrate current, I_{SUBS};

(b) V_{DSB} would be altered by changes in surface-state occupancy, D_S;

(c) substrate current is in turn altered by D_S, and thus

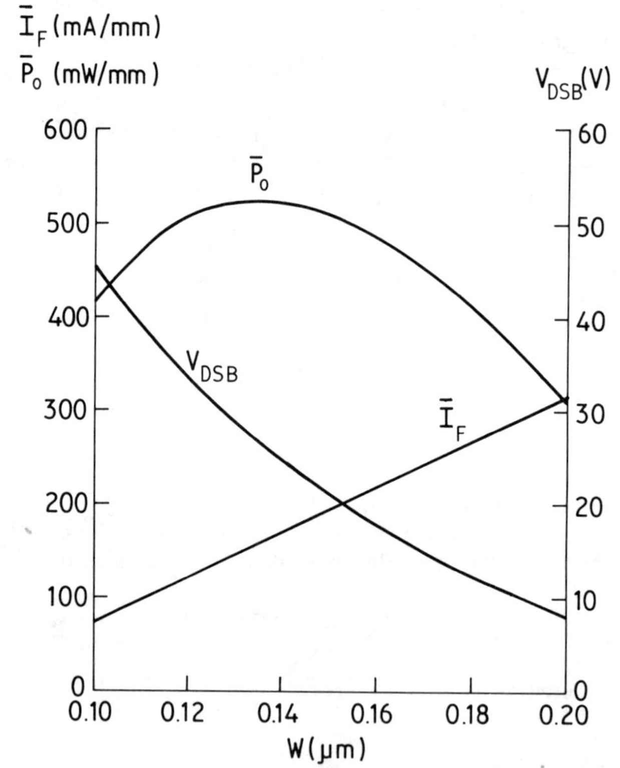

Fig. 6.59 Calculated dependence upon channel layer thickness (under the gate) of drain-source initial avalanche voltage, open channel current and output power density for a 0.7 μm GaAs power FET element.

(d) $\bar{P}_O$ will be altered by both I_{SUBS} and D_S. The peak power point may be raised or lowered, and will be displaced laterally. Often, the effect is to move the maximum to higher $\bar{I}_F$ values.

Nevertheless, the values in Fig. 6.59 are not too far removed from practical values. Figure 6.60, for instance, shows practical data from very carefully controlled experiments [6.25] in which electron-beam lithography was used to define the entire multi-finger power FET structure, and purpose-grown and carefully profiled layers were used (the profiling included deep level transient spectroscopy). The solid lines are theoretical plots based on prior knowledge of the material and structure; no adjustable parameters have been used. It is seen that the curves are a reasonable fit to the experimental data, the scatter in which is almost certainly due to D_S and I_{SUBS} variations from chip-to-chip (all chips are from the same layer).

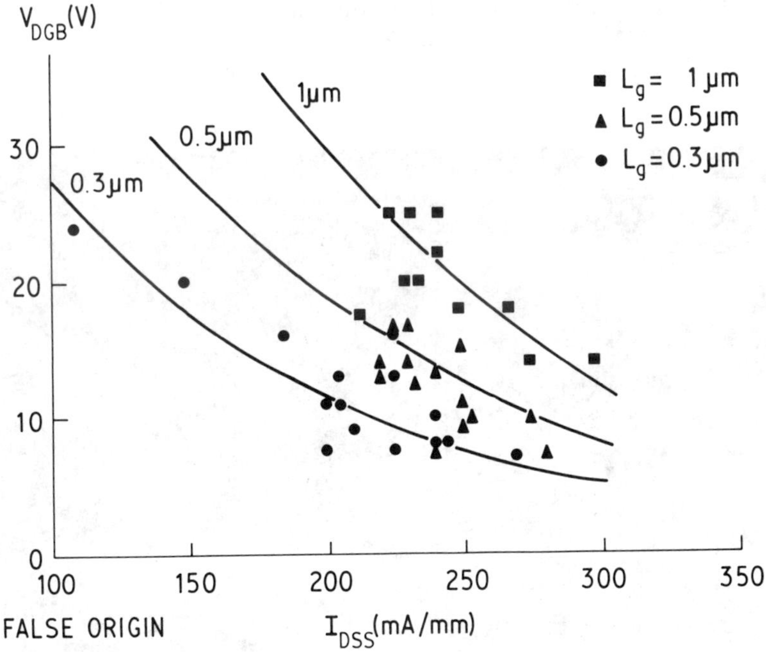

Fig. 6.60 Experimental gate-drain voltage for initial avalanche in 1 µm, 0.5 µm and 0.3 µm MESFETs. The curves are theoretical predictions based on a knowledge of the FET material and structure factors without any adjustable parameters or fitting factors.

Given the restrictions underlying Fig. 6.59, several general deductions can be made.

(a) V_{DSB} decreases, roughly in an inverse fashion, as the layer doping-thickness product NW increases.

(b) $\bar{I}_F$ increases as NW increases.

(c) There is an optimum value of NW at which $\bar{P}_O$ is a maximum. $\bar{P}_O$ falls quite rapidly on either side of this fairly narrow maximum.

As Fig. 6.60 shows, when care is taken, some or all of these trends can be seen in practical devices. On many occasions, however, anomalies exist, especially in V_{DSB} and $\bar{P}_O$. We will cite two. The first is the rate-dependence of V_{DSB} [6.26], in which V_{DSB} has a value which apparently depends on the speed of the bias pulse used to measure it. Secondly, in Fig. 6.61 is shown the initial breakdown characteristic [6.25] of a 0.8 μm gate length FET which had $\bar{I}_{DSS} \approx 250$ mA mm^{-1} (bearing in mind that $\bar{I}_{DSS}$ is a reflection of the NW product upon which V_{DSB} depends). The device has 82 V applied to the drain, without ill effect, which is many times the value expected from eqn (6.54). A very high density of slow surface states on this particular device is responsible for this behavior; close inspection of the photograph shows small steps in the

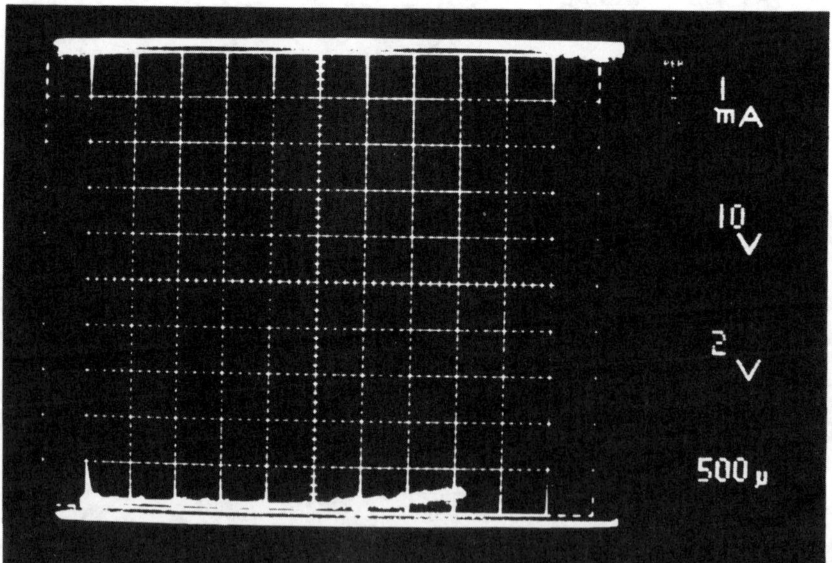

Fig. 6.61 Slowly applied drain sweep showing 82 V applied between drain and source of a 0.8 μm gate length GaAs MESFET.

characteristic where the curve tracer sweep was incrementally increased. At each stopping point the surface states slowly charged, thus progressively deferring the onset of initial avalanche. As a microwave power device, the FET was useless, because the surface states cannot respond in the same way at microwave frequencies.

Experimental facts such as these mean that eqns (6.54) and (6.56) must be employed with caution. Nonetheless, they provide a rough guide to the numbers which can be expected from appropriate technology. The very great and very severe difficulty with the GaAs power MESFET is that the power performance depends to first order upon the occupied surface-state density, and upon the N-layer to buffer-substrate interface potential, both of which are difficult to control. Most of the time D_S drifts to a value which is unfavorable to output power. Occasionally, devices are fabricated in which D_S drift and other uncontrolled factors conspire to give welcome but largely irreproducible high values of $\bar{P}_O$ which are almost exclusively the only values reported in the literature. MMIC engineering, however, is not about one-off best results, but about what can be achieved routinely.

References

6.1 Ladbrooke, P. H. Some effects of wave propagation in the gate of a microwave MESFET, *Electron. Lett.*, **14**(1), 5, 21–22 January 1978.

6.2 Barton, T. M. and Ladbrooke, P. H. The role of the device surface in the high voltage behaviour of the GaAs MESFET, *Solid-State Electron.*, **29**(8), 807–813, August 1986.

6.3 Wilson, K. and Bridge, J. P. unpublished work carried out at GEC Research, Wembley.

6.4 Ladbrooke, P. H. Power limiting due to impact ionization in GaAs MESFETs, *Electron. Lett.*, **17**(10), 338–339, 14 May 1981.

6.5 Willing, H. A., Rauscher, C. and De Santis, P. A technique for predicting large signal performance of a GaAs MESFET, *IEEE MTT-S Internat. Microwave Symp. Digest*, 132–134, Ottawa, 27–29 June 1978.

6.6 Pucel, R. A., Haus, H. A. and Statz, H. Signal and noise properties of gallium arsenide field-effect transistors, *Advances in Electronics and Electron Physics*, **38**, 195–265, Academic Press, New York, 1975.

6.7 Sone, J. and Takayama, Y. A small-signal analytical theory for GaAs field-effect transistors at large drain voltages, *IEEE Trans. Electron Devices*, **ED-25**(3), 329–337, March 1978.

6.8 Abeles, J. H., Leheny, R. F., Chang, G. K. and Allen, S. J. Experimental measurement of a high-field dipole in GaAs field-effect transistors, *Appl. Phys. Lett.*, **49**(20), 1387–1389, 17 November 1986.

6.9 Turner, B. in Howes, M. J. and Morgan, D. V. (eds), *Gallium Arsenide—Materials, Devices and Circuits*, 361–427, Wiley, Chichester, 1985.

6.10 Camacho-Penalosa, C. and Aitchison, C. S. Modelling frequency dependence of output impedance of a microwave MESFET at low frequencies, *Electron. Lett.*, **21**(12), 528–529, 6 June 1985.

6.11 Ladbrooke, P. H. and Blight, S. R. Drift and unpredictability in the dc and microwave characteristics of GaAs MESFETs due to surface states, *JEDEC Workshop on GaAs Reliability*, paper I-1, Grenelefe, FL, 27 October 1986.
6.12 Barton, T. M., Snowden, C. M., Richardson, J. R. and Ladbrooke, P. H. Narrow pulse measurement of drain characteristics of GaAs MESFETs, *Electron. Lett.*, **23**(13), 686–687, 18 June 1987.
6.13 Caddick, C. J., unpublished work carried out at GEC Research, Wembley.
6.14 Graffeuil, J., Hadjoub, Z., Fortea, J. P. and Pouysegur, M. Analysis of capacitance and transconductance frequency dispersions in MESFETs for surface characterization, *Solid-State Electronics*, **29**(10), 1087–1097, October 1986.
6.15 Ladbrooke, P. H. and Blight, S. R. Low-field, low-frequency dispersion of transconductance in GaAs MESFETs with implications for other rate-dependent anomalies, *IEEE Trans. Electron Devices*, **ED-35**(3), 257–267, March 1988.
6.16 Wolf, P. Microwave properties of Schottky-barrier field-effect transistors, *IBM J. Res. Develop.*, **14**(2), 125–141, March 1970.
6.17 Strid, E. W. 26 GHz wafer probing for MMIC development and manufacture, *Microwave Journal*, **29**(8), 71–82, August 1986.
6.18 Barna, A. *VHSIC Technologies and trade-offs*, 10–13, Wiley, New York, 1981.
6.19 Delagebeaudeuf, D., Chevrier, J., Laviron, M. and Delescluse, P. A new relationship between the Fukui coefficient and optimal current value for low-noise operation of field-effect transistors, *IEEE Electron Device Lett.*, **EDL-6**(9), 444–445, September 1985.
6.20 Brewitt-Taylor, C. R., Robson, P. N. and Sitch, J. E. New estimate of the minimum noise figure of a MESFET, *Electron. Lett.*, **14**(25), 818–820, 7 December 1978.
6.21 Fukui, H. Optimal noise figure of microwave GaAs MESFETs, *IEEE Trans. Electron Devices*, **ED-26**(7), 1032–1037, July 1979.
6.22 Blight, S. R. *Surface and Bulk Traps in Materials and Devices for GaAs Integrated Circuits*, Ph.D. Thesis, Cardiff, University of Wales Institute of Science and Technology, March 1987.
6.23 Croydon, W. F. and Parker, E. H. C. *Dielectric Films on Gallium Arsenide*, 109–114, Gordon and Breach, London, 1981.
6.24 Wemple, S. H., Flahive, P. G., Allyn, C. L., Schlosser, W. O. and Iglesais, D. E. Design principles for source-via power GaAs FETs, *Inst. Phys. Conf. Ser. No. 63, Int. Symp. GaAs and Related Compounds*, 437–442, Oiso, Japan, 20–23 September 1981.
6.25 Ladbrooke, P. H., Zwicknagl, P. and Lowe, C. Unpublished experimental work done at Cornell University, Ithaca, New York, 1983.
6.26 Ladbrooke, P. H. and Martin, A. L. Material and structure factors affecting the large-signal operation of GaAs MESFETs, *Proc. Conf. Semi-Insulating III-V Materials*, 313–320, Nottingham, 14–16 April 1980.

6.A Appendix to Chapter 6: Effective drift velocity versus electric field in a GaAs MESFET

There are five factors which determine the effective velocity-field characteristic in GaAs MESFETs. They are:

(i) the velocity-field characteristic for long, undoped GaAs samples (where by "long" is meant that the sample should be very many times longer than a mean-free path for scattering);

(ii) the manner in which doping affects the basic $v(E)$ curve through ionized- and neutral-impurity scattering;

(iii) modifications to transport in the longitudinal, or x, direction brought about by electric fields in the transverse, or y, direction;

(iv) temporal factors, involving such issues as the distance traveled by an electron in one mean-free time between scattering events, and velocity overshoot; and

(v) lattice temperature.

6.A.1 $v(E)$ in long samples of undoped bulk GaAs

Figure 6.A.1 shows the generally accepted form of the velocity-field characteristic of undoped bulk GaAs. For low electric fields, an increase in electric field strength brings about a proportional increase in drift velocity with a drift mobility of $\mu_0 \approx 7800\ \mathrm{cm^2\ Vs^{-1}}$. As the electric field is progressively increased, the average energy acquired by the Γ-valley electrons from the field increases until, at an energy of approximately 0.1 eV,

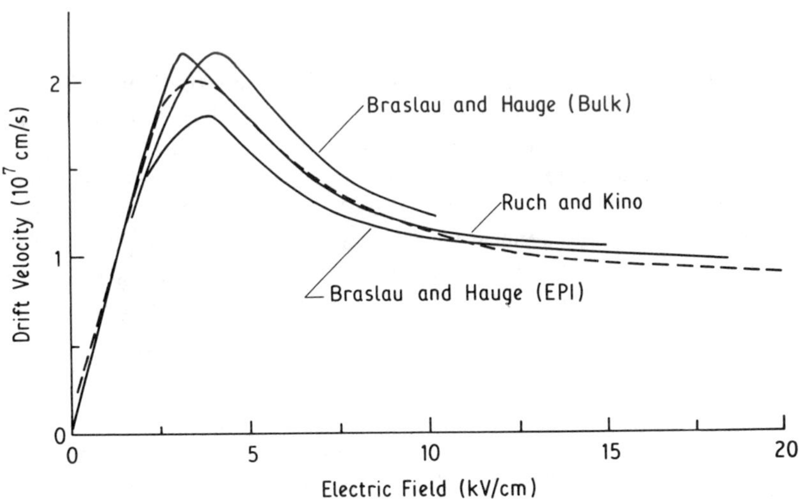

Fig. 6.A.1 Electron drift velocity versus electric field for undoped bulk GaAs (after [6.A.9]).

corresponding to an electron temperature of around 1100 K†, the rate at which electrons lose energy to polar optical phonons ceases to increase with electron energy so the carriers go on accelerating in the field until they have sufficient energy to transfer to the upper valleys [6.A.1]. The next set of conduction-band valleys above the Γ valley is the set of valleys designated L along the $\langle 111 \rangle$ directions [6.A.2], in which the drift velocity of electrons is essentially 10^7 cm s^{-1}. As electrons progressively transfer from high-mobility, high-velocity states in the Γ valley to low-velocity states in the L valleys, the drift velocity in the direction of the field, averaged over all electrons, falls to give rise to the familiar $v(E)$ characteristic.

Evidently, the runaway of electrons in the central (Γ) conduction-band valley beginning at electron temperatures [6.A.3] around 1100 K is a factor of major importance in determining the $v(E)$ characteristic of GaAs. If the polar optical mode of lattice vibration were more efficient in removing energy from the electron system above this temperature, the energy of electrons in the Γ valley would not run away but would stabilize, basically in accordance with the simple single-electron energy balance equation:

$$B(T_e) = q\mu(T_e)E^2$$

where the left hand side of the equation is the energy lost per second by an electron due to collisions with the lattice, and the right hand side is the energy acquired per second by an electron from the field.

Thus, if $B(T_e)$ were proportional to T_e and of the form

$$B(T_e) = \frac{T_e - T_0}{\tau_{\mathscr{E}}}$$

(where T_0 is the lattice temperature and $\tau_{\mathscr{E}}$ is an energy relaxation time, assumed constant), then electrons in the central valley would not have sufficient energy to transfer to the upper valleys until an electric field of around 10 kV cm^{-1} were applied to the sample. Early estimates put the electron energy required for transfer at about 0.32 eV to 0.36 eV, whereas nowadays 0.28 eV to 0.29 eV is the generally accepted value. It is principally the time during which electrons in the Γ valley are in the runaway state, with energies between 0.1 eV and 0.29 eV, which leads to the possibility of electron velocity overshoot (Section 6.A.4).

†This value of electron temperature is from [6.A.3], which discarded the factor of 3/2 in the definition of electron temperature given in eqn (5.A.5). Accounting for the 3/2 would give a value of approximately 750 K for the electron temperature T_e which is equivalent to an electron energy of 0.1 eV.

6.A.2 Ionized impurity scattering

In Fig. 6.A.2 is shown the effect of doping upon the velocity-field characteristic [6.A.4]. Forces exerted upon electrons moving through the electrostatic potential field associated with an ionized impurity centre deflect the electron without affecting the electron energy, i.e. electron momentum tends to be randomized by ionized impurities, but the energy is unchanged. Ionized impurity scattering is thus elastic in nature.

The tendency to randomize momentum results in a net reduction in the component of electron velocity in the direction of the applied field. Hence the electron mobility is reduced, at low electric fields, to an extent which is doping dependent (Fig. 6.A.3, after [6.A.5, 6.A.6]). Because the total electron energy is unchanged, however, the field at which the electrons will acquire sufficient energy to runaway is also unchanged, so that electron transfer still occurs at a field in the neighborhood of 3.5 kV cm^{-1}. These two facts, i.e. reduced low-field mobility and unchanged threshold field, conspire to reduce the peak electron drift velocity, as shown in Fig. 6.A.2. The ultimate saturated drift velocity at field strengths of a few tens of kV cm^{-1} is unchanged because v_{sat} is governed by deformation potential scattering and by energy loss to polar optical phonons in the L valleys. It is thus unaffected by elastic scattering by ionized and neutral impurities.

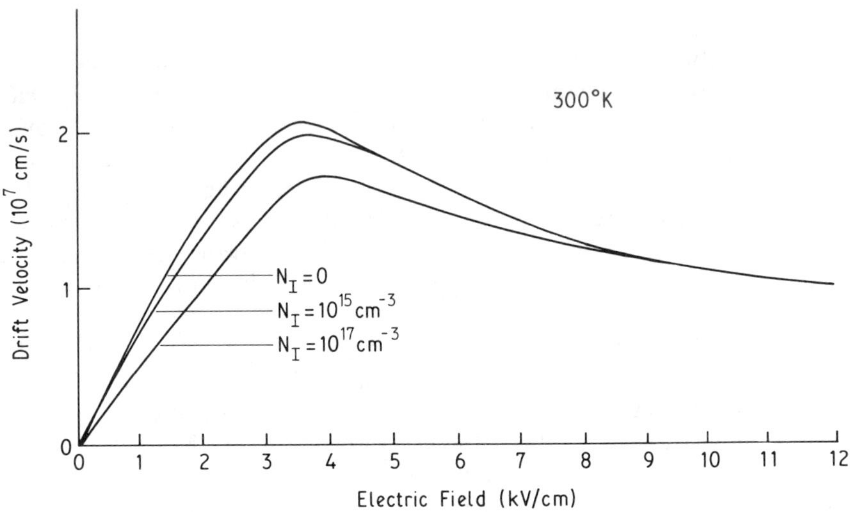

Fig. 6.A.2 Effect of doping upon the velocity-field characteristic of GaAs (after [6.A.4]).

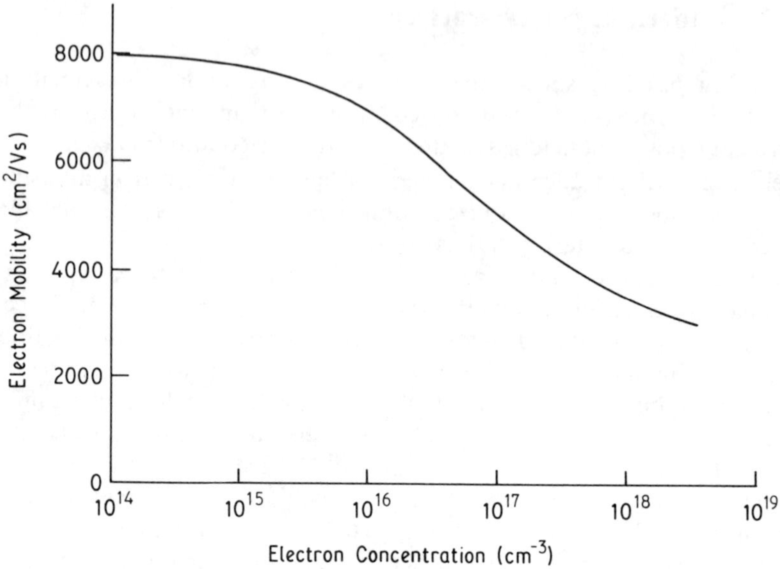

Fig. 6.A.3 Dependence of low-field electron mobility upon N-type doping concentration in GaAs (after [6.A.5] and [6.A.6]).

6.A.3 Effect of transverse electric fields

In equilibrium, some electrons in the channel of a GaAs MESFET occupy regions of the channel in which there are y-directed fields, i.e. fields which are transverse to the longitudinal or x direction (Fig. 6.A.4). Although these fields can, by themselves, give rise to no electron heating, Brennan and Hess have shown how, in the presence of an x-directed field, they may affect the energy of the electron distribution. Furthermore, Fig. 6.8 illustrates how the channel electrons in an operating FET experience a transverse electric field originating from the positive ionic charge in the depletion region extension into the gate-drain space.

Figure 6.A.5 shows the calculated degradation of the longitudinal $v_x(E_x)$ characteristic which follows from the presence of even moderate y-directed fields [6.A.8]. The calculations which led to this figure explicitly assumed that the electrons are transported along the direction of the net field $E = (E_x^2 + E_y^2)^{1/2}$, which makes the implicit assumption that the transverse y-directed field is fully effective in heating the electrons. In contrast, Brennan's and Hess's work may be interpreted to show that the transverse field which is effective in imparting energy to the electron system is dependent upon the

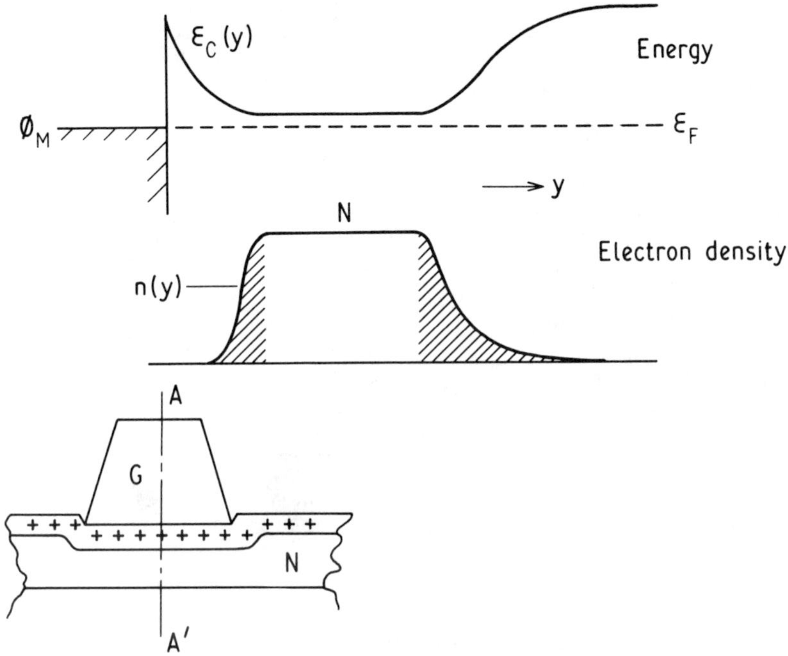

Fig. 6.A.4 Showing (shaded) the channel electrons which experience transverse or y-directed fields in equilibrium.

magnitude of the longitudinal field E_x as well as the transverse field itself, E_y. A typical value for the effective transverse field is about one-third of the actual transverse field. Thus, from Fig. 6.A.5, serious degradation of the longitudinal velocity-field characteristic can be expected for transverse fields in the channel of only 10 kV cm^{-1}; the low-field mobility is degraded, the peak velocity is much reduced, and the region of negative differential velocity tends to disappear. The ultimate saturated drift velocity is, however, unchanged.

6.A.4 Temporal factors, and velocity overshoot

Above an electron energy of approximately 0.1 eV, the longitudinal polar optical mode of lattice vibration becomes ineffective in removing energy from the electron distribution in the Γ valley of GaAs; the mean rate at which electron energy is lost to polar optical phonons actually decreases slightly, so the polar optical mode is unable to remove energy from the electrons at the same rate at which it is gained from the field. In fields greater than 3.5 kV cm^{-1}, therefore, the electrons simply go on accelerating until they

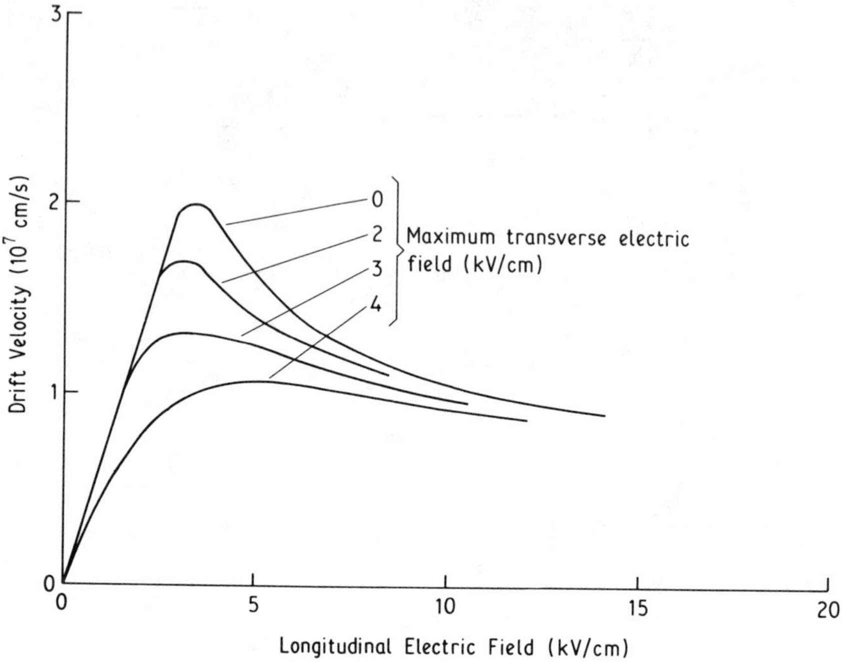

Fig. 6.A.5 Degradation in longitudinal $v(E)$ characteristic due to an effective transverse electric field $\boldsymbol{E}_y$ (after [6.A.8]).

attain an energy sufficient for transfer to the next-highest conduction-band valleys (i.e. the L valleys). The energy required for transfer is equal to the valley separation, minus the energy of a polar optical phonon, i.e. approximately 0.29 eV − 0.036 eV ≈ 0.25 eV. During the time electrons accelerate from 0.1 eV to 0.29 eV in the Γ valley, they may attain velocities in the direction of the field well in excess of the peak velocity in Fig. 6.A.1 (the limit set by the maximum group velocity is about 8×10^7 cm s^{-1}), i.e. the electron drift velocity may transiently overshoot the values which apply for long samples in the steady state.

Simple momentum and energy conservation equations can be used to make order of magnitude estimates for the times and distances for which overshoot may persist. In the case of a sample subjected to a longitudinal field of a few kV cm^{-1}, we typically find

$$\tau_{\text{overshoot}} \approx 1 \text{ psec}$$

$$x_{\text{overshoot}} \approx 0.5 \ \mu\text{m}$$

Transverse electric fields dramatically affect these answers. In the first place, a higher net field allows the electron to acquire the energy difference of 0.19 eV in a shorter time. Secondly, the electron is no longer accelerated purely along the x direction in the process of acquiring this energy, so that its component of velocity along the longitudinal channel direction also decreases. The foregoing two factors lead approximately to

$$x_{\text{overshoot}} \approx \frac{1.75}{E_x[1+(E_y/E_x)^2]} \ (\mu\text{m})$$

where E_x is the longitudinal field in kV cm^{-1} and E_y is the effective transverse field which, from Section 6.A.3, may typically be between about one-third of the actual field value when the transverse field is electrostatic in nature, and the full transverse field value where transport in the transverse field occurs. With $E_x = 3.5$ kV cm^{-1}, an effective field of $E_y \approx 10$ kV cm^{-1} reduces the overshoot distance to

$$x_{\text{overshoot}} \approx 0.05\ \mu\text{m}$$

Since transverse field effects of at least this magnitude occur in GaAs MESFETs at typical working voltages and currents (e.g. $V_{DS} = 3$ V, $I_D = 0.5I_{DSS}$ for a 0.5 μm gate length FET), we conclude that the overshoot distance will often be an order of magnitude smaller than the carrier transit length $(L+X)$ in the channel of the FET. On these grounds, velocity overshoot can be neglected in real devices.

6.A.5 Lattice temperature

Figure 6.A.6 shows the calculated effect of lattice temperature upon the velocity-field characteristic of bulk GaAs [6.A.4]. The low-field mobility, the peak velocity and the saturated drift velocity are all reduced.

Deviations from the 300 K velocity-field characteristic, which cannot be neglected, can therefore be expected in operating devices, particularly in power GaAs FETs where the channel temperature may rise to 150 °C or greater.

6.A.6 Conclusion as to the effective $v(E)$ characteristic

The combined influences of doping, transverse fields, scattering rate and lattice (i.e. channel) temperature, all effective simultaneously, lead to a velocity-field relationship for longitudinal (i.e. x-directed) transport in a FET channel which is quite different from that for bulk, undoped, GaAs. Because

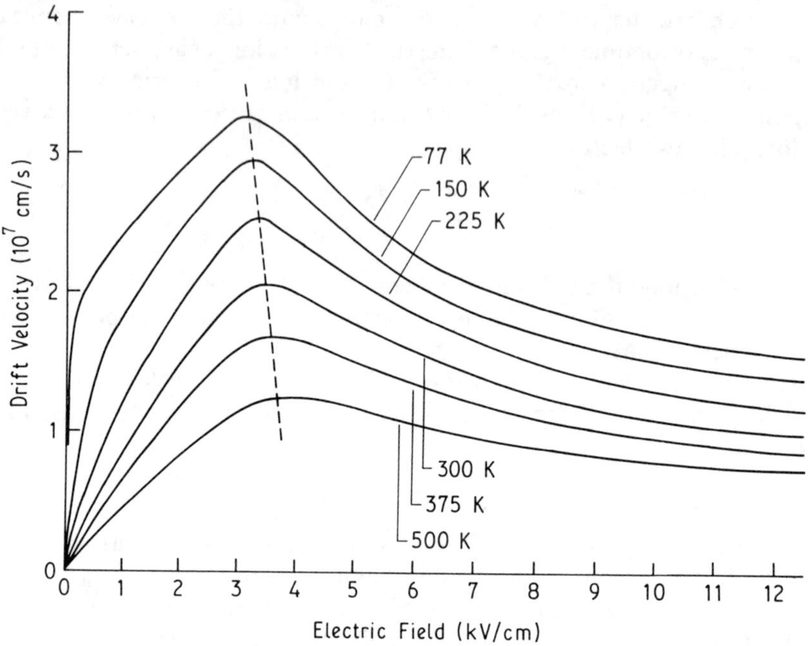

Fig. 6.A.6 Effect of lattice temperature upon the velocity-field characteristic of bulk GaAs (after [6.A.4]).

velocity overshoot can be neglected in real devices in which there are transverse fields, $v(E)$ may, for all practical purposes, be assumed to be an instantaneous function with a form as shown in Fig. 6.A.7 at 300 K. The low-field mobility is reduced by (a) increased doping and (b) higher transverse (i.e. y-directed) fields.

The saturated drift velocity ($E \geqslant 10$ kV cm^{-1}) is unchanged at around 10^7 cm s^{-1}. v_{sat} is, however, temperature dependent; it reduces with increasing lattice temperature.

In summary

Mobility—effect of doping and impurities:

$$\mu_0 \approx 2500\text{–}3500 \text{ cm}^2 \text{ V}^{-1} \text{ s}^{-1}$$

at typical channel doping levels.

Mobility—effect of transverse fields: as a rough guide, E_y may reduce μ_0 further to about one-third of the as-doped value.

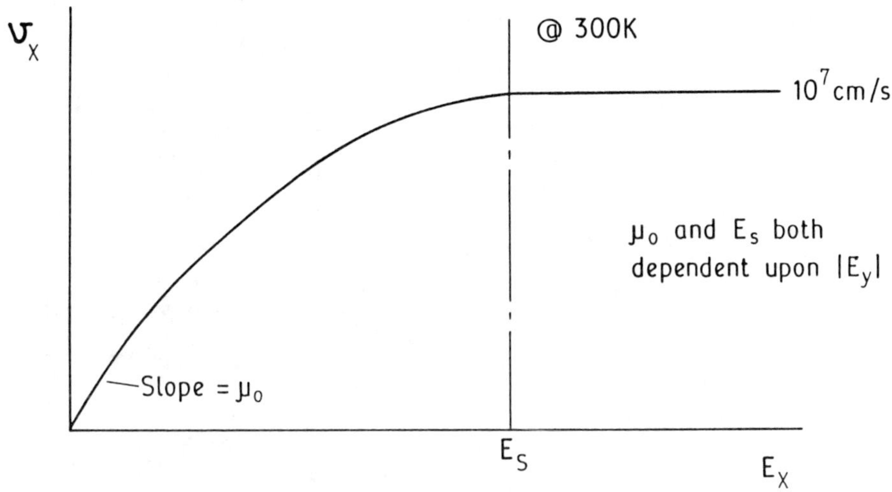

Fig. 6.A.7 Effective velocity-field characteristic for longitudinal transport of electrons in a GaAs FET channel.

Mobility—effect of lattice temperature:

$$d\mu_o/dT_0 \approx -0.4\% \text{ per K} \quad (\text{at a field of } 1 \text{ kV cm}^{-1})$$

Saturated drift velocity—effect of lattice temperature:

$$(dv_{sat}/dT_0) \approx -0.2\% \text{ per K} \quad (\text{at a field of } 12 \text{ kV cm}^{-1})$$

(v_{sat} is unaffected by doping and transverse fields).

Thus, very importantly, the saturated drift velocity is expected to be that of bulk GaAs, i.e. close to 10^7 cm s^{-1} at 300 K.

Experimental confirmation of this expectation can be provided from the equations in this book in two ways. The first involves rearranging the reverse modelling equation for gate length, eqn (8.2), to give

$$v_{sat} = \frac{g_{m0}L_G}{2(C_{gc}+C_{gd})}\left[\frac{3}{2}+\frac{\epsilon Z_G}{C_{gd}}\right]$$

Applying this expression to the various manufacturers' data of Tables 8.1 and 8.2 of Chapter 8 yields Fig. 6.A.8. There is no clear trend with gate length but the scatter increases at smaller gate lengths, which is probably a reflection of the percentage uncertainty in L_G increasing as L_G decreases. The mean value of saturated drift velocity evident in this figure is 0.91×10^7 cm s^{-1}.

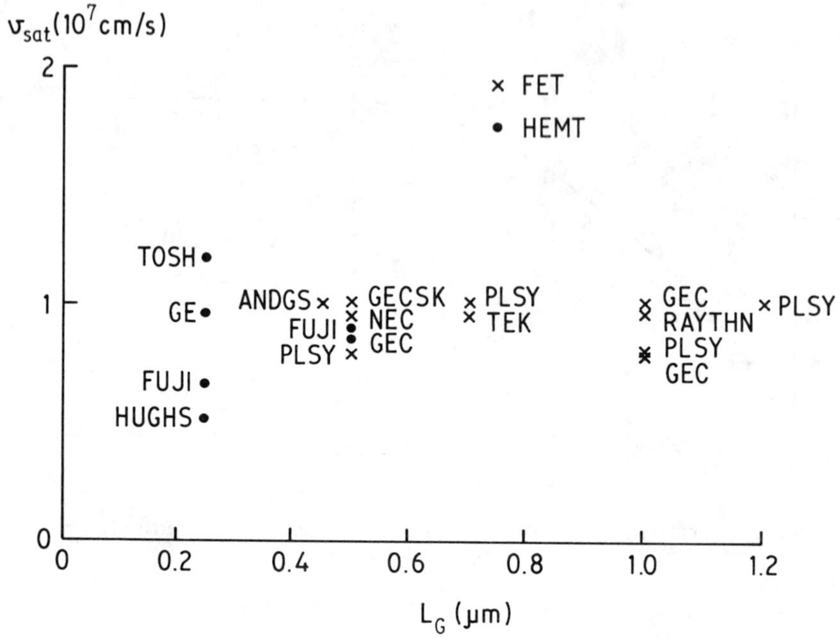

Fig. 6.A.8 Saturated drift velocity obtained by analysis of scattering parameter measurements on a selection of practical GaAs MESFETs and HEMTs.

The second method involves the expressions for τ_{gm}, C_{gc} and C_{gd} from Table 6.1 of Chapter 6. We have

$$\tau_{gm} = \frac{1}{v_{sat}}\left(\frac{X}{2} - \frac{2d}{1+2X/L_G}\right)$$

From eqn (6.10) of Chapter 6, the expression for C_I may be used to eliminate d to give

$$\tau_{gm} = \frac{1}{v_{sat}}\left[\frac{X}{2} - \frac{2\epsilon L_G Z_G}{C_I(1+2X/L_G)}\right]$$

Now

$$C_{gd} = \frac{2\epsilon Z_G}{(1+2X/L_G)}$$

So

$$\tau_{gm} = \frac{L_G}{v_{sat}}\left(\frac{X}{2L_G} - \frac{C_{gd}}{C_I}\right)$$

Rearranging the expression for C_{gd} to give (X/L_G), and eliminating $(X/2L_G)$,

$$v_{sat}=\frac{L_G}{2\tau_{gm}}\left(\frac{\epsilon Z_G}{C_{gd}}-\frac{2C_{gd}}{C_I}-\frac{1}{2}\right) \tag{6.A.1}$$

C_I is unfortunately not a measurable quantity, so we must replace it by parameters that are. Manipulating the equations for C_{gc} and C_{gd} of Table 6.1, allows us to write:

$$C_I=C_{gc}+C_{gd}-\frac{\epsilon Z_G L_G}{2d}\left(\frac{\epsilon Z_G}{C_{gd}}-\frac{1}{2}\right)$$

Using the equation for g_{m0} to eliminate d,

$$C_I=C_{gc}+C_{gd}-\frac{g_{m0}L_G}{2v_{sat}}\left(\frac{\epsilon Z_G}{C_{gd}}-\frac{1}{2}\right)$$

One of two strategies may now be employed: either a value of v_{sat} can be found between eqn (6.A.1) and the equation for C_I by iteration or, because C_I appears in a small correction term in the basic equation for v_{sat}, we may with little error set $v_{sat}=10^7$ cm s^{-1} in the expression for C_I and use the resultant value of C_I in eqn (6.A.1) (which amounts to accepting the first estimate in the iteration method).

A variation on this second theme uses g_{m0} in place of C_I to eliminate d from the expression for τ_{gm}, and C_{gd} to eliminate X as before. Straightforward algebraic manipulation then gives

$$v_{sat}=\frac{L_G\left(\dfrac{\epsilon Z_G}{C_{gd}}-\dfrac{1}{2}\right)}{2\left(\tau_{gm}+\dfrac{C_{gd}}{g_{m0}}\right)} \tag{6.A.2}$$

In all these expressions for v_{sat}, the experimental values of C_{gc} and C_{gd} should strictly be corrected for geometric capacitance.

In Table 6.A.1 are shown the results of applying these methods to an ion-implanted foundry device having gate dimensions of $L_G=0.5$ μm, $Z_G=250$ μm. The non-iterated option was used in calculating v_{sat} from eqn (6.A.1) so that, where v_{sat} deviates markedly from 10^7 cm s^{-1}, the value calculated from eqn (6.A.2) must be considered the more accurate of the two. The mean value of all the v_{sat} entries in Table 6.A.1, as calculated from eqn (6.A.2), is 0.96×10^7 cm s^{-1}.

It is important to emphasize that two conditions must be satisfied if these methods are to yield values for v_{sat} which can be believed.

(i) Reliable data must be available for C_{gc}, C_{gd} and, particularly, τ_{gm}. The

Table 6.A.1 v_{sat} evaluated from τ_{gm} for a foundry device nominal gate dimensions: $L_G = 0.5$ μm; $Z_G = 250$ μm

V_{SG} [V]	V_{DS} [V]	C_{gc} [pF]	C_{gd} [pF]	τ_{gm} [ps]	g_{m0} [mS]	v_{sat} by (6.A.1) (cm s^{-1})	v_{sat} by (6.A.2) (cm s^{-1})
	2.0	0.227	0.031	1.323	40.4	0.33×10^7	0.60×10^7
	2.5	0.271	0.026	1.780	38.2	0.65	0.72
	3.0	0.273	0.023	2.026	36.6	0.80	0.84
	3.5	0.284	0.021	1.972	36.4	1.03	1.02
	4.0	0.287	0.019	2.654	36.0	0.94	0.95
0.2	4.5	0.293	0.017	2.966	33.0	1.04	1.04
	5.0	0.303	0.016	3.366	32.2	1.03	1.02
	5.5	0.312	0.014	3.816	31.1	1.14	1.13
	6.0	0.317	0.013	4.077	30.2	1.20	1.19
	2.0	0.247	0.030	1.430	37.7	0.44	0.61
	2.5	0.259	0.026	1.394	37.4	0.81	0.86
	3.0	0.261	0.023	1.671	36.3	0.95	0.97
	3.5	0.262	0.022	1.913	35.1	0.93	0.94
0.4	4.0	0.268	0.020	2.272	33.7	0.98	0.98
	4.5	0.275	0.018	2.781	32.2	0.99	0.99
	5.0	0.281	0.016	3.236	30.8	1.06	1.05
	5.5	0.291	0.015	3.439	30.4	1.12	1.11
	6.0	0.297	0.014	3.886	29.4	1.11	1.10
	2.0	0.241	0.028	1.046	37.9	0.78	0.87
	2.5	0.237	0.026	1.345	36.0	0.79	0.87
	3.0	0.239	0.024	1.491	34.7	0.79	0.87
	3.5	0.245	0.022	1.691	33.9	1.02	1.03
0.6	4.0	0.252	0.021	1.965	33.2	0.99	1.00
	4.5	0.258	0.019	2.392	32.0	1.02	1.02
	5.0	0.267	0.018	2.567	31.3	1.07	1.05
	5.5	0.271	0.016	3.147	29.8	1.08	1.07
	6.0	0.279	0.015	3.534	29.1	1.08	1.07

Note that $C_{gs\ geom} = 0.023$ pF; $C_{gd\ geom} = 0.002$ pF and $V_P \approx 2.5$ V.

best way to ensure the acceptability of such data is to make microwave measurements over a whole range of bias points and to require that the results exhibit the form of Fig. 6.39, especially Fig. 6.39(c). Data in which the τ_{gm} curves are interleaved or twisted should be regarded as suspect.

(ii) Precise knowledge of the geometric contributions to the capacitances must be available so that the measured values of capacitance can be corrected to yield C_{gc} and C_{gd} for insertion into eqns (6.A.1) and (6.A.2).

We therefore conclude, somewhat disappointingly perhaps, that there is no velocity overshoot in FETs and that, for all practical purposes, the velocity which governs all the important aspects of GaAs FET behavior is a sedate 10^7 cm s^{-1}.

References

6.A.1 Maloney, T. J. Polar mode scattering in ballistic transport GaAs devices, *IEEE Electron. Dev. Lett.*, **EDL-1**(4), 54, April 1980.

6.A.2 Littlejohn, M. A., Hauser, J. R. and Glisson, T. H. Velocity-field characteristics of GaAs with Γ_6^C—L_6^C—X_6^C conduction-band ordering, *J. Appl. Phys.*, **48**(11), 4587–4590, November 1977.

6.A.3 Huang, R. S. and Ladbrooke, P. H. The physics of excess electron velocity in submicron-channel FETs, *J. Appl. Phys.*, **48**(11), 4791–4798, November 1977.

6.A.4 Ruch, J. G. and Fawcett, W. Temperature dependence of the transport properties of gallium arsenide determined by a Monte Carlo method, *J. Appl. Phys.*, **41**(9), 3843–3849, August 1970.

6.A.5 Walukiewicz, W., Lagowski, J., Jastrzebski, L., Lichtensteiger, M. and Gatos, H. C. Electron mobility and free-carrier absorption in GaAs: determination of compensation ratio, *J. Appl. Phys.*, **50**(2), 899–908, February 1979.

6.A.6 Walukiewicz, W., Lagowski, J. and Gatos, H. C. Electron mobility in *N*-type GaAs at 77 K; determination of the compensation ratio, *J. Appl. Phys.*, **53**(1), 769–770, January 1982.

6.A.7 Brennan, K. and Hess, K. A theory of enhanced impact ionization due to the gate field and mobility degradation in the inversion layer of MOSFETs, *IEEE Trans. Electron. Dev. Lett.*, **EDL-7**(2), 86–88, February 1986.

6.A.8 Ladbrooke, P. H. Some possible consequences of a transverse doping gradient in bulk-effect oscillators, *J. Phys. D: Appl. Phys.*, **3**, 437–442, March 1970.

6.A.9 Fawcett, W., Boardman, A. D. and Swain, S. Monte Carlo determination of electron transport properties in gallium arsenide, *J. Phys. Chem. Solids*, **31**(9), 1963–1990, September 1970.

Chapter 7

High electron mobility transistors

7.1 Introduction

The high electron mobility transistor, or HEMT, is a field-effect transistor having current carriers (electrons) densely confined in a thin sheet of order 100 Å thickness. Borrowing terminology from Drude's theory of metals, in which the density of free electrons is very large (approximately 10^{22} cm^{-3}), the thin sheet of current-carrying electrons in a HEMT is referred to as a two-dimensional electron gas, or 2-DEG, which suggests another acronym by which the device is also known: the TEGFET. As will become clear in Sections 7.2 and 7.3, the HEMT consists of epitaxially grown layers of compound semiconductors which have different bandgap energy, so the interface between them constitutes a heterojunction. Only the wider band gap material is N-type doped, leading to the alternative acronyms of MODFET (meaning modulation doped FET) and SDHT (meaning selectively doped heterojunction transistor). All these shortened forms refer to exactly the same device.

For application to MMICs, the most attractive feature of the HEMT is its low noise. Despite the reservations expressed in the introduction to Section 6.12, there always will be a benefit in reducing the noise figure of an MMIC. At millimetric frequencies particularly [7.1], the HEMT promises a lower noise figure (in dB) than gain (also expressed in dB), and to have a higher gain than noise figure is distinctly useful. Thus, on this basis, the HEMT is a strong candidate for millimetre-wave integration. In discrete form, a HEMT structure has given greater than 6 dB gain in a circuit at 94 GHz [7.2], thus demonstrating the feasibility of amplification all the way up to 100 GHz.

At lower frequencies, for example X- and J-bands, the HEMT at present finds application mainly as a discrete device in low-noise hybrid amplifiers, where on-circuit tuning to realize the minimum noise figure can be carried out with relative ease. Noise figures from 0.7 dB to 1 dB with 10 dB associated gain are commonplace from 0.5 μm gate length devices at X-band (around 10 GHz).

Much of the work of Chapter 6 on the physical basis of equivalent circuits applies to the HEMT, with the possible addition of resistors across C_{gc} and C_{gd} to account for gate leakage current which is sometimes observed, but whose origin is not wholly clear. That work must be supplemented by expressions for the HEMT equivalent circuit elements, in terms of device structure, where readily possible (Section 7.5). Prior to that, a short discussion will be given of the physical basis and structure of the device (Section 7.3) in terms of the different band structure available from $Al_xGa_{1-x}As$ alloys, depending upon the aluminium fraction, x ($0 \leqslant x \leqslant 1$). Although other material systems are being actively worked upon (e.g. $Ga_{0.85}In_{0.15}As$ strained layers with $Al_{0.3}Ga_{0.7}As$ confinement layers; and $Ga_{0.47}In_{0.53}As$ lattice–matched to $Al_{0.48}In_{0.52}As$ confinement layers grown on InP), the present chapter will address the $Al_xGa_{1-x}As$ system only.

7.2 Energy band line-up

If aluminium (Al) is part-substituted for gallium (Ga) in a growing crystalline layer, an alloy semiconductor results which has a lattice constant almost identical to that of GaAs. Consequently, lattice-matched layers of $Al_xGa_{1-x}As$ can be grown on GaAs substrates, and further layers of GaAs can similarly be grown lattice-matched over the $Al_xGa_{1-x}As$. Molecular beam epitaxy (MBE), and organo-metallic vapor-phase epitaxy (OMVPE) or metal–organic chemical vapor deposition (MOCVD), may all be used for the growth of such layered structures with interfaces sharp on the scale of a few monolayers (i.e. $\leqslant$10 Å). Al may be substituted for Ga anywhere in the fractional range $0 \leqslant x \leqslant 1$, giving GaAs at one extreme ($x = 0$), and aluminium arsenide (AlAs) at the other ($x = 1$). x is the molecular proportion of Al to Ga, known simply as the mole fraction.

As a function of x, the energies of the conduction-band minima in $Al_xGa_{1-x}As$, relative to the top of the valence band at the Γ point, are as in Fig. 7.1 [7.3]. For GaAs, the energy separation between the valence band edge and the lowest of the conduction-band minima (the Γ minimum, located at a wave vector k of (000)), is 1.42 eV at 300 K, i.e. the bandgap of GaAs is 1.42 eV. At a mole fraction of 0.3, the Γ minimum is still the lowest, but the bandgap increases to approximately 1.8 eV. For mole fractions $x \geqslant 0.45$, the Γ minimum is no longer the lowest, but the X minima are. Thus, for example, for $x = 0.8$, the bandgap of $Al_{0.8}Ga_{0.2}As$ is about 2.1 eV. For AlAs, the bandgap is approximately 2.17 eV. It is thus clear that in the vicinity of an interface

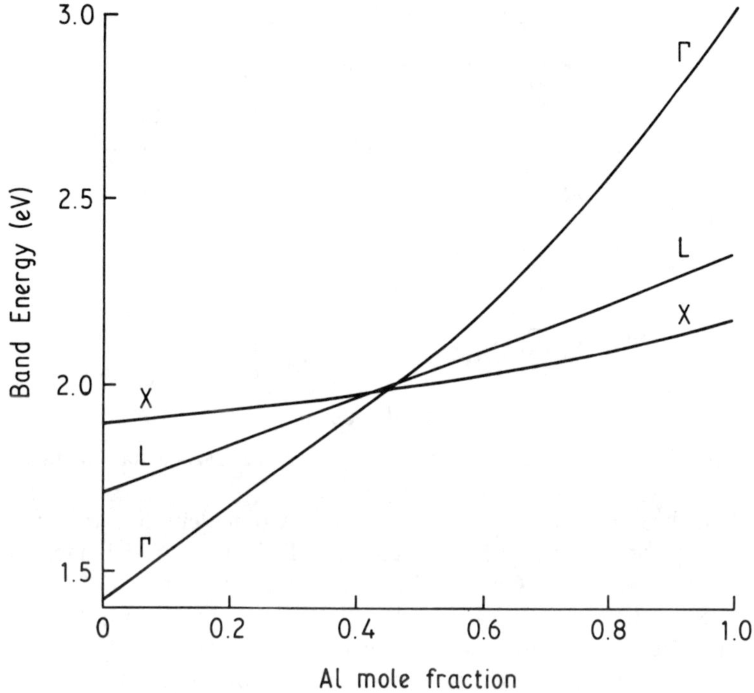

Fig. 7.1 Variation in Γ, L and X conduction band valley energies with Aluminium mole fraction x in $Al_xGa_{1-x}As$.

between GaAs and $Al_xGa_{1-x}As$ there will be a difference in bandgap energy† that will in general occur partly in the form of a discontinuity in the valence band, $\Delta\mathscr{E}_v$, and partly as a discontinuity in the conduction band, $\Delta\mathscr{E}_c$, as in Fig. 7.2.

The underlying physical principles governing the manner in which the bands line-up and determine $\Delta\mathscr{E}_c$ and $\Delta\mathscr{E}_v$, is an extensive subject [7.4]. For present purposes, the conduction-band discontinuity for $x \leqslant 0.45$ is principally of interest, and may be shown to be given approximately [7.3] by

$$\Delta\mathscr{E}_c = 0.81x \qquad x \leqslant 0.45 \tag{7.1}$$

Thus, for example, for $x = 0.3$, $\Delta\mathscr{E}_c \approx 0.24$ eV. It may be shown that the maximum conduction-band discontinuity occurs for an Al mole fraction of

†We here make the common assumptions that the interface is sharp on an atomic scale, and that the bulk bandstructure of both materials exists or applies right up to the interface. The valence band edge at k = (000) is denoted by $\mathscr{E}_v$ and the lowest of the conduction band minima is denoted by $\mathscr{E}_c$.

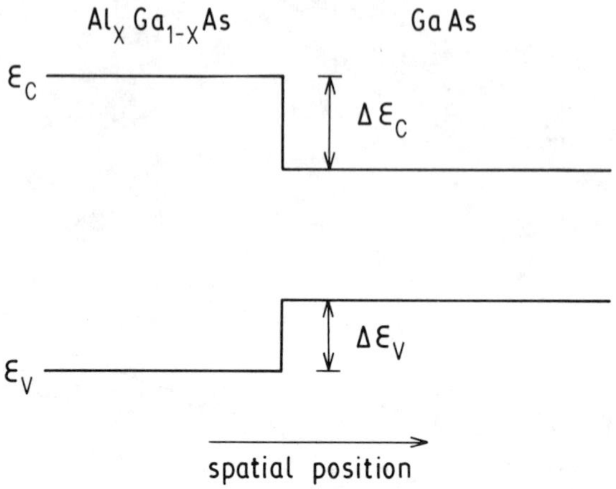

Fig. 7.2 Energy band offsets at an $Al_xGa_{1-x}As$–GaAs heterojunction.

0.45, and has a value of about 0.36 eV. Coincidentally, this value is immediately comparable with the intervalley (Γ–L) separation in GaAs which means that, if electrons in GaAs in the vicinity of an interface with $Al_xGa_{1-x}As$ become energetic enough to transfer to the L valleys in GaAs, they are also energetic enough to surmount the energy barrier $\Delta\mathcal{E}_c$ and be scattered into the $Al_xGa_{1-x}As$. For the time being, however, the important point is that a step in the conduction band edge of 0.2 eV to 0.3 eV can be realized by appropriate choice of Al mole fraction, x.

7.3 Physical basis and structure

Consider a series of thin, alternating, layers of GaAs and $Al_xGa_{1-x}As$ as in Fig. 7.3(a). Suppose during growth of the layers a flux of dopant (e.g. Si) is directed at the epitaxially growing surface whenever Al is, so that the $Al_xGa_{1-x}As$ layers, and only these layers, become N-type doped. If we suppose for a moment that the electrons associated with the dopant atoms remain in the immediate locality of these donor atoms, then we would have the schematic representation of Fig. 7.3(b). Recognizing, however, that the binding energy of electrons to the dopant atoms might be roughly 0.01 eV, and that the thermal energy of an electron at 300 K is 0.026 eV, the electrons will not be bound but will be free to move throughout the crystal in all directions (i.e. randomly) with velocities governed by their thermal energy. Sooner or later, if the $Al_xGa_{1-x}As$ layers are thin enough, virtually all electrons will pass across an interface and enter a GaAs layer, whereupon the energy $\Delta\mathcal{E}_c$ they possess above the conduction-band edge in the GaAs layer will be lost by

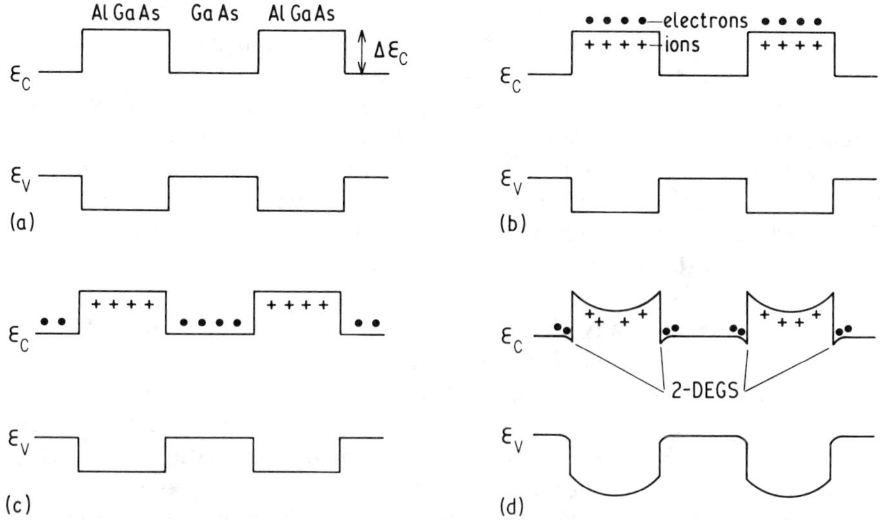

Fig. 7.3 Energy band variation through an $Al_xGa_{1-x}As$–GaAs superlattice: (a) undoped; (b) with N-type (silicon) dopant confined to the wider bandgap material (assuming, hypothetically, that the free electrons associated with the Si dopant remain close to the donor ions); (c) assuming transfer of the free electrons and (hypothetically) uniform distribution in the GaAs; (d) accounting for space-charge effects.

phonon emission. If, on the other hand, the N–$Al_xGa_{1-x}As$ layers are thicker (several hundred Ångströms), then only the hundred Ångströms or so nearest the heterointerface (for typical doping densities) transfers its electrons to the GaAs; the rest remains undepleted. Once electrons are in the potential well, i.e. the GaAs, they have no way of acquiring sufficient energy to escape back into the $Al_xGa_{1-x}As$ in equilibrium (i.e. with no external electrical or optical excitation), and so they remain confined (Fig. 7.3(c)). Because the electrons are nevertheless free to move, however, they are attracted to the heterojunction interfaces by the electrostatic fields arising on the ionized impurity atoms in the $Al_xGa_{1-x}As$ (Fig. 7.3(d)). The electrons form thin sheets of charge on the GaAs side of the interfaces. Electric field arising on the positively charged dopant ions in the $Al_xGa_{1-x}As$ terminates on these sheets of electron charge. Integrating Poisson's equation twice, given the distribution of positive and negative charges in Fig. 7.3(d), gives rise to a spatial dependence of potential shown by the band bending in Fig. 7.3(d).

The final result of the selective doping, or modulation doping, of the heterojunction layers is thus a spatial separation of the donor ions and the free electrons. Consequently, the scattering of the electrons by ionized impurities is much reduced compared with electrons in uniformly doped bulk material for two reasons. First, the electrons are now remote from the electrostatic fields surrounding all the ions which deflect the electrons' trajectories (i.e. scatter them). Secondly, the electrons tend to screen one another from such fields. Because ionized impurity scattering has its greatest effect at low fields, we might expect a major improvement in the low-field mobility of electrons compared with electrons in bulk N-type material. At room temperature, low-field mobilities of 6000–7000 $cm^2\ V^{-1}\ s^{-1}$ can be achieved, compared with 2500–3500 $cm^2\ V^{-1}\ s^{-1}$ for bulk material, depending upon doping level. Furthermore, if the lattice is cooled to reduce the number of phonons and hence the phonon scattering rate, exceedingly high values of low-field mobility can be recorded (typically, 2×10^4–$7 \times 10^4\ cm^2\ V^{-1}\ s^{-1}$ at 77 K and up to $10^6\ cm^2\ V^{-1}\ s^{-1}$ at nominally 4.3 K). Values of sheet carrier concentration in the range $5 \times 10^{11}\ cm^2$ to $10^{12}\ cm^{-2}$ are commonly obtained in the 2-DEG.

Close inspection of the conduction-band edge variation in the vicinity of a 2-DEG in Fig. 7.3(d) will show that the potential variation is roughly triangular in shape. Furthermore, the width of this triangular potential well is only about 100 Å, so the electrons in the 2-DEG might be expected to occupy quantized levels in the well, i.e. the problem is strictly quantum mechanical in nature [7.5].

Although HEMTs using multiple 2-DEGs have been constructed, we shall confine our attention to field-effect transistors using only one 2-DEG. We thus consider growing a layer structure having only one selectively-doped N–$Al_xGa_{1-x}As$ to π-GaAs interface†. Starting with a semi-insulating GaAs substrate, we may grow either

(a) a π-GaAs layer first, followed by an N-doped $Al_xGa_{1-x}As$ layer (Fig. 7.4(a)) or

(b) an N-doped $Al_xGa_{1-x}As$ layer first, followed by a π-GaAs layer (Fig. 7.4(b)).

The first is referred to as a conventional HEMT structure (CHEMT or simply HEMT), whereas the second is termed an inverted HEMT structure (or IHEMT). Generally, higher mobilities have been obtained for conventional HEMT layers, a fact which has steered attention towards this particular structure somewhat at the expense of the IHEMT. To reflect this dominance, the following discussion will be restricted to the conventional HEMT.

As indicated in Fig. 7.4, field-effect transistors may be fabricated from the basic layers by alloying source and drain regions so as to provide ohmic

†The notation π-GaAs means GaAs that is very lightly p-type, e.g. $N_A \leqslant 10^{14}\ cm^{-3}$. Such lightly doped material typically results from nominally undoped growth of GaAs in an MBE system.

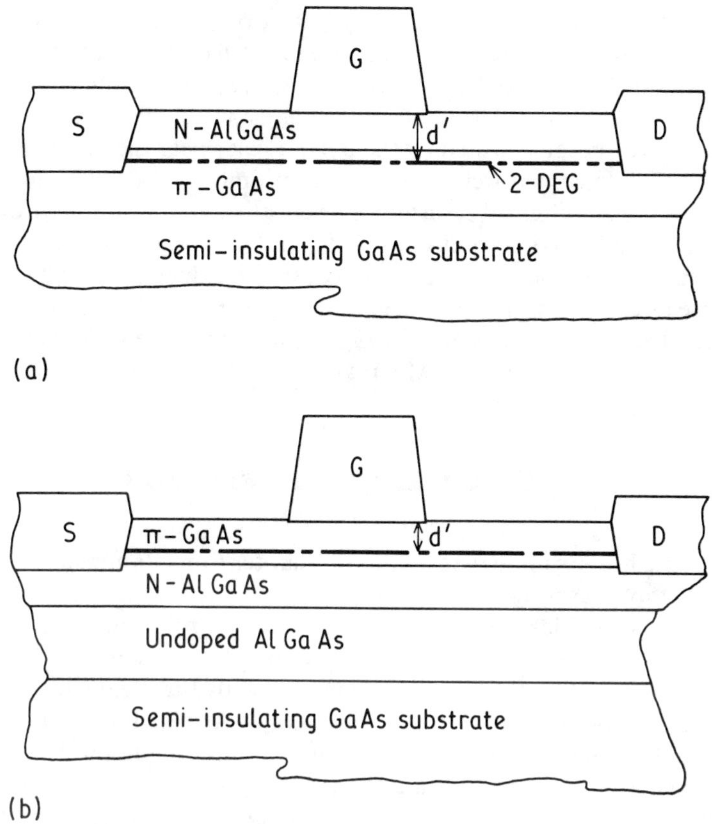

Fig. 7.4 Basic cross-sectional diagrams of (a) conventional HEMT (CHEMT) and (b) inverted HEMT (IHEMT).

contact to the 2-DEG with, between the ohmics, a Schottky gate just as in the bulk GaAs FET. In use, the device is biased exactly as a MESFET. In the case of the HEMT, the application of negative bias to the gate at first depletes the otherwise undepleted N–$Al_xGa_{1-x}As$ layer of electrons. When the depletion edge enters the 100 Å or so nearest the interface which is already depleted by virtue of having transferred its electrons to form the 2-DEG, further application of negative gate bias starts to deplete the 2-DEG itself. Thus the channel of current-carrying electrons, which the 2-DEG represents, can be modulated by a signal applied to the gate electrode. Once channel modulation occurs in this way, the device exhibits properties of gain and amplification up until the point where sufficient negative gate bias has been applied to deplete the 2-DEG fully, when the device is said to be pinched off.

All the foregoing has very obvious and immediate parallels with the bulk

GaAs MESFET. The most distinguishing feature of the HEMT compared with the FET is that conduction between source and drain occurs in a relatively well-confined channel at a fixed distance from the gate metal (i.e. the distance d' in Fig. 7.4(a)). Neglecting for the moment the almost inevitable deconfinement of electrons as they become heated by the intense fields near the drain-end of the gate, the well-defined and well-confined channel means that the transconductance $g_{m0} = \epsilon v_{sat} Z_G / d'$ can be sustained even at quite small values of drain current, I_D. Such behavior is unlike the MESFET, where at low values of I_D the depletion edge enters the tail of the profile causing d' to increase rapidly, and g_{m0} to collapse or compress, as in Fig. 6.33(a). From Section 6.12 and eqn 6.50, we can expect that the ability to maintain g_{m0} at low current levels will lead to lower values of noise figure for the HEMT than for the FET, as is observed.

7.4 Practical HEMT structures

Drawing heavily upon the findings of Chapter 6 and its appendix, we may anticipate those aspects of the underlying physics which are of major importance or significance to HEMT operation at room temperature. The discussion will be conducted with reference to a practical HEMT structure, which has two additional features compared with the elementary form of Fig. 7.4(a): an undoped spacer layer of $Al_xGa_{1-x}As$ which further separates the 2-DEG from the electrostatic potential of the ionized impurity centers in the N-doped $Al_xGa_{1-x}As$, and an overlying layer of heavily N-type GaAs usually referred to as the capping layer, or simply the N^+ cap (Fig. 7.5). The gate is located in a recess etched in the cap. The initial purpose of the cap appears to have been to obtain ohmic contacts of low resistance to the 2-DEG, although it has also a major bearing on the gain and noise achievable through f_T and R_S, and also upon the output power per millimetre of gatewidth.

With reference to Fig. 6.36, the high doping-thickness product of the cap and N–$Al_xGa_{1-x}As$ layers, and the high electron mobility in the 2-DEG, all contribute to keeping the source resistance, R_S, small. For instance, if

$$N_{cap} = 10^{18}\ \text{cm}^{-3} \qquad \mu_{0cap} = 2500\ \text{cm}^2\ \text{V}^{-1}\ \text{s}^{-1}$$

$$Y_{cap} = 750\ \text{Å}$$

$$N = 10^{18}\ \text{cm}^{-3} \qquad \mu_{0AlGaAs} = 1800\ \text{cm}^2\ \text{V}^{-1}\ \text{s}^{-1}$$

$$Y_A = 500\ \text{Å}$$

$$n_{s0} = 6.10^{11}\ \text{cm}^{-2} \qquad \mu_{02-DEG} = 7000\ \text{cm}^2\ \text{V}^{-1}\ \text{s}^{-1}$$

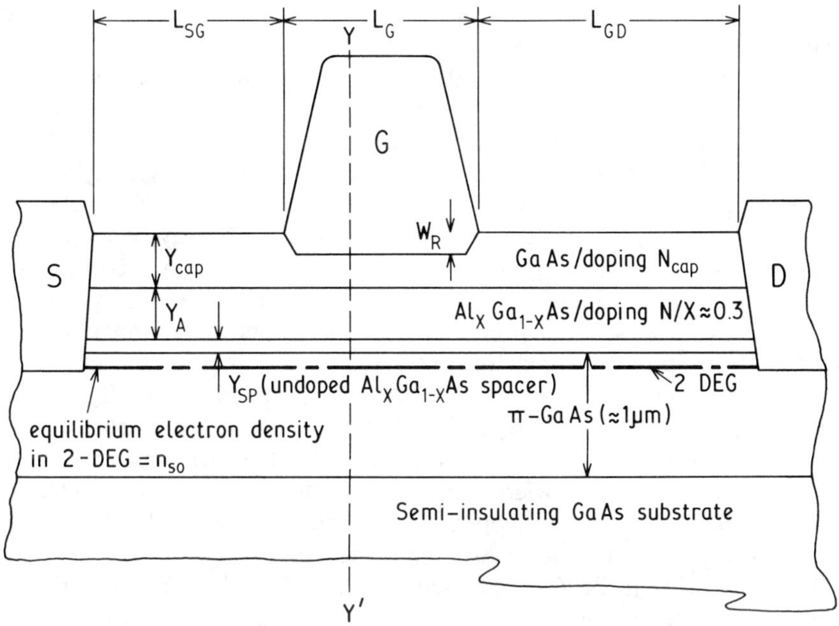

Fig. 7.5 Structural and material parameters for calculating principal equivalent circuit elements of a HEMT.

then the resistances of the three regions for $L_{SG} = 0.5$ μm and $Z_G = 150$ μm are

cap layer: 2.03 Ω (This allows for a surface potential of 0.8 V)

$Al_xGa_{1-x}As$ layer: 2.89 Ω

2-DEG: 4.96 Ω

Thus the high electron mobility in the 2-DEG makes only a minor contribution to keeping R_{SG} small; paralleling the above three resistances gives

$$R_{SG} = 0.96\ \Omega$$

For $\rho_c \approx 0.2\ \Omega$ mm, $R_{contact} \approx 1.33\ \Omega$, so that

$$R_S = R_{contact} + R_{SG} \approx 2.3\ \Omega$$

Referring next to Fig. 6.9, the region labeled region I, over which charge exchange with the depletion edge occurs by dielectric relaxation, will be charged by high-mobility electrons in the HEMT, which is a benefit. Over this region, the channel remains well-defined and well-confined. Over the regions marked II and III in Fig. 6.9, however, electrons in the HEMT, as in the FET, are accelerated in the channel fields $E_x(x)$ to energies which allow electron transfer to occur to the L valleys, variously taken to be 0.28–0.36 eV above the Γ conduction-band minimum. From eqn (7.1) for $x = 0.3$ we have $\Delta\mathscr{E}_c \approx$ 0.24 eV, so the electrons in regions II–III clearly have enough energy to enter the $Al_xGa_{1-x}As$ if they happen to be scattered in the direction of the interface, or, if they remain in the GaAs, to be well above the energy of the 2-DEG triangular well. In other words, the 2-DEG property is lost in regions II and III; the electrons become deconfined and occupy bulk $Al_xGa_{1-x}As$ or bulk GaAs electron states. We expect, therefore, these electrons to move through regions II–III at the saturated drift velocity for bulk GaAs, i.e. 10^7 cm s^{-1} (see the appendix to Chapter 6) or, worse, at the saturated drift velocity for bulk $Al_xGa_{1-x}As$, i.e. down to about 4×10^6 cm s^{-1}. Since the noise theory of Section 6.12 was developed on the assumption that the device is biased so that regions II and III are strictly limited in extent, this issue of electron deconfinement should not jeopardize any benefit the modulation-doped structure may bring to noise figure reduction.

Our interest is in devices operated under high-field conditions at room temperature. On the basis of the foregoing brief discussion, therefore, we may expect no major improvements in device performance due to improvements in electron transport; velocity overshoot is not expected to play any major role. Neither do we expect the quantum–mechanical nature of 2-DEG available states to be particularly important. The primary attribute of the HEMT structure is the excellent channel confinement over most of the 2-DEG in the shadow of the gate, which allows the transconductance to be sustained as the drain current is reduced (by application of gate bias) to relatively low values. The fact that the electron drift mobility is high over region I of the channel helps to keep f_T high, and hence the power gain high. Finally, the source resistance tends to be low, which is a coincidental result of the high value of doping-thickness product of typical HEMT layers.

As with the FET, we are concerned mainly with the microwave characteristics of the HEMT, not the dc behavior.

7.5 Principal equivalent circuit elements

If we neglect the fact that gate leakage current in some HEMTs requires a resistance in parallel with C_{gc} (and perhaps another resistor in parallel with C_{gd}), the equivalent circuit for the HEMT is otherwise the same as that for the

GaAs FET (Fig. 6.44). Expressions for the three important elements g_{m0}, C_{gc} and C_{gd} are available, subject to some conditions which may be understood by reference to Fig. 7.6. The figure shows the potential variation along a plane such as Y–Y′ in Fig. 7.5. We restrict ourselves to the case where the following conditions apply.

(i) The gate recess is shallower than the N^+ cap (i.e. $W_R \leqslant Y_{cap}$).

(ii) The applied gate voltage (which is negative with respect to the source, i.e. $V_{GS'} \leqslant 0$ or, equivalently, $V_{S'G} \geqslant 0$) is large enough for the associated gate depletion edge to have reached the 2-DEG after fully depleting the N^+ cap and N–$Al_xGa_{1-x}As$ in the shadow of the gate†, (i.e. $n_s \leqslant n_{s0}$).

(iii) The gate voltage is not so large as to cause pinch off (i.e. $n_s > 0$).

Summation of the potentials across the various regions in Fig. 7.6 gives, after some re-arrangement,

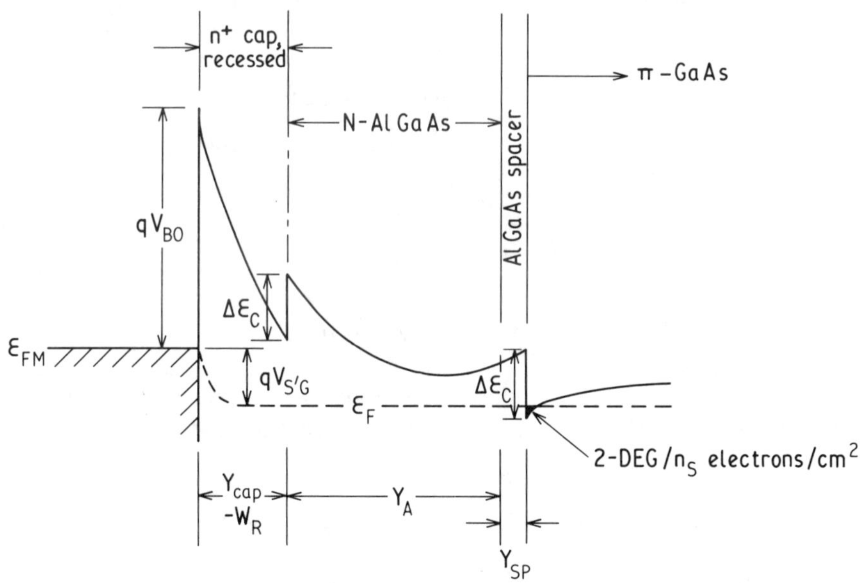

Fig. 7.6 HEMT energy band diagram along a plane such as Y–Y′ in Fig. 7.5.

†When the gate voltage is not large enough to have depleted fully the N–$Al_xGa_{1-x}As$, the electrons in the $Al_xGa_{1-x}As$ contribute to the source-drain current. The phraseology "parallel conduction" is used to describe this condition.

$$n_s = \frac{-(\epsilon/q)(V_{S'G}+V_{BO})+0.5N_{cap}(Y_{cap}-W_R)^2+NY_A(Y_{cap}-W_R)+\dfrac{\epsilon N Y_A{}^2}{2\epsilon_{A1}}}{Y_{cap}-W_R+80\text{ Å}+\dfrac{\epsilon}{\epsilon_{A1}}(Y_A+Y_{SP})} \tag{7.2}$$

In eqn (7.2), we have assumed that the charge centroid of the 2-DEG is offset by 80 Å from the heterointerface. ϵ_{A1} is the permittivity of $Al_xGa_{1-x}As$, given by

$$\epsilon_{A1} = \epsilon - \epsilon_0 \times 3x$$

As before, ϵ is the permittivity of GaAs.

We shall use eqn (7.2) as a check that conditions (ii) and (iii) are satisfied in any calculation involving equivalent circuit elements. When these conditions are in fact satisfied, it may be shown, by methods similar to those employed in Sections 6.3.2 and 6.3.3, that

$$g_{m0} \approx \frac{\epsilon v_{sat} Z_G}{d} \tag{7.3}$$

$$C_{gc} \approx \frac{\epsilon Z_G}{d}\left[L_G + \frac{\epsilon(V_{D'S'}+V_{S'G}+V_{BO})}{q(N_{cap}Y_{cap}+NY_A)-(1.6q\epsilon N_{cap})^{1/2}-qn_s}\right] \tag{7.4}$$

$$C_{gd} \approx \frac{\epsilon^2 Z_G{}^2 L_G}{2dC_{gc}-1.5\epsilon Z_G L_G} \tag{7.5}$$

$$d = Y_{cap} - W_R + 80\text{ Å} + \frac{\epsilon}{\epsilon_{A1}}(Y_A + Y_{SP}) \tag{7.6}$$

To illustrate the use of these equations, we consider a HEMT for millimetre wave operation with the following structure and bias:

$L_G = 0.25\ \mu m$	$N_{cap} = 2\times10^{18}\ cm^{-3}$	$N = 10^{18}\ cm^{-3}$	
$Z_G = 100\ \mu m$	$Y_{cap} = 750$ Å	$Y_A = 300$ Å	$x = 0.3$
$W_R = 700$ Å	$n_{s0} = 7\times10^{11}\ cm^{-2}$	$Y_{SP} = 30$ Å	
$V_{BO} = 0.8$ V	$V_{S'G} = 0$ V	$V_{D'S} = 2.5$ V	

As a first step, eqn (7.2) is used to check that $n_s \leqslant n_{s0}$ at the stated gate bias, so eqns (7.4–7.6) are valid. From eqn (7.2),

$$n_s = 1.72\times10^{11}\ cm^{-2}\ (<n_{s0}) \tag{7.7}$$

Hence we may proceed to calculate the equivalent circuit elements:

$$g_{m0} = 23.7 \text{ mS}$$

$$C_{gc} = 0.0636 \text{ pF}$$

$$C_{gd} = 0.0178 \text{ pF}$$

Most of the other equivalent circuit elements can be calculated using the methods of Chapter 6 (including $\mathscr{L}_g$, R_g, R_i, R_s, R_d, C_o and the geometric capacitances). Estimation of r_o, however, requires a treatment of deconfined 2-DEG current, or substrate current, of rather greater sophistication than the FET substrate current treatment given in Section 6.6. (It follows, of course, that eqns (7.3–7.5) neglect deconfined current.)

Returning to the conditions (i) and (ii) just preceding eqn (7.2), equations which remove these restrictions can be written to generalize eqns (7.2–7.6) to devices where the gate recess is deeper than the cap layer, and to operation where parallel conduction occurs (a regime which is without much interest for low-noise applications). The need to account for deconfined current in regions II–III will remain.

Despite the foregoing reservations, the equations given here have proved useful in HEMT design, and in understanding the microwave characteristics of such devices in terms of their structural features. Experience to date tentatively suggests that the equations relate best to devices in which the π-GaAs layer is in fact P-type doped to maybe 10^{15} cm^{-3}–10^{16} cm^{-3}, such doping helping to reduce the degree of electron deconfinement from the 2-DEG. Comparison with a numerical implementation of equations in which deconfinement is accounted for [7.6] has partly led to this view.

7.6 HEMT noise

Delagebeaudeuf *et al.* [7.7] have presented some evidence that eqn (6.50), or equivalently eqn (6.52), i.e.

$$NF_{\min} \approx 1 + 2\omega \frac{C_{gc}}{g_{m0}} \left(\frac{R_s + R_g}{R_i} \right)^{1/2} \tag{6.52}$$

applies to HEMTs as well as to FETs. In a practical determination of R_s, R_g and R_i, there is redundancy in the way the total input resistance obtained from S-parameter measurements can be divided among these three resistive elements. The resulting uncertainty in the values of R_s, R_g and R_i makes proof that eqn (6.52) applies a little difficult. It is useful as a guide, however, and may be used in conjunction with eqns (7.2–7.4) to estimate the minimum noise figure of a HEMT structure. To give an example, suppose the 0.25 μm device previously considered has

$$\rho_c = 0.1\ \Omega\ \text{mm} \quad \text{and} \quad L_{SG} = 0.5\ \mu\text{m}$$

The resistances of the various layers in the source-gate space are then†:

Cap layer: 1.53 Ω

$Al_xGa_{1-x}As$ layer: 8.68 Ω

2-DEG: 6.38 Ω

Hence we find $R_{SG} = 1.08\ \Omega$ and $R_s = 2.08\ \Omega$. Assuming a gold gate metallization with a device plan view as in Fig. 6.1(a) we have $m = 2$ and, for a metallization thickness of $h = 0.25\ \mu$m, eqn (6.43) gives

$$R_g \approx 3.2\ \Omega$$

Using eqn (7.7), the drain current is (neglecting substrate current)

$$I_D \approx q n_s v_{sat} Z_G \approx 2.75\ \text{mA}$$

From eqn (6.7), with $\mu_{2\text{-DEG}} = 7000\ \text{cm}^2\ \text{V}^{-1}\ \text{s}^{-1}$, R_i is then estimated as

$$R_i \approx 13\ \Omega$$

Finally, from eqn (6.52), the 0.25 μm HEMT in question could be expected to have a minimum noise figure of $NF_{min} \approx 1.55$ dB at 20 GHz and $NF_{min} \approx 0.85$ dB at 10 GHz.

7.7 Prospects for HEMT integration

As already stated, one interpretation of the benefits to be derived from the HEMT is that of channel electron confinement. The 2-DEG allows the product $g_{m0}R_i^{1/2}$ in eqn (6.52) to be higher than in a FET, thus lowering the noise. This property commends the HEMT as the transistor for millimetre wave integration. The power performance of HEMTs has also recently been shown to be better than that of typical GaAs MESFETs at millimetric frequencies [7.2]. As yet, however, little has been announced in the way of HEMT integration, although some reports do exist [7.9]. Important questions relating to failure mechanisms, mean-time-to-failure, and high

†Here we assume a surface depletion of 240 Å in the N^+ cap, a mobility of 2000 $\text{cm}^2\ \text{V}^{-1}\ \text{s}^{-1}$ in the cap; a 100 Å depletion of the $Al_xGa_{1-x}As$ adjacent to the 2-DEG; a mobility in the undepleted $Al_xGa_{1-x}As$ of 1800 $\text{cm}^2\ \text{V}^{-1}\ \text{s}^{-1}$; and a mobility of 7000 $\text{cm}^2\ \text{V}^{-1}\ \text{s}^{-1}$ in the 2-DEG.

power microwave pulse recovery (relevant to transmit–receive applications) will, in the fulness of time, be answered, and will determine the long-term prospects for HEMT integration in analog applications.

References

7.1 Berenz, J. J. High electron mobility transistors, *IEEE MTT-S Int. Microwave Symp.* Newsletter, 43–52, Summer 1984.

7.2 Smith, P. M., Chao, P. C., Duh, K. H. G., Lester, L. F., Lee, B. R. and Ballingall, J. M. Advances in HEMT technology and applications, *IEEE MTT-S Int. Microwave Symp. Dig.*, 749–752, Las Vegas, 9–11 June 1987.

7.3 Hill, A. J. and Ladbrooke, P. H. Dependence of conduction-band discontinuity on aluminium mole fraction in GaAs/AlGaAs heterojunctions, *Electron. Lett.*, **22**(4), 218–220, 13 February 1986.

7.4 Duggan, G. A critical review of heterojunction band offsets, *J. Vac. Sci. Technol. B.*, **3**(4), 1224–1230, July/August 1985.

7.5 Störmer, H. L., Gossard, A. C. and Wiegmann, W. Observation of inter-subband scattering in a 2-dimensional electron system, *Solid State Comm.*, **41**(10), 707–709, March 1982.

7.6 Hill, A. J. and Ladbrooke, P. H. Unpublished work carried out as an extension of Hill, A. J. *An Analytic Model for High Electron Mobility Transistors*, Ph.D Thesis, Durban, University of Natal, 1986.

7.7 Delagebeaudeuf, D., Chevrier, J., Laviron, M. and Delescuse, P. A new relationship between the Fukui coefficient and optimal current value for low-noise operation of field-effect transistors, *IEEE Electron: Device Lett.*, **EDL-6**(9), 444–445, September 1985.

7.8 Nishimoto, C., La Rue, R., Bandy, S., Day, M., Eckstein, J., Webb, C., Yuen, C. and Zdasiuk, G. A 2-20 GHz, high gain, monolithic HEMT distributed amplifier, *IEEE Microwave and Millimetre-Wave Monolithic Circuits Symp. Dig.*, 109–113, Las Vegas, 8–9 June 1987.

7.9 Berenz, J., Yen, H. C., Esfandiari, R., Nakano, K., Sato, T., Velebir, J. and Ip, K. 44 GHz low noise amplifier, *IEEE Microwave and Millimetre-Wave Monolithic Circuits Symp. Dig.*, 15–18, Las Vegas, 8–9 June 1987.

Chapter 8

Reverse modeling GaAs MESFETs and HEMTs†

8.1 Introduction

In MESFET-based circuits, one of the principal yield-limiting elements may be the MESFET itself. Contemporary microwave instrumentation easily possesses sufficient accuracy and precision to resolve differences in scattering parameters between nominally identical FETs which, in turn, can be translated into differences in equivalent circuit values by the use of a computer-based routine for fitting equivalent circuits to measured S-parameters. Whether or not such differences in equivalent circuit element values matter to the overall behavior of an MMIC can be ascertained by simulating the entire network with one of the proprietary network analysis programs such as SUPERCOMPACT© or TOUCHSTONE© but, even when that is done, there is little but experience and intuition to guide the technologist to the reasons why the FET equivalent circuits are all different in the first place. What is wanted is a set of rules which identify, one by one, the material and device structural causes of FET equivalent circuit differences, and hence MMIC differences, i.e. we ideally require a set of equations which translate the differences in equivalent circuits into differences in doping density, gate length and gate recess depth so that the technologist can identify those features requiring tighter process control. A procedure of this kind is just the reverse of the customary procedure known as modeling, wherein the equivalent circuit elements are calculated or forecast from a knowledge of the material and structural make-up of the transistor. For this reason, the task of deducing materials-related and device-structural related values from measured S-parameters and fitted equivalent circuits is here termed "reverse modeling".

8.2 Reverse modeling for gate length

We start with the expression for C_{gc} from Table 6.1:

$$C_{gc} = \frac{\epsilon Z_G L_G}{d}\left(1 + \frac{X}{2L_G} - \frac{2d}{L_G + 2X}\right)$$

†This chapter follows closely the text of 'Reverse modelling of GaAs MESFETs and HEMTs' (Ladbrooke, P. H.) reproduced with permission from the *GEC Journal of Research*, *Vol.* **6**, *(No. 1)*, 1988 (pp. 1–9).

Using the expressions for C_{gd} and g_{m0} from Table 6.1, this may be rewritten as

$$C_{gc} = \frac{g_{m0}}{v_{sat}}\left(L_G + \frac{X}{2}\right) - C_{gd}$$

so that

$$L_G + \frac{X}{2} = \frac{v_{sat}}{g_{m0}}(C_{gc} + C_{gd}) \tag{8.1}$$

We note in passing that this relationship has major significance to the current cut-off, or transition, frequency f_T. We shall return to this issue immediately after deriving the desired reverse modeling relationship for L_G.

Now,

$$C_{gd} = \frac{2\epsilon Z_G}{1 + 2X/L_G} \tag{6.15}$$

so that

$$X = L_G\left(\frac{\epsilon Z_G}{C_{gd}} - \frac{1}{2}\right)$$

Using this relationship to eliminate X from the expression for $(L_G + X/2)$ gives

$$L_G = \frac{2v_{sat}(C_{gc} + C_{gd})}{g_{m0}(3/2 + \epsilon Z_G/C_{gd})} \tag{8.2}$$

This expression is the reverse modeling equation for gate length in its simplest form. It allows the metallurgical gate length to be deduced from a measurement of the FET S-parameters after reducing them to equivalent circuit values. Note that the gate width, Z_G, must be known.

Before studying the applications of the foregoing equations, we return to the question of f_T, the current cut-off frequency. Referring to the very much simplified equivalent circuit of Fig. 8.1, f_T is defined as the frequency at which $|i_d/i_g|$ falls to unity under short-circuit output conditions. We have

$$i_d = g_{m0}v$$

$$i_g = j\omega(C_{gc} + C_{gd})v$$

so

$$\left|\frac{i_d}{i_g}\right| = \frac{g_{m0}}{\omega(C_{gc} + C_{gd})}$$

Hence, by definition,

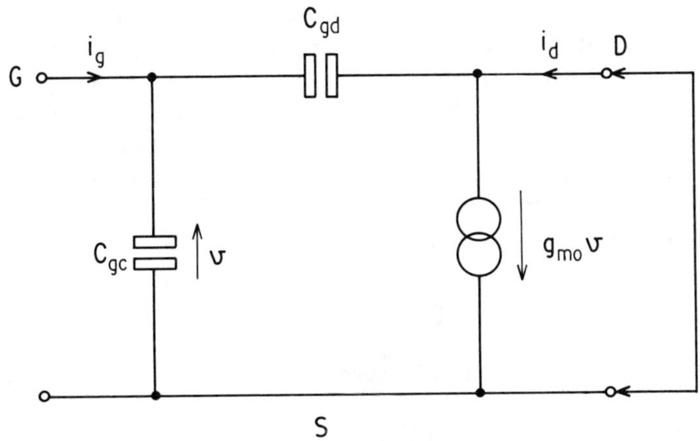

Fig. 8.1 Basic equivalent circuit elements needed for reverse modeling of GaAs FETs and HEMTs.

$$f_T = \frac{g_{m0}}{2\pi(C_{gc} + C_{gd})} \tag{8.3}$$

Substituting for $g_{m0}/(C_{gc}+C_{gd})$ in eqn (8.1), we have

$$f_T = \frac{v_{sat}}{2\pi(L_G + X/2)} \tag{8.4}$$

There are some very important observations to be made from this equation.

(a) If we write, as is commonly done,

$$f_T = \frac{1}{2\pi\tau} \tag{8.5}$$

then,

$$\tau = \frac{L_G + X/2}{v_{sat}}$$

τ is therefore the net charging time of the total depletion region; it is not the transit time through the channel as is so frequently assumed.

(b) The elimination of R_i in arriving at Fig. 8.1 implies that the length of region I in Fig. 6.7 must be vanishingly small. In that case the entire depletion edge would be charged and discharged by velocity saturated electrons, including the part of length L_G. This is expressed by the term L_G/v_{sat} in the relationship for τ.

(c) The fact that the conditions discussed in (b) strictly do not exist in practical FETs means that eqn (8.4) is by no means the best basis upon which to estimate practically occurring values of v_{sat} from experimentally derived values of f_T. Somewhat surprisingly though, eqn (8.4) does appear to offer very useful estimates of v_{sat} when f_T is known, and similarly useful estimates of f_T when v_{sat} can be regarded as known. The methods given in the appendix to Chapter 6 offer an alternative basis for determining v_{sat} from measured data.

(d) Eqn (8.4) can be shown, in fact, to apply in the limit $L_G \ll X$, i.e. it applies strictly in the limit of vanishingly small gate length. f_T reaches exactly the same limit in FETs as in bipolar transistors.

We now return to the issue of reverse modeling for the metallurgical gate length, eqn (8.2).

Measured values of C_{gc} and C_{gd} will normally consist in part of geometric capacitance associated with the electrodes. In contrast, eqn (8.2) requires the space-charge capacitances associated with the depletion layer only. Consequently, the geometric components of capacitance must be subtracted from the measured values of gate-channel and gate-drain capacitance to yield C_{gc} and C_{gd} for insertion into eqn (8.2). Typically, such a correction changes the capacitances by about 10%.

For rapid appraisal work, the corrections for geometric capacitance can be ignored and, to compensate, C_{gd} can be omitted from the numerator to give the following working expression:

$$L_G \approx \frac{2v_{sat}C_{gc}}{g_{m0}(3/2+\epsilon Z_G/C_{gd})} \qquad (8.6)$$

To illustrate the equivalence of eqns (8.2) and (8.6) in a practical case, we quote the following measured equivalent circuit elements for a foundry-type ion implanted 0.5 μm FET, which has a gate width of $Z_G = 250$ μm:

$$C_{gc} = 0.23 \text{ pF}$$

$$C_{gd} = 0.032 \text{ pF}$$

$$g_{m0} = 44 \text{ mS}$$

Taking $v_{sat} = 10^7$ cm sec^{-1}, eqn (8.6) then gives

$$L_G = 0.44\ \mu\text{m}$$

The best estimates for the geometric capacitances of the particular transistor in question are

$$C_{gs\ geom} = 0.023 \text{ pF}$$

$$C_{gd\ geom} = 0.002 \text{ pF}$$

which, upon subtraction from the uncorrected capacitances, yield the following values for the space-charge capacitances for insertion into eqn (8.2) (and for completeness the value of the intrinsic transconductance is repeated also):

$$C_{gc} = 0.207 \text{ pF}$$

$$C_{gd} = 0.03 \text{ pF}$$

$$g_{m0} = 44 \text{ mS}$$

Equation (8.2) then gives

$$L_G = 0.44\ \mu\text{m}$$

which is exactly the result obtained previously. Note that a small margin of conservatism has been allowed in this FET with regard to gate length. In a foundry device, it is sound practice to avoid oversize gates as reference to the example of Fig. 11.10 will show.

To put eqn (8.6) to a more comprehensive test, Table 8.1 lists equivalent circuit data from a number of FETs from independent manufacturers. The table also gives the reverse modeled gate length calculated from eqn (8.6). All results are plotted in Fig. 8.2, which assesses the deduced value of L_G against

Table 8.1 Reverse modeled gate length for various FETs

Company	*Nominal* L_G/Z_G (μm)	C_{gc} (pF)	C_{gd} (pF)	g_{m0} (mS)	L_G (μm) (eqn 8.6)	*Reference*
Anadigics	0.45/200	0.18	0.014	27.6	0.42	Sample FET data: C. Huang, private communication
GEC	0.5/250	0.23	0.032	44	0.44	GEC "SK" foundry FET
GEC	1.0/300	0.285	0.033	25	0.90	Developmental device
GEC	1.0/300	0.17	0.024	10.5	1.10	Developmental device
NEC	0.5/260	0.221	0.025	35	0.48	NEC 70090
Plessey	0.5/900	0.522	0.09	72	0.55	Type Z584/0.5
Plessey	0.7/900	0.9	0.045	72	0.66	Type GAT6/3
Plessey	1.0/900	1.053	0.117	80.1	1.10	Type Z584/1
Plessey	1.2/900	0.855	0.072	54	1.08	Plessey SOFET
Raytheon	1.0/2000	1.9	0.12	115	0.97	Power FET: GaAs IC Symposium 1983
Tektronix	0.7/300	0.331	0.02	30.2	0.68	IEEE Trans MTT-30, July 1982

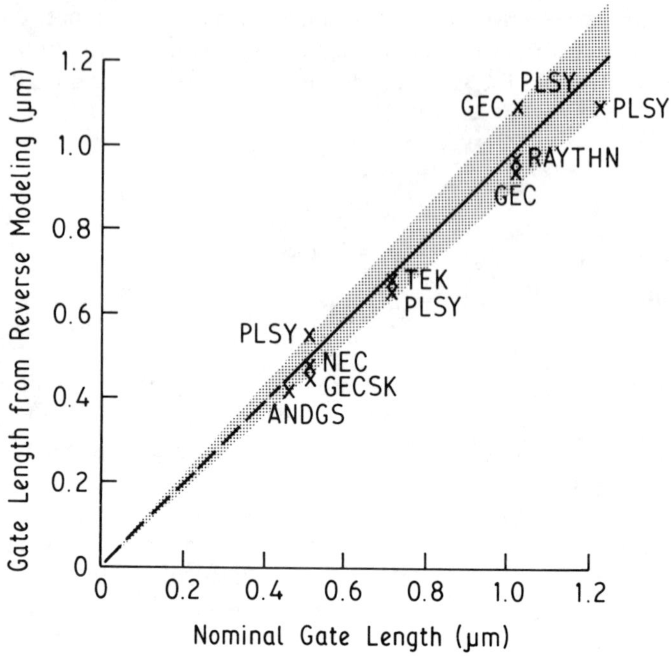

Fig. 8.2 Test of equation for reverse modeling metallurgical gate length using GaAs FET data from various manufacturers. Shaded zone indicates $\pm 10\%$ error bounds.

the nominal value. An arbitrary error band of 10% has also been shown. Bearing in mind that the error between the reverse modeled and nominal gate lengths is an accumulation of errors in the factors listed below, the agreement is satisfactory. The factors are:

(a) the ability to fabricate gates to a nominal length,
(b) measurement of *S*-parameters,
(c) fitting of equivalent circuits to *S*-parameters,
(d) errors in equation (8.6).

Note that in none of these FETs is the channel doping profile known, and that there is no need to know it in order to calculate the gate length by reverse modeling. Neither is it necessary for the assumption to be fulfilled that the surface depletion depth equal the gate recess depth. As the next section will illustrate, it is merely necessary that the gate recess depth is comparable with the surface depletion depth.

8.3 Errors in eqns (8.2) and (8.6)

The assumption implicit in the equations of Table 6.1, that the surface depletion depth equals the recess depth, clearly breaks down for a planar FET with no gate recess. Physically, in this limit, the depletion layer extension into the gate-drain space is poorly screened from surface charge so that the gate-drain and gate-channel capacitances are surface-charge dependent to an extent which cannot be ignored. The use of eqns (8.2) or (8.6) in this limit results in an underestimate for L_G. As a guide to the magnitude of the errors which can be expected, we consider two FETs, one with a shallow gate recess and one with no recess at all. Using the more comprehensive set of forward- and reverse-modeling equations which account for surface charge and channel doping, and from which the equations in this book are obtained as a simplified subset, C_{gc}, C_{gd}, and g_{m0} can all be calculated from the transistor's material and structural specification, then the apparent value of L_G can be calculated from eqns (8.2) or (8.6). We consider two FETs, one device having a recess depth of $W_R = 0$, the other $W_R = 500$ Å, and both with

$$N = 3.5 \times 10^{17} \text{ cm}^{-3}$$

$$L_G = 0.5 \text{ μm}$$

$$Z_G = 200 \text{ μm}$$

$$\phi_s = 0.7 \text{ eV}$$

Figure 8.3 shows the error expected as a function of gate bias at a drain bias of 3.5 V. The principal finding is that devices with no gate recess, or perhaps with recesses which are very shallow compared with the surface depletion depth, cannot be reverse modeled using eqns (8.2) or (8.6) near open channel operating conditions. Providing the pinch-off voltage is large enough to allow it, planar FETs should instead be measured at a gate bias which depletes the channel to a depth equal to about twice the surface depletion depth. Since many GaAs FETs have gate recesses, however, this is not a serious restriction. The more elaborate reverse modeling scheme, from which the present analysis derives, is not subject to this restriction in the same way. It yields the correct gate length, together with other material and device structural information, even for devices without a gate recess operated at, or near, $V_{GS} = 0$ V.

8.4 Extension to HEMTs

With the increasing interest being shown in HEMTs, a diagnostic capability of reverse modeling to determine metallurgical gate length is

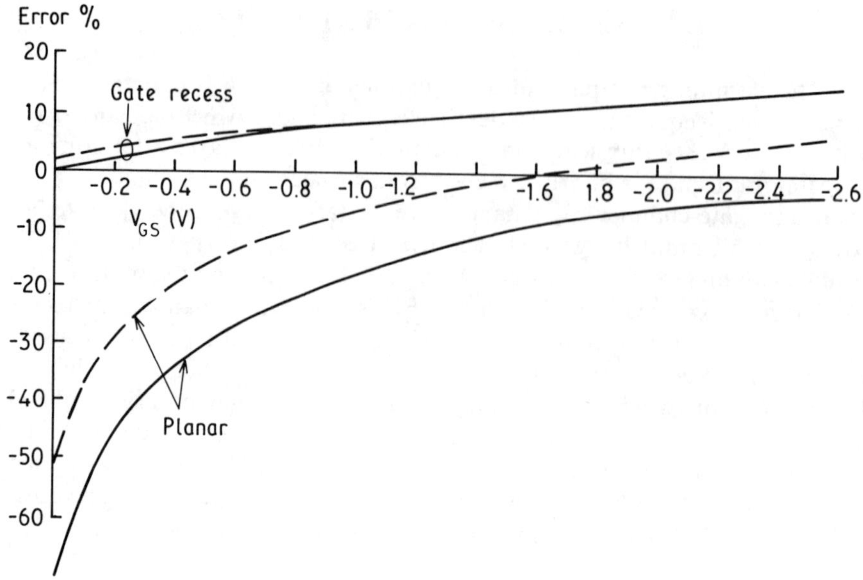

Fig. 8.3 Typical errors in lowest-order equations for reverse modeling of metallurgical gate length taking a 0.5 μm FET as an example: --- simplified equation without correction for geometric capacitances; ——— full equation with correction for geometric capacitances.

valuable from several points of view. HEMTs exhibit lower noise than MESFETs. As is well known, however, the noise figure is proportional to gate length, so a proper identification of contributions to the measured noise figure is helped, among other things, by a knowledge of the actual gate length of the transistor under study.

To see why the same reverse modeling equation should apply to HEMTs, we first review the physical processes which determine the speed of response of the GaAs FET to small signals. Referring to Fig. 6.9, the net charging time of the depletion region is determined by the time taken for the charge to change in the depletion edge in response to a small change in gate-source voltage. In turn, this depends upon

(i) the depth of depletion, d,
(ii) the extension of the space charge layer towards the drain, X, and
(iii) the rate at which electrons can move into and out of the depletion edge (i.e. the electron mobility and velocity).

Turning now to Fig. 8.4, much the same applies to the HEMT, except

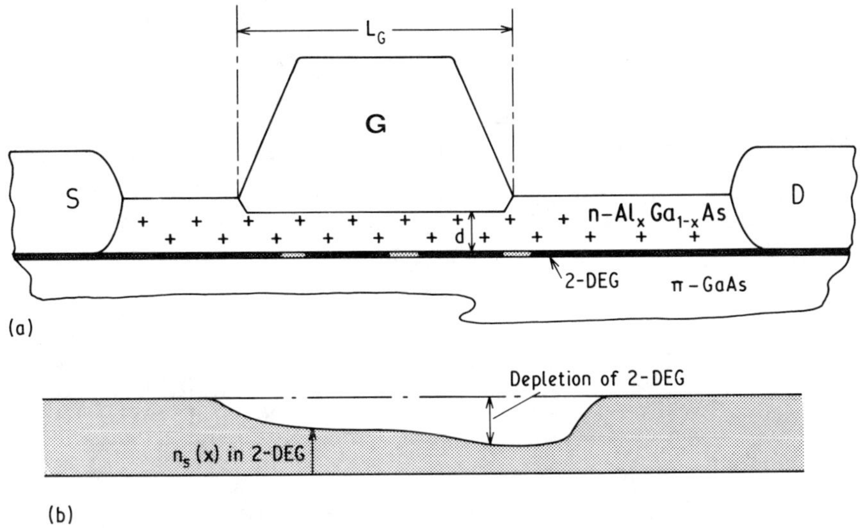

Fig. 8.4 Cross-section of HEMT showing form of 2-DEG depletion relevant to reverse modeling.

that the charge changes now occur in the two-dimensional electron gas (2-DEG), and at the right-hand edge of the depletion in the n-$Al_xGa_{1-x}As$. (We are here using a lower case n, rather than N, to emphasize the fact that there are free electrons in the $Al_xGa_{1-x}As$ which must respond to the field changes associated with the movement of the depletion edge.) Hence d in (i) above becomes the gate metal-to-2-DEG separation. Because d does not appear explicitly in eqns (8.2) or (8.6), this fact is of no direct consequence. What is important, however, is that the field solutions should be similar in the FET and the HEMT so that, for given values of d, X and L_G, the relationships:

$$C_{gc} = f(d, X, L_G, \ldots)$$

$$C_{gd} = f(d, X, L_G, \ldots)$$

$$g_{m0} = f(d, v_{sat}, \ldots)$$

are of the same functional form for HEMTs as for FETs. It turns out that the field solutions are sufficiently similar for the same reverse modeling equations to be workable, as Table 8.2 and Fig. 8.5 show. Just as for the FET, these results are obtained with $v_{sat} = 10^7$ cm sec^{-1}.

Table 8.2 Reverse modeled gate lengths for various HEMTs

Company	*Nominal* L_G/Z_G(μm)	C_{gc} (pF)	C_{gd} (pF)	g_{m0} (mS)	L_G(μm) (eqn 8.6)	*Reference*
Fujitsu	0.25/200	0.18	0.045	60	0.30	*IEEE MTT-S Dig.* 1021, 1987
Fujitsu	0.5/200	0.31	0.037	60	0.49	Ibid.
GE	0.25/150	0.271 +0.011	0.015	85.7	0.25	*Electron. Dev. Lett.*, **EDL-6**(10), October 1985
GEC	0.5/120	0.156	0.022	29.3	0.50	Developmental device
Hughes	0.25/60	0.098	0.014	22.9	0.33	*IEEE MTT-S Dig.* 1016, 1987
Toshiba	0.25/200	0.18	0.012	54	0.20	Ibid., 1012

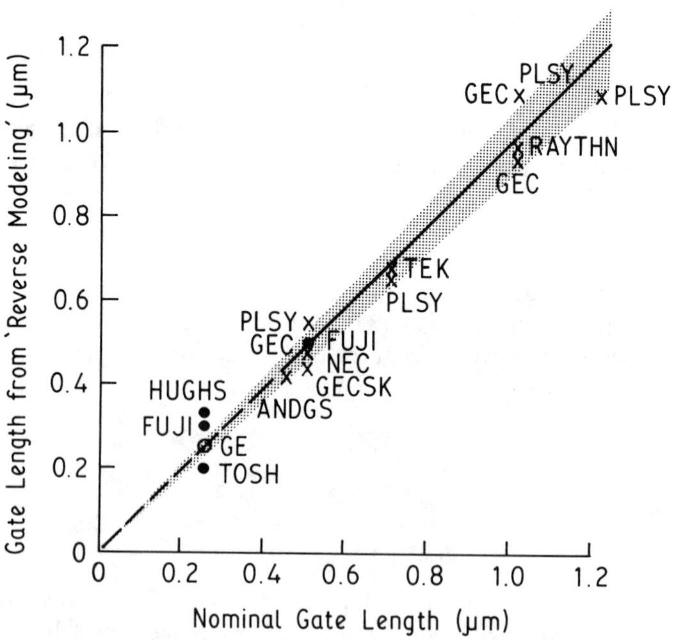

Fig. 8.5 Test of reverse modeling equation for HEMTs from various manufacturers, as well as FETs. Shaded zone defines ±10% error bounds.

8.5 General comments

It is interesting to compare the values of the metallurgical gate length deduced here with the values obtained from the very simple expression:

$$L_G \approx v_{sat} C_{gc} / g_{m0} = L_{elec} \tag{8.7}$$

which is easily obtained from eqns (6.10) and (6.18). The symbol L_{elec} is used as a reminder that the equation gives what is often referred to as an effective electrical gate length. L_{elec} is calculated in Table 8.3 for all the devices so far considered in Tables 8.1 and 8.2. The general lack of agreement between L_{elec} and the metallurgical gate length, as obtained by reverse modeling, demonstrates the importance of including the space charge layer extension, X, in analyses involving C_{gc}, C_{gd}, f_T, and all related analyses. To have the dimensions of the depletion layer correctly represented is therefore important.

Figures 8.2 and 8.5 offer little or no evidence to support the idea that velocity overshoot occurs in FETs and HEMTs with gate lengths down to 0.25 μm. At least part of the reason for this result is that the space charge layer extension, X, is comparable with the gate length itself; the spatial extent of the overall region with which charge exchange is taking place is perhaps 1.5 to 2.5 times the gate length, as inserting practical values for C_{gd} into eqn (6.15) will

Table 8.3 Effective electrical gate length for devices in Tables 8.1 and 8.2

Device type	*Company*	*Nominal* L_G/Z_G(μm)	L_{elec} (μm)
FET	Anadigics	0.45/200	0.65
FET	GEC	0.5/250	0.52
FET	GEC	1.0/300	1.14
FET	GEC	1.0/300	1.62
FET	NEC	0.5/260	0.63
FET	Plessey	0.5/900	0.73
FET	Plessey	0.7/900	1.25
FET	Plessey	1.0/900	1.31
FET	Plessey	1.2/900	1.58
FET	Raytheon	1.0/2000	1.65
FET	Tektronix	0.7/300	1.10
HEMT	Fujitsu	0.25/200	0.30
HEMT	Fujitsu	0.5/200	0.52
HEMT	GE	0.25/150	0.33
HEMT	GEC	0.5/120	0.53
HEMT	Hughes	0.25/60	0.43
HEMT	Toshiba	0.25/200	0.33

verify. The total transit distance is therefore many times the likely overshoot distance (see appendix to Chapter 6) in the presence of transverse electric fields as exist in real devices.

Given the form of eqns (8.2) and (8.6), any errors in C_{gc} and g_{m0}, and to some extent in C_{gd}, lead to compound errors in L_G. High quality S-parameter measurements are essential, with consistent calibrations that place the reference planes at well-defined and repeatable positions on the chip. A Cascade microwave prober and a Hewlett Packard 8510 network analyzer provide the necessary quality of measurement. Such a facility makes readily possible the measurement of S-parameters over a complete range of bias. Subsequent conversion to equivalent circuit form provides a stringent test of the reverse modeling eqns (8.2) and (8.6). In Fig. 8.6 is shown an example of three FETs measured and reverse modeled over a range of bias. Figure 8.6(a) shows two devices off the same wafer, with a clear difference in gate length between them. Even though there is some scatter in the plots, it is possible to discern a systematic change in the reverse modeled value of L_G with increasing drain bias. There is no systematic change with gate bias. We conclude the deduced value of L_G is in this regard independent of the bias point, which is as it should be. Experience has shown, however, that operation too near pinch off sometimes leads to erroneously large values of L_G obtained from eqns (8.2) or (8.6), and so is to be avoided. The apparent increase in gate length with increasing drain bias is 10% per volt or less, and stems directly from the decrease in transconductance with increasing drain bias observed in many GaAs MESFETs. The tilt in the larger gate length device in Fig. 8.6(a) is among the worst seen throughout this work. Commonly, discrepancies like those evident in the shorter gate length FET in Fig. 8.6(a), and in Fig. 8.6(b), are obtained. Bearing in mind that the errors and uncertainties contributing to Figure 8.6 include errors in S-parameter measurements and errors in the routines for reducing S-parameters to equivalent circuits, as well as errors in the reverse modeling equations themselves, the consistency achieved in L_G in Fig. 8.6 for any given transistor is good enough to render the technique extremely useful as a diagnostic tool.

As already mentioned, the work described in this book derives from a more comprehensive set of equations for both forward and reverse modeling the GaAs FET. Although these are based upon the assumption of an equivalent uniform doping in the channel, they allow values for the doping density, surface charge and gate length to be recovered by reverse modeling. The second of these, the occupied surface-state density, is important because it is difficult to control, yet it affects the microwave behavior of a FET, dramatically so in the case of a power FET. Further, the gate length is recovered without the large error evident in Fig. 8.3 for devices without a gate recess operated at zero gate bias. It would appear possible in principle to

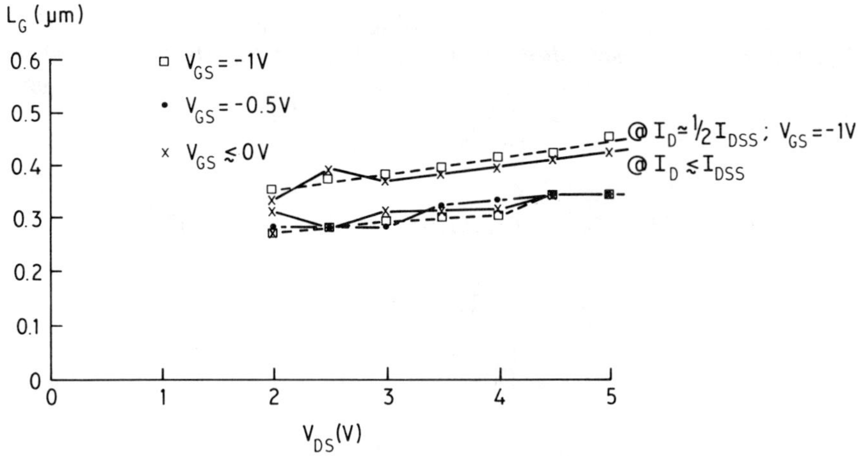

(a)

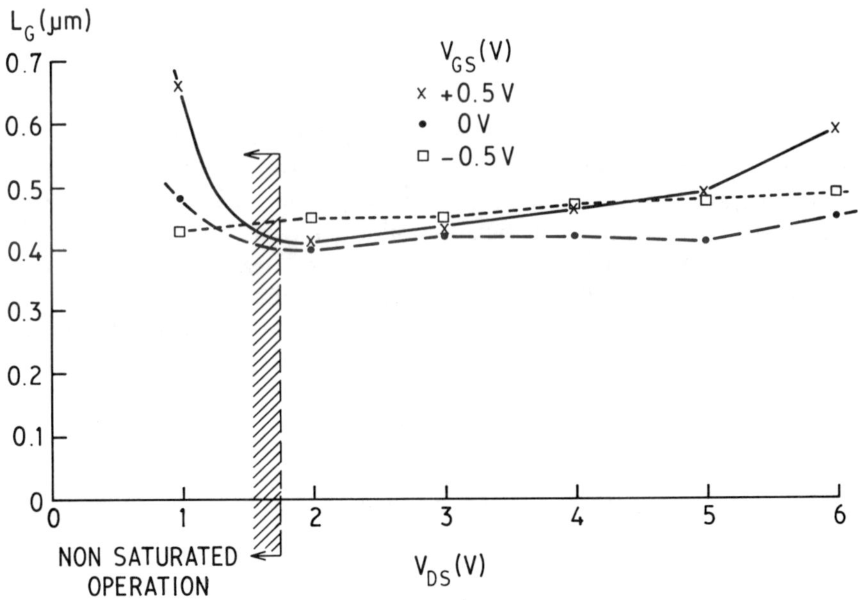

(b)

Fig. 8.6 Test of validity of reverse modeling by assessing independence of reverse modeled metallurgical gate length from bias conditions: (a) showing clear difference in gate length of two FETs from the same wafer; (b) for a foundry device.

extend the method to arbitrary channel doping profiles while retaining a compact formalism, thereby preserving all the advantages of physical insight offered, and rapid execution even on a small computing facility. We shall consider a simple method of reverse modeling for channel doping in the next section.

8.6 Reverse modeling for channel doping

Equations (6.18) and (6.8) of Table 6.1 are

$$g_{m0} = \frac{\epsilon v_{sat} Z_G}{d}$$

$$d = \left[\frac{2\epsilon(V_{S'G} + V_{BO})}{qN}\right]^{1/2}$$

N is assumed to be uniform, say of value N_u. By simple rearrangement of these two equations,

$$N_u = \frac{2(V_{S'G} + V_{BO})}{q\epsilon}\left(\frac{g_{m0}}{v_{sat} Z_G}\right)^2 \tag{8.8}$$

When N is non-uniform, this reverse modeling expression does not give the actual doping density but the equivalent uniform doping density, N_u. N_u will in general be a function of the gate bias $V_{S'G}$, reflecting the fact that, as the depletion edge moves through the non-uniform free-electron distribution $n(y)$ (see Fig. 6.27), the doping averaged over the depletion, N_u, will vary also. Note that because N_u is the value of equivalent doping that would give rise to the same value of d at a given $V_{S'G}$, N_u is not the ordinary average but is given by

$$N_u = \frac{2}{d^2}\int_0^d \mathrm{d}y \int_0^y \mathrm{d}y' N(y')$$

Consider our example FET from Fig. 6.25 with

$$g_{m0} = 31.5 \text{ mS} \quad \text{at} \quad V_{S'G} = 0.5 \text{ V (i.e. } V_{GS'} = -0.5 \text{ V)}$$

$$Z_G = 200\ \mu\text{m} \quad V_{BO} = 0.8 \text{ V}$$

We then find

$$N_u = 3.5 \times 10^{17} \text{ cm}^{-3}$$

As far as this simple (and incomplete) argument goes, an interesting property of N_u is that its maximum value can never be greater than the maximum in the real profile, $N(y)$. Conversely, the actual doping in the channel of the above FET would be at least $3.5 \times 10^{17}\ \mathrm{cm}^{-3}$, and almost certainly greater at some point in the profile if the profile were non-uniform.

In FETs where substrate current contributes to g_{m0} (see Section 6.6), the reverse modeled doping density is too large by the factor:

$$(\mathrm{d}I_{SUBS}/\mathrm{d}V_{S'G})^2$$

For example, if we taken the same FET as previously but acknowledge the flow of substrate current governed by an N-layer-to-substrate barrier energy of $\phi_B = 0.7$ eV, we could expect, from Fig. 6.31:

$$g_{m0} = 40\ \mathrm{mS} \quad \text{at} \quad V_{D'S'} = 3\ \mathrm{V}$$

Equation (8.8) would then have us believe that

$$N_u = 5.65 \times 10^{17}\ \mathrm{cm}^{-3}$$

whereas the actual doping density is $3.5 \times 10^{17}\ \mathrm{cm}^{-3}$. In this circumstance, therefore, the actual channel doping is somewhat lower than the apparent value, N_u. For these reasons, extreme caution needs to be exercised before drawing firm conclusions as to channel doping using this method. Self-consistent corrections for substrate current can in principle be made, but they are too extensive for their inclusion here. Nevertheless, if careful checks are made against test structures having known doping levels, it is possible to gain a feel, by experience, for those devices where eqn (8.8) provides a useful estimate.

8.7 Conclusion

A technique has been described which allows the doping and metallurgical gate length of a GaAs MESFET or HEMT to be deduced from microwave equivalent circuit data. Data from a number of manufacturers' devices have been analyzed. For gate lengths in the 0.5 μm to 1.0 μm range, the deduced gate length and the nominal value agree to within 10%. For 0.25 μm HEMTs, the agreement is typically to within 0.05 μm. The scheme presented here is a simplified version of a more comprehensive set of reverse modeling algorithms which allow information regarding occupied surface-state density and channel doping level to be deduced, as well as gate length. These techniques of reverse modeling have proved useful in diagnosing out-of-specification MMICs, process control monitoring, understanding MMIC yield figures, interpreting HEMT noise and in gathering intelligence about published and commercial devices and circuits using FETs or HEMTs.

Chapter 9

Design limits

9.1 Introduction

One of the most important phases in integrated circuit design is the early determination of whether a particular performance specification is feasible and, if so, what are the broad characteristics of the active and passive elements required to achieve it. In other words, it is important to have a set of guidelines, or coarse rules, to assess the feasibility of, and possible solutions to, any proposed MMIC specification. Diagrammatically, such a process of initial constraint can be represented as in Fig. 9.1, step 2. Without this step, unwarranted time and effort might otherwise be expended in attempting to find designs or even practical solutions where none is intrinsically possible. Where a solution is in fact possible, the design proper can subsequently be carried out using very much more comprehensive rules as at step 3 in Fig. 9.1,

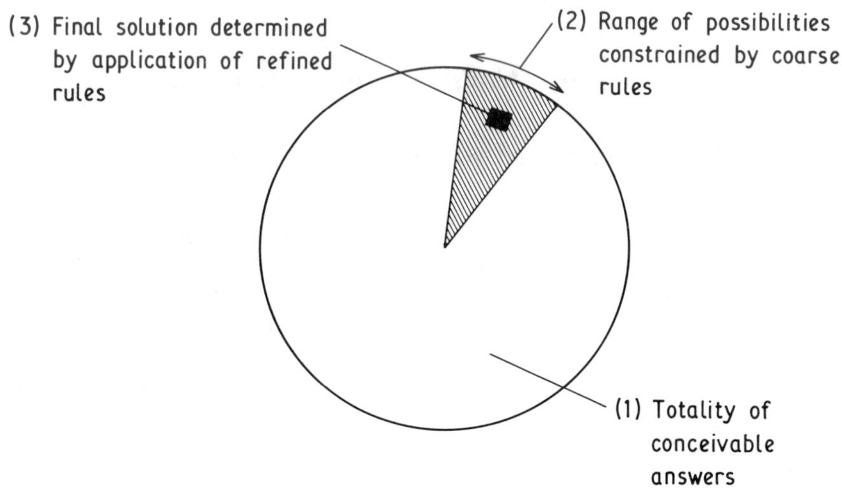

Fig. 9.1 Steps in constraining the solution possibilities.

which may frequently involve software such as TOUCHSTONE© or SUPERCOMPACT©.

In this chapter we attempt to establish some coarse constraints imposed principally by the active elements in an MMIC. For small-signal circuits such rules principally concern the following limits of FETs and HEMTs

(a) stability (in the sense of tendency to oscillate),

(b) frequency response (f_T and f_{max}, and their physical significance),

(c) power gain,

(d) input and output reflection coefficient over a specified bandwidth (involving impedance level, Z), and

(e) minimum noise figure.

For large-signal (power) circuits, the first four of the foregoing criteria are important and need to be supplemented by general rules giving the possible values of

(f) maximum sustainable drain voltage,

(g) maximum drain current,

(h) output power P_O,

(i) $P_O Z f^2$ i.e. the compromises possible between output power, impedance level and frequency of operation.

We consider the small-signal criteria first.

9.2 Limits to small-signal behavior

9.2.1 Gain, frequency and voltage, and gate length

In Fig. 9.2 is shown the simplest possible equivalent circuit† which makes provision for the inescapable properties of the intrinsic transistor, namely‡; gate strip resistance R_g, gate-channel capacitance C_{gc}, channel resistance R_i and transconductance, g_{m0}. It can be argued that R_g can in principle be made arbitrarily small by increasing the height and conductivity of the gate metal without limit. To relate to practical cases, however, we choose to include R_g and to give it a finite value.

†With no feedback elements we conveniently avoid, for the time being, the question of stability.

‡For simplicity, we here ignore the intrinsic component of gate inductance, $\mathscr{L}_g$.

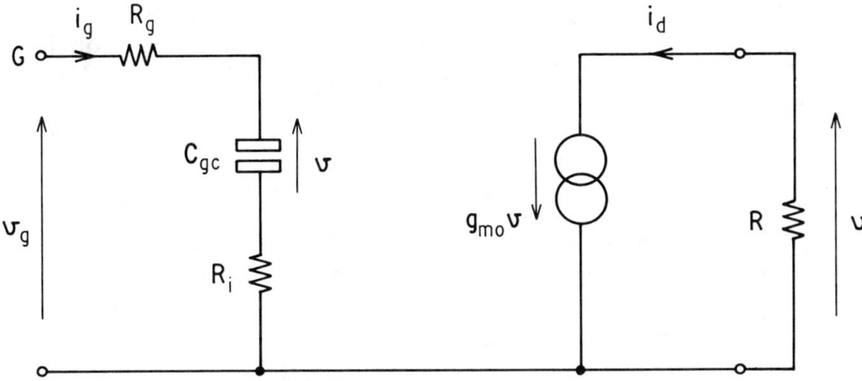

Fig. 9.2 Unilateral equivalent circuit for first-order GaAs FET gain calculations.

From Fig. 9.2, the current gain is

$$G_i = \left|\frac{i_d}{i_g}\right| = \frac{g_{m0}}{\omega C_{gc}}$$

Because the current cut-off frequency, or transition frequency, f_T, is by definition the frequency at which G_i falls to unity, we have

$$f_T = g_{m0}/2\pi C_{gc} \tag{9.1}$$

and so

$$G_i = f_T/f \tag{9.2}$$

The voltage gain is readily determined as

$$G_v = \left|\frac{v_o}{v_g}\right| = \frac{g_{m0}R}{[1+\omega^2 C_{gc}{}^2(R_g+R_i)^2]^{1/2}}$$

So that, in the limit $[\omega C_{gc}(R_g+r_i)]^2 \gg 1$,

$$G_v \approx \frac{g_{m0}R}{\omega C_{gc}(R_g+R_i)} = G_i\,\frac{R}{(R_g+R_i)} = \left(\frac{f_T}{f}\right)\frac{R}{(R_g+R_i)} \tag{9.3}$$

The power gain is

$$G_P = \frac{\text{power delivered to load}}{\text{power input to FET}} = \frac{g_{m0}{}^2 R}{\omega^2 C_{gc}{}^2(R_g+R_i)}$$

$$= \left(\frac{f_T}{f}\right)^2 \frac{R}{(R_g+R_i)} \tag{9.4}$$

More realistically, if the FET output resistance r_o had been included across the generator in Fig. 9.2, the maximum power gain would be obtained when $R = r_o$; the power delivered to the load R would then be $(g_{m0}v)^2 R/4 = (g_{m0}v)^2 r_o/4$ so that

$$G_P = \left(\frac{f_T}{f}\right)^2 \frac{r_o}{4(R_g + R_i)} \tag{9.5}$$

From these equations it follows that the higher the frequency at which G_P falls to unity, the higher will be the power gain at any lower frequency, f. The unity power gain frequency is also referred to as the maximum frequency of oscillation, f_{max}, so that, from eqn (9.5),

$$f_{max} = \frac{f_T}{2}\left(\frac{r_o}{R_g + R_i}\right)^{1/2} \tag{9.6}$$

Although this equation needs further refinement to apply to real transistors (see the appendix to this chapter), it correctly identifies the following facts:

(i) f_{max}, not f_T per se, is the frequency figure of merit which is most relevant to MMICs where power gain matters;

(ii) low input resistance is necessary for good power gain.

In the previous chapter it was shown that (eqn 8.4)

$$f_T = \frac{v_{sat}}{2\pi(L_G + X/2)} \tag{9.7}$$

Eliminating the space-charge layer extension X using the equations of Table 6.1,

$$(V_{D'G} + V_{B0})f_T = \frac{qNd}{\epsilon}\left(\frac{v_{sat}}{2\pi} - L_G f_T\right) \tag{9.8}$$

Equation (9.8) expresses the compromises possible between operating voltage and frequency. For example, suppose we tried to configure a transistor to deliver 10 dB gain into a load of 50 Ω at 94 GHz. Such a transistor might have $L_G = 0.125$ μm, $R_g = 1$ Ω and $R_i = 4$ Ω so that, from eqn (9.4), an f_T value of:

$$f_T = 94 \text{ GHz}$$

is required. If $N = 2 \times 10^{18}$ cm^{-3} and $d = 300$ Å, eqn (9.8) shows that the maximum gate-drain voltage at which the device could be operated would be (assuming $V_{B0} = 0.8$ V),

$$V_{D'G} \approx 2.9 \text{ V}$$

Alternatively, if the designer of millimeter-wave monolithic integrated circuits (M^3IC) could be content with 5 dB gain, the transistor would need an f_T of (from eqn (9.4))

$$f_T \approx 53 \text{ GHz}$$

and so, from eqn (9.8)

$$V_{D'G} \leqslant 13.9 \text{ V}$$

These results represent the limit of what is possible: in practical devices and circuits, the position is less ideal and lower numbers would be expected. Impact ionization may impose a lower limit on $V_{D'G}$ in any case.

From the right-hand side of eqn (9.8) the maximum allowable gate length of an active element to give any desired f_T value can be determined. Thus, for example, for an f_T of 53 GHz, we must have

$$L_G < \frac{v_{sat}}{2\pi f_T} = 0.3 \text{ µm}$$

Thus, we could contemplate achieving ≈5 dB gain at 94 GHz with a 0.25 µm device but, continuing with the same values of N and d as previously, from eqn (9.8) with $f_T \approx 53$ GHz such a device would of necessity work at

$$V_{D'G} \leqslant 3.4 \text{ V}$$

9.2.2 Effect of gatewidth, Z_G

Even for a FET or HEMT active element that is fixed in cross-section, as in present-day foundry processes, the gatewidth Z_G is a variable open to the MMIC designer. It therefore becomes important to understand the effect of gatewidth on the various electrical parameters with which a circuit engineer would normally be concerned.

Two cases need to be carefully and clearly distinguished, as in Fig. 9.3. In case (a), different total gatewidths Z_G are realized by the parallel integration of individual gatestrips, each of width Z_{GU} (the subscript "U" referring to the width of each gate unit). Parallel connection of gate strips of width Z_{GU} is illustrated in Fig. 9.3(a). In case (b), the width of the gate strip itself, Z_{GU}, is changed to realize a different (total) Z_G value (Fig. 9.3(b)).

If we neglect the different parasitic element values that will inevitably be associated with different pad sizes and electrode cross-overs in parallel-integrated device structures, and concentrate on the remaining equivalent circuit elements, we find that each of the elements C_{gsa}, C_{gc}, R_i, R_s, C_{gd}, C_{gda}, g_{m0}, r_o, C_{ds1} and R_d has the same numerical value provided the total Z_G is the same in cases (a) and (b) of Fig. 9.3. (Refer to Fig. 6.43 for identification of

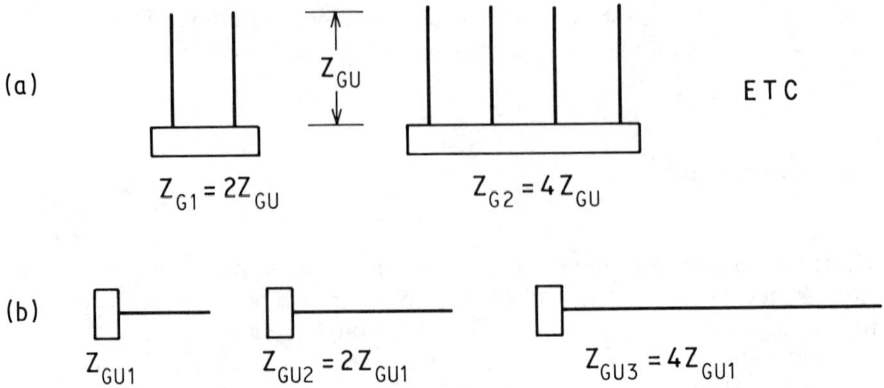

Fig. 9.3 Ways of achieving different total gatewidth: (a) paralleling gate fingers; (b) extending gate fingers.

these elements). Only $\mathscr{L}_g$ and R_g are different (and we will continue to neglect $\mathscr{L}_g$ in this discussion). So far as R_g is concerned, we have for case (a),

$$R_g = \frac{\rho Z_G}{3m^2 h L_G} = \frac{\rho Z_{GU}}{3mhL_G}$$

where m is the number of strips connected in parallel. For case (b),

$$R_g = \frac{\rho Z_{GU}}{3hL_G} = \frac{\rho Z_G}{3hL_G}$$

If we can assume that the source and load impedances can be adjusted, or scaled, as Z_S/m and Z_L/m respectively when, in case (a), FETs of m fingers are integrated, then case (a) is directly equivalent to a parallel connection of m FETs as in Fig. 9.4. If each transistor (and its associated Z_S and Z_L) is identical to every other of the m FETs, then the gain and noise figure will be the same for all transistors, and furthermore will be unaltered by the fact of connecting all the FETs in parallel. In other words, the overall gain and noise figure of a FET are decided principally by the per unit properties of a single strip of width Z_{GU}; increasing Z_G by paralleling such strips merely lowers the impedance level at which the prescribed values of G_P and NF_{min} are obtained. We now return to case (b).

To illustrate two important points concerning the effect of Z_{GU} on the gain properties, we consider Figs 9.5 and 9.6, which are derived from practical data for a 0.5 μm inverted HEMT having $Z_{GU}=60$ μm. The measured equivalent circuit has been stripped of pad capacitance and re-scaled to other per-unit strip widths of $Z_{GU}=120$ μm, 240 μm and 300 μm. In this form the data are applicable to active elements for integration in a HEMT-based

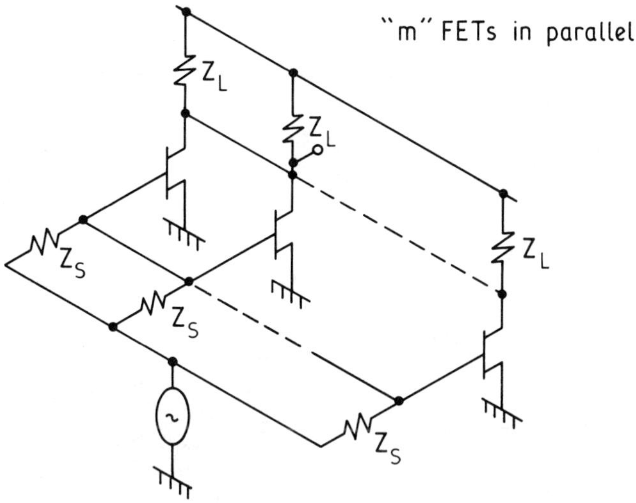

Fig. 9.4 Parallel connection of identical small FETs, equivalent to Fig. 9.3(a).

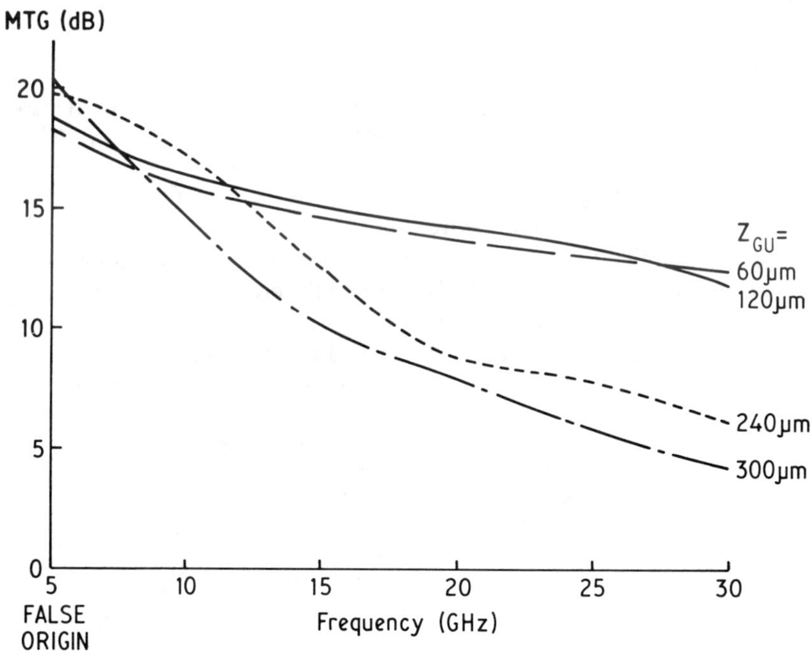

Fig. 9.5 Variation of expected circuit gain for various per-unit-gatewidth MMIC IHEMTs obtained by scaling practical data from a device with $L_G = 0.5$ μm and $Z_G = 60$ μm.

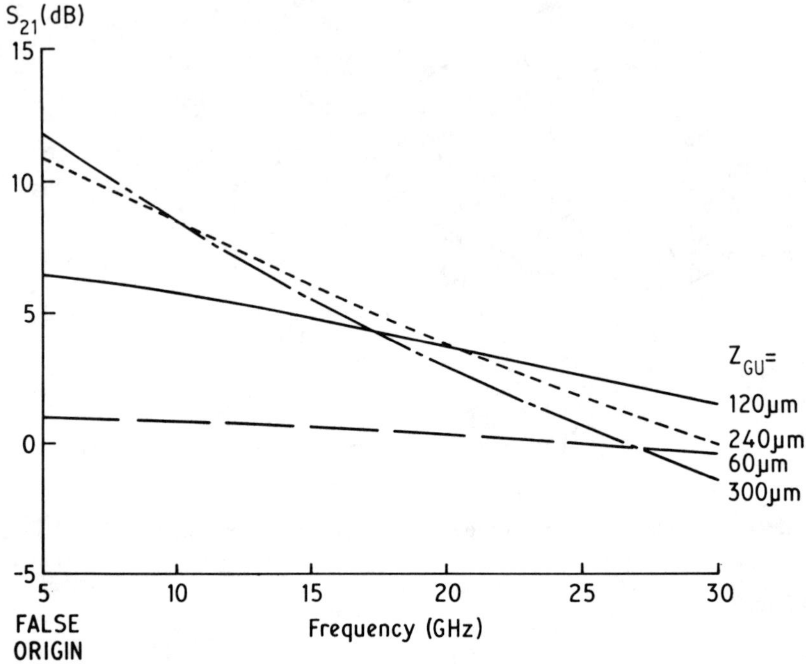

Fig. 9.6 Variation of expected s_{21} for various per unit gatewidth MMIC IHEMTs obtained by scaling practical data from a device with $L_G = 0.5$ µm and $Z_G = 60$ µm.

MMIC because, in integration, the gate and drain pads can be very small as they are not required to support wire bonds.

Figure 9.5 shows the expected circuit gain based on a calculation of the maximum tunable gain (MTG). (MTG will be covered in Section 9.2.4). The first important point, demonstrated in Table 9.1, is that the per unit gatewidth for best gain is, relatively speaking, larger at low frequencies and smaller at high frequencies. At 12 GHz, $Z_{GU} = 120$ µm would typically be used.

Table 9.1 Per unit gatewidth for best gain

f(GHz)	Z_{GU} (µm)
6	240 or 300
12	60, 120 or 240
18	60 or 120
>30	60

The second important point is illustrated by Fig. 9.6, which shows s_{21} for the self-same devices (of different Z_{GU}) as in Fig. 9.5. Comparison of the two figures clearly shows that s_{21} cannot, by itself, be used as a guide to circuit gain. Notably, the s_{21} of the 60 μm gate strip device is very low over the range 15–20 GHz, whereas the expected circuit gain over this frequency range is very much better than the 240 μm and 300 μm devices. A simple account of the behavior of s_{21} can be given as follows.

Consider Fig. 9.7, which shows a transmission line of characteristic impedance Z_0 supporting a wave of amplitude v^+ incident on a load Z_L, and a reflected wave of amplitude v^-. We have

$$v^- = \left(\frac{Z_L - Z_0}{Z_L + Z_0}\right) v^+ = \Gamma v^+$$

where Γ is the reflection coefficient. Also,

$$v_L = v^+ + v^- = \frac{2Z_L}{Z_L + Z_0} v^+ \tag{9.9}$$

When Z_L consists of the FET, as in an S parameter measurement, we have (from Fig. 9.8)

$$Z_L = (R_i + R_g) + \frac{1}{j\omega C_{gc}}$$

and

$$v = \frac{v_L}{j\omega C_{gc} Z_L} \tag{9.10}$$

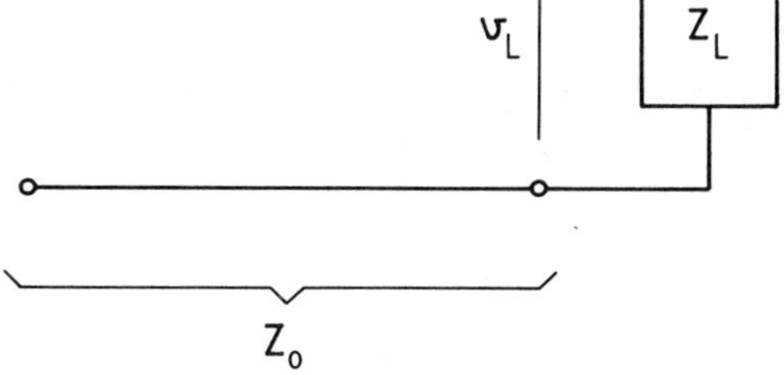

Fig. 9.7 Waves on a transmission line terminated in a load $Z_L = Z_0$.

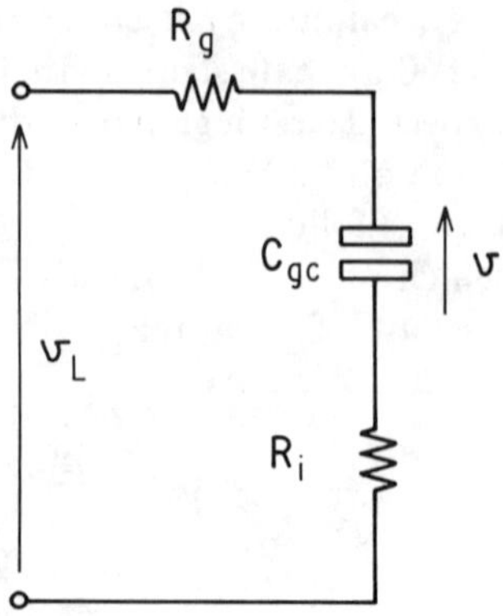

Fig. 9.8 Input part of FET equivalent circuit as the load Z_L on a transmission line, where $Z_L = (R_i + R_g) + 1/j\omega C_{gc}$.

Combining eqns (9.9) and (9.10), we obtain

$$v = \frac{2v^+}{j\omega C_{gc}(Z_L + Z_0)}$$

or

$$v = \frac{2v^+}{1 + j\omega C_{gc}(R_i + R_g + Z_o)}$$

The complete (but simplified) network for s_{21} measurement of the FET is shown in Fig. 9.9. Ignoring any reflected waves in the output circuit, s_{21} is by definition given by

$$|s_{21}| = \left|\frac{v_o}{v^+}\right| = \left|\frac{2g_{m0}Z_0}{1 + j\omega C_{gc}(R_i + R_g + Z_0)}\right|$$

Hence

$$|s_{21}| = \frac{2g_{m0}Z_0}{[1 + \omega^2 C_{gc}{}^2(R_i + R_g + Z_0)^2]^{1/2}} \tag{9.11}$$

To illustrate the consequences of different gatewidths upon $|s_{21}|$, consider an IHEMT with the following values:

$$\bar{g}_{m0} = 210 \text{ mS mm}^{-1}$$

$$R_g = 0 \text{ (see below)}$$

$$\bar{R}_i = 1\ \Omega\text{ mm}$$

$$\bar{C}_{gc} = 1.67 \text{ pF mm}^{-1}$$

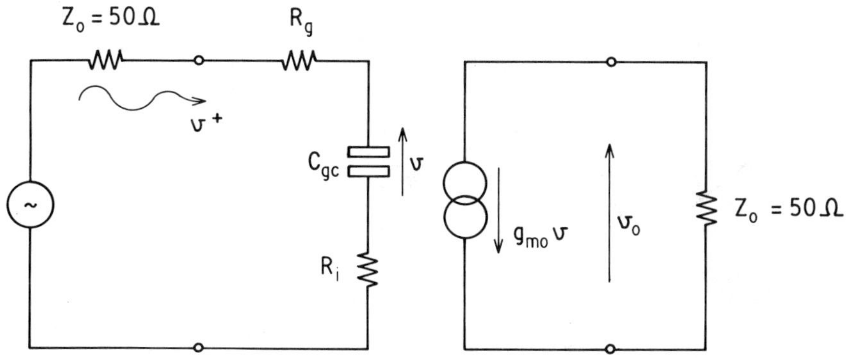

Fig. 9.9 Simplified network for s_{21} measurement of a FET.

Recalling that $g_{m0} = \bar{g}_{m0} \times Z_G$, etc., devices of different Z_G then have $|s_{21}|$ as calculated from eqn (9.11) and plotted in Fig. 9.10. Note that, having set $R_g = 0$, the answers apply to either configuration (a) or (b) in Fig. 9.3. Physically, as can very simply be confirmed by inserting values into eqn (9.11), the $|s_{21}|$ of narrow gatewidth devices is low due to a combination of impedance mismatch and lower transconductance than wider devices. That leaves aside, however, the question of the improvement possible in power gain if the source and load impedances connected to the FET can be adjusted.

We finally arrive at three important conclusions regarding total gatewidth, Z_G, and per unit gatewidth, Z_{GU}.

(i) The expected circuit gain, or maximum tunable gain (MTG), of a FET or HEMT operated in a system where the source and load impedances can be adjusted is governed by the per unit gatewidth, Z_{GU}, rather than the total gatewidth, Z_G. Figure 9.5 gives an example. The differences in this figure stem from the fact that R_g and $\mathscr{L}_g$ increase as Z_{GU} increases.

(ii) Differences in $|s_{21}|$ for devices of different total gatewidth Z_G, as exemplified by Fig. 9.6, arise mainly from limited voltage drive to a 50 Ω load (i.e. the term $g_{m0}Z_0$ in eqn (9.11)) and, to some extent, impedance mismatch. R_g does not play a first-order role, as evidenced by the fact that Fig. 9.10 accounts broadly for the differences appearing in Fig. 9.6 despite the simple nature of the derivation (of Fig. 9.10).

(iii) s_{21} is not a useful guide to expected circuit gain (or MTG).

In the expression for minimum noise figure (eqn (6.52)),

$$NF_{\min} \approx 1 + 2\omega \frac{C_{gc}}{g_{m0}} \left(\frac{R_s + R_g}{R_i} \right)^{1/2} \qquad (9.12)$$

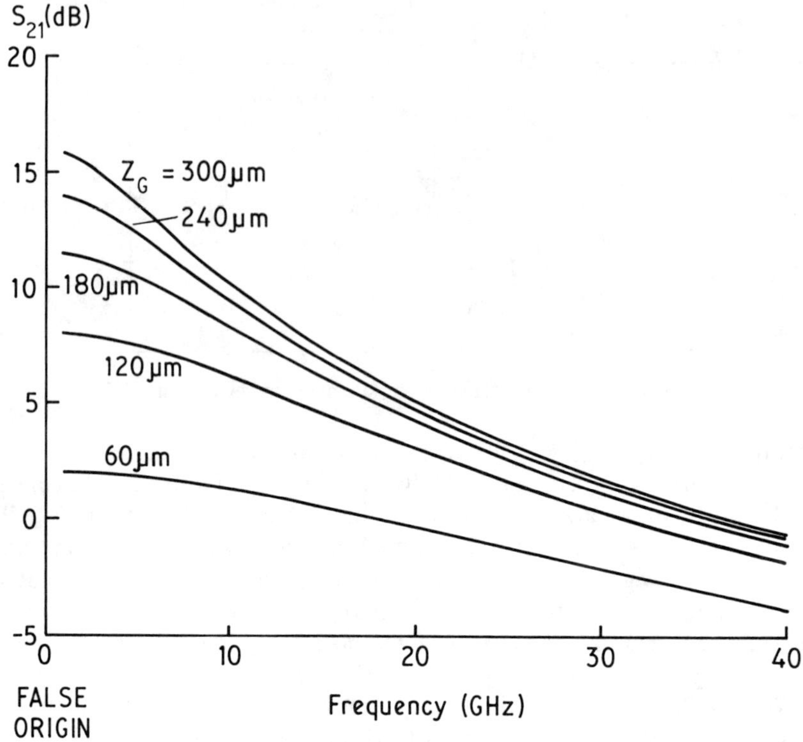

Fig. 9.10 Results of simple analysis for s_{21}. IMPORTANT NOTE: $R_g=0$ in these plots.

we note that C_{gc} and g_{m0} scale as Z_G, and that R_s and R_i scale as Z_G^{-1} regardless of whether case (a) or (b) of Fig. 9.3 applies, whereas R_g scales as $Z_G/m^2 = Z_{GU}/m$. Hence, assuming all other factors to be unaltered, the lowest noise is achieved by having as many gate strips as practicable to realize a given total gatewidth $Z_G = mZ_{GU}$. Clearly, once R_g has been rendered small relative to R_s in eqn (9.12), further reduction in R_g by increasing the number m of gatestrips is pointless. In practice, $m=4$ is a typical limit for small-signal low-noise devices.

9.2.3 Input and output reflection coefficient

Frequently, the input termination to an MMIC, i.e. the impedance of the source or generator from which the MMIC is fed, will be chosen so as to maximize the transfer of signal power from the generator to the chip, in which event the reflection coefficient at the input needs to be as small as possible. In other cases, the input may be systematically mismatched over the frequency band of interest to achieve gain slope compensation.

Whereas an excellent match can normally be achieved at a spot frequency, the reflection coefficient over a broader band of frequencies is not so easy to anticipate. Some guidelines are required.

We consider the problem of matching the input of a FET in an MMIC represented by the simplified network of Fig. 9.11. The FET in question may be the first device in an MMIC, in which case we will be concerned with an input matching network; or it may be an intermediate FET on the chip, in which event we are concerned with the properties of an interstage matching network. Because an artificial transmission line is generally considered to offer a constant characteristic impedance over a usefully broad band, we shall first consider a matching scheme which embodies the FET into an artificial transmission line as in Fig. 9.12, where $C = C_2$. (For simplicity, we assume $R_2 \approx 0$.)

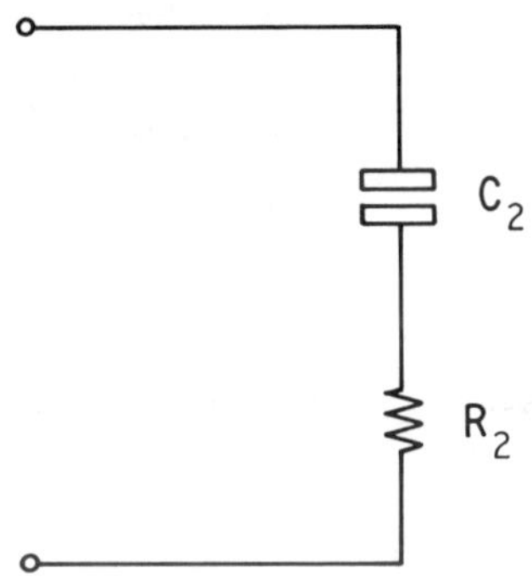

Fig. 9.11 Approximate input equivalent network of FET to be matched.

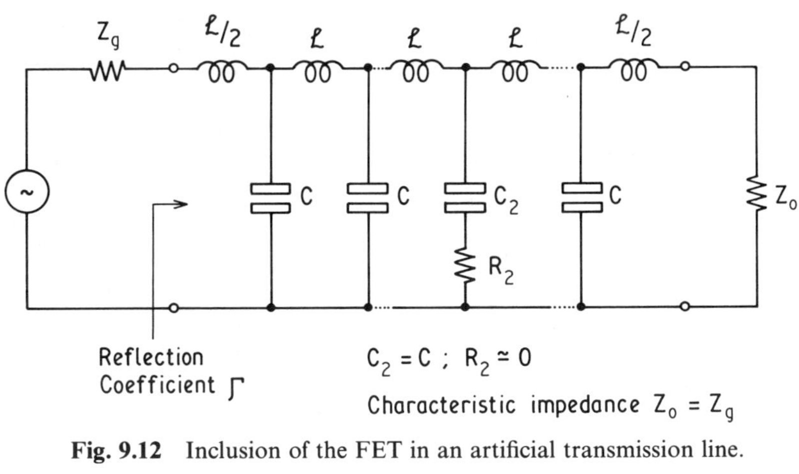

Fig. 9.12 Inclusion of the FET in an artificial transmission line.

With this broadbanding scheme the magnitude of the reflection coefficient, $|\Gamma|$, can be made small by choosing the series inductance per section, $\mathscr{L}$, such that the characteristic impedance of the line equals the impedance of the source or generator, Z_g, i.e. $Z_0 = Z_g$. Excitation of a FET incorporated into an artificial transmission line in this way occurs as the voltage wave passes before being absorbed in the matched terminating impedance at the right-hand end of the line. There is nevertheless a limited bandwidth available before the reflection coefficient Γ starts to increase to non-negligible values, as we shall next show.

The standard theory of artificial lines [9.1] gives the characteristic impedance, Z_0, and the cut-off frequency, f_C, as

$$Z_0 = \left(\frac{\mathscr{L}}{C}\right)^{1/2} = \left(\frac{\mathscr{L}}{C_2}\right)^{1/2}$$

$$f_C = \frac{1}{\pi(\mathscr{L}C)^{1/2}}$$

Combining these equations, we have

$$f_C = \frac{1}{\pi Z_o C} = \frac{1}{\pi Z_0 C_2} \tag{9.13}$$

Now it may be shown that the impedances Z_{01} and Z_{02}, respectively the impedances at the mid-series and mid-shunt points, points 1 and 2 in Fig. 9.13, are given by [9.1]:

$$Z_{01} = Z_0(1 - x^2)^{1/2} \qquad Z_{02} = \frac{Z_0}{(1 - x^2)^{1/2}}$$

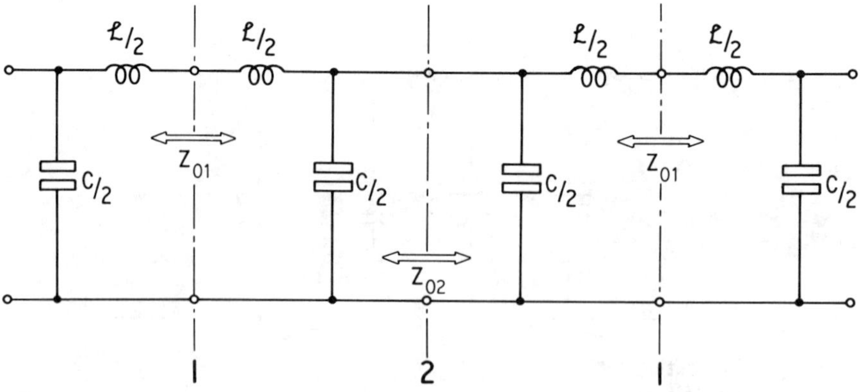

Fig. 9.13 Mid-series and mid-shunt impedances of an artificial transmission line.

where

$$x = f/f_C = \pi f Z_0 C = \pi f Z_0 C_2$$

Regardless of whether the line is fed from the generator at a mid-series or mid-shunt point, the reflection coefficient is:

$$\Gamma = \left.\frac{Z_{0i} - Z_0}{Z_{0i} + Z_0}\right|_{i=1 \text{ or } 2} = \frac{1-(1-x^2)^{1/2}}{1+(1-x^2)^{1/2}}$$

i.e.

$$|\Gamma| = \frac{1-[1-(f/f_C)^2]^{1/2}}{1+[1-(f/f_C)^2]^{1/2}} \tag{9.14}$$

A plot of this reflection coefficient is given in Fig. 9.14. It shows that a voltage standing wave ratio of less than 2:1, or equivalently a return loss of

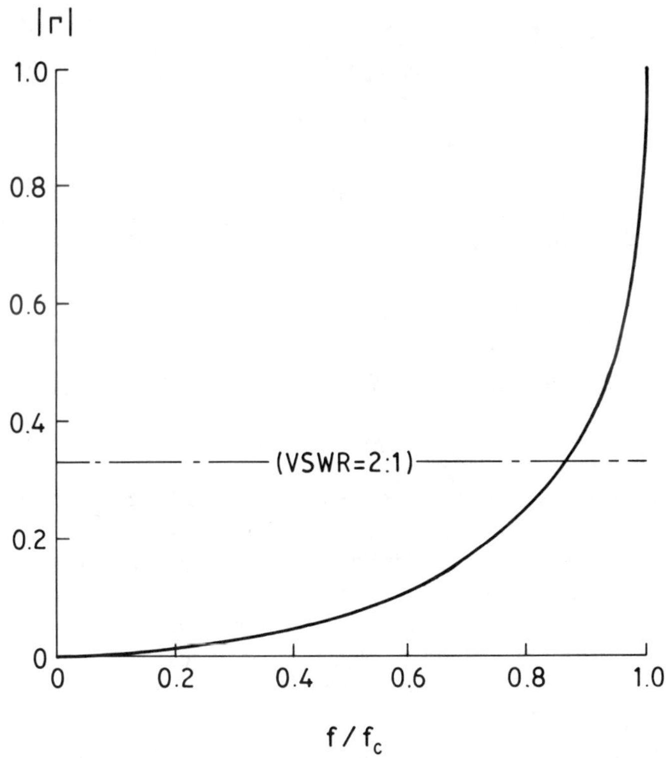

Fig. 9.14 Reflection coefficient versus frequency for an artificial line that is matched to a generator at frequencies approaching zero.

10 dB or less, can be sustained up to approximately $f = 0.86\, f_C$. Eliminating f_C from eqn (9.14) using eqn (9.13) furthermore shows that

(i) for a given $|\Gamma|$, higher operating frequency f can be traded for a lower impedance level, Z_0;
(ii) for a given impedance level, Z_0, $|\Gamma|$ is lower at lower frequencies, f;
(iii) at a given frequency, f, $|\Gamma|$ is lower for a lower Z_0.

To see what this means in practice, suppose we wished to design a distributed matching network to achieve a VSWR of less than 2:1 up to 18 GHz for a power FET having $C_2 = 1$ pF. The impedance level required can be found by the following steps. For a VSWR $\leqslant$ 2:1 at 18 GHz,

$$f_C = \frac{18 \text{ GHz}}{0.86} = 21 \text{ GHz}$$

i.e.

$$(\pi Z_0 C_2)^{-1} = 21 \text{ GHz}$$

$$Z_0 = \frac{1}{\pi \times 1 \text{ pF} \times 21 \text{ GHz}} = 15\ \Omega$$

Alternatively, if it were desired to use this FET in a 50 Ω system, the cut-off frequency would become

$$f_C = \frac{1}{\pi \times 50\ \Omega \times 1 \text{ pF}} = 6.4 \text{ GHz}$$

The highest frequency at which a VSWR of 2:1 could be achieved would then be

$$f = 0.86\, f_C = 5.5 \text{ GHz}$$

Hence a distributed matching network of the type under consideration offers no possibility of matching a fairly large power FET element to 50 Ω above about 5 GHz. The MMIC designer then has two choices. One is to consider breaking-up the FET element into sub-elements having smaller capacitance C_2, these smaller FET elements each forming one of the capacitances in Fig. 9.12. Such a move then provides the basis of a distributed power amplifier (see Section 10.5). The other is to consider another type of matching network.

Before leaving the artificial line, it is interesting to reflect upon the nature of the match it apparently provides, and the nature of the excitation of the FET (Fig. 9.15). Evidently, little power may be absorbed by the FET as the wave

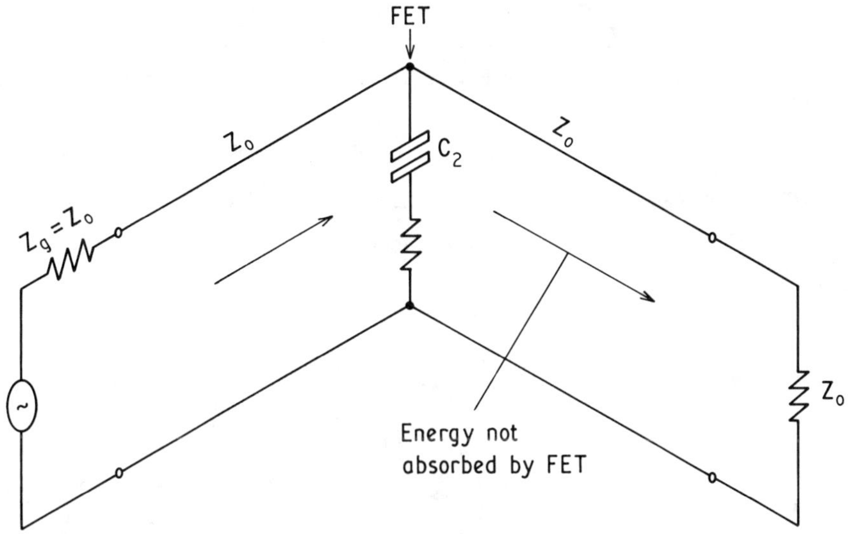

Fig. 9.15 Illustrating the power dumping property of an artificial line matching scheme.

from the generator propagates along the line, meaning that the FET "sees" an impedance quite different from the conjugate of its own impedance (as would be required for maximum transfer of signal power from the wave to the FET). In such circumstances, much if not most of the energy in the wave must either be reflected back toward the generator, or be transmitted onwards towards the load. Referring to Fig. 9.14, for $f \leqslant 0.5 f_C$ the predominant effect is that of transmission, the signal ultimately being absorbed in the terminating impedance Z_0 at the right hand end of the line (Fig. 9.15). The artificial line used in a distributed amplifier thus contrives to create a good match by dumping signal power not absorbed by the FET. As such, it is just one method of creating a good match by signal dumping. A directional coupler (such as a Lange coupler) is another method.

The second type of matching network we shall consider is one where no provision for power dumping is made, so the objective is to ensure as much of the incident signal power as possible is absorbed by the FET. Any which is not will be reflected back towards the generator, thereby contributing to a finite value of reflection coefficient.

In Fig. 9.16 is shown a matching network interposed between the generator and the FET. The network is assumed to be an impedance transformer with the properties of

(i) transforming the impedance $|Z_2|$ connected at port 2 by the magnitude factor N^2 to give a magnitude $N^2|Z_2|$ at port 1, and additionally

(ii) introducing a phase factor $\exp(j\theta)$ so that the transformed impedance at port 1 is

$$Z_1 = N^2 \exp(j\theta) Z_2$$

subject to the restriction that the real part of Z_1, $\mathrm{Re}(Z_1)$ should satisfy

$$\mathrm{Re}(Z_1) > 0$$

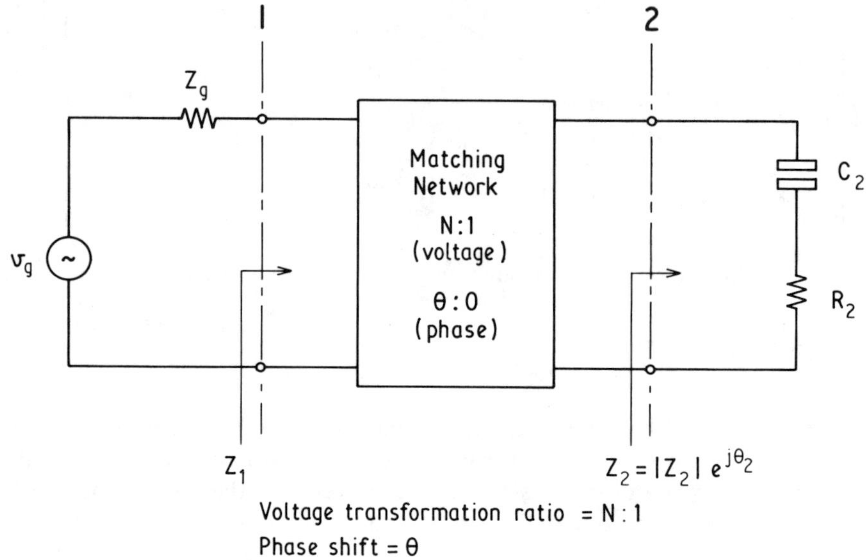

Fig. 9.16 Idealized matching transformer.

Assuming the generator impedance Z_g to be real, we then have for the reflection coefficient at the plane 1 in Fig. 9.16:

$$\Gamma = \frac{Z_g - Z_1}{Z_g + Z_1} = \frac{Z_g - N^2|Z_2|\exp[j(\theta+\theta_2)]}{Z_g + N^2|Z_2|\exp[j(\theta+\theta_2)]}$$

After expanding $\exp[j(\theta+\theta_2)]$ as $[\cos(\theta+\theta_2)+j\sin(\theta+\theta_2)]$ and taking the modulus, we find

$$|\Gamma|^2 = \frac{A - \cos(\theta+\theta_2)}{A + \cos(\theta+\theta_2)}$$

with

$$A = \frac{Z_g^{\,2} + N^4|Z_2|^2}{2Z_g N^2|Z_2|}$$

Given that $\cos(\theta+\theta_2)\leqslant 1$, and given also $|\Gamma|\geqslant 0$, we have that A must be greater than or equal to 1. The condition $A=1$ leads to the minimum value of $|\Gamma|$ and is satisfied by the choice (first matching condition):

$$N=(Z_g/|Z_2|)^{1/2}$$

We then have

$$|\Gamma|^2=\frac{1-\cos(\theta+\theta_2)}{1+\cos(\theta+\theta_2)}$$

In this expression we are free to choose θ in the interests of minimizing the reflection coefficient, subject only to the constraint $\mathrm{Re}(Z_1)>0$, or equivalently to $\cos(\theta+\theta_2)>0$. The phase factor θ_2 is that associated with the FET (i.e. C_2 and R_2 in series):

$$\theta_2=\tan^{-1}[1/(\omega C_2 R_2)]$$

For present purposes, we shall assume the phase factor θ to be linearly proportional to frequency, as might be afforded by a length of non-dispersive line, so,

$$\theta=\mathrm{K}\omega$$

We then have

$$|\Gamma|^2=\frac{1-\cos\{\mathrm{K}\omega-\tan^{-1}[1/(\omega C_2R_2)]\}}{1+\cos\{\mathrm{K}\omega-\tan^{-1}[1/(\omega C_2R_2)]\}} \tag{9.15}$$

When we are concerned with a band of frequencies ω_L to ω_H, we may clearly choose K so as to give $|\Gamma|=0$ at one particular frequency. We shall now prove that, if we so adjust K that $|\Gamma|=0$ at the highest frequency in the band, ω_H, then the restriction that $\cos(\theta+\theta_2)$ must be greater than zero will be satisfied. Thus consider the choice (second matching condition):

$$\mathrm{K}=(1/\omega_H)\tan^{-1}[1/(\omega_H C_2 r_2)] \tag{9.16}$$

then, at any $\omega\leqslant\omega_H$,

$$\cos\{\mathrm{K}\omega-\tan^{-1}[1/(\omega C_2R_2)]\}=\cos\left\{\frac{\omega}{\omega_H}\tan^{-1}[1/(\omega_H C_2R_2)]\right.$$
$$\left.-\tan^{-1}[1/(\omega C_2R_2)]\right\}$$

Since $\tan^{-1}(1/\omega C_2R_2)\leqslant\pi/2$, and since also

$$\tan^{-1}[1/(\omega_H C_2R_2)]<\tan^{-1}[1/(\omega C_2R_2)] \tag{9.17}$$

we have

$$[\theta+\theta_2]=\frac{\omega}{\omega_H}\tan^{-1}[1/(\omega_H C_2 R_2)]-\tan^{-1}[1/(\omega C_2 R_2)]$$

$$<\left(\frac{\omega}{\omega_H}-1\right)\tan^{-1}[1/(\omega C_2 R_2)]<\left(\frac{\omega}{\omega_H}-1\right)\frac{\pi}{2}$$

Thus $\theta+\theta_2$ lies in the range $-\pi/2\leqslant\theta+\theta_2\leqslant 0$, so that $\cos(\theta+\theta_2)\geqslant 0$, as required. Consequently, with $|\Gamma|$ adjusted to be zero at ω_H, the highest frequency in the band, the condition in eqn (9.16) may be substituted into eqn (9.15) to yield an expression from which the reflection coefficient at any frequency $\omega\leqslant\omega_H$ may be determined, that is

$$|\Gamma|^2=\frac{1-\cos\left\{\dfrac{\omega}{\omega_H}\tan^{-1}[1/(\omega_H C_2 R_2)]-\tan^{-1}[1/(\omega C_2 R_2)]\right\}}{1+\cos\left\{\dfrac{\omega}{\omega_H}\tan^{-1}[1/(\omega_H C_2 R_2)]-\tan^{-1}[1/(\omega C_2 R_2)]\right\}} \tag{9.18}$$

Although this expression is algebraically exact, its use is rarely warranted for quick first appraisals. Making use of the inequality in eqn (9.17), we may write

$$|\Gamma|^2\geqslant\frac{1-\cos\left\{\left(\dfrac{\omega}{\omega_H}-1\right)\tan^{-1}[1/(\omega C_2 R_2)]\right\}}{1+\cos\left\{\left(\dfrac{\omega}{\omega_H}-1\right)\tan^{-1}[1/(\omega C_2 R_2)]\right\}}$$

This expression may be further condensed by substituting

$$\Delta\omega=(\omega_H-\omega)$$

$$\cos(-\Delta\omega/\omega_H)=\cos(\Delta\omega/\omega_H)$$

$$\text{and } Q=(1/\omega C_2 R_2)$$

Then,

$$|\Gamma|^2\geqslant\frac{1-\cos\left(\dfrac{\Delta\omega}{\omega_H}\tan^{-1}Q\right)}{1+\cos\left(\dfrac{\Delta\omega}{\omega_H}\tan^{-1}Q\right)} \tag{9.19}$$

Values of $|\Gamma|$ for various values of quality factor, Q, are plotted in Fig. 9.17. What this figure tells us is that, even when we achieve an exact match at some point in the frequency band of interest, at some other point in the band the reflection coefficient will be at least as large as the values given by the curves of Fig. 9.17. The wider the band of interest, the worse the reflection coefficient somewhere in the band will be. Similarly, the larger the Q of the FET, the larger will be the reflection coefficient at the worst point in the band.

By way of illustration, consider the task of designing a 6–18 GHz power MMIC around a power FET having $C_2=1$ pF and $R_2=3\ \Omega$. Taking the mid-band point of 12 GHz for the purposes of estimating Q, we have

$$Q=4.4$$

Then with $\Delta\omega/\omega_H=12$ GHz/18 GHz$=0.67$, we estimate $|\Gamma|$ as

$$|\Gamma|\geqslant 0.48$$

(By comparison, the exact value from eqn (9.18) is $|\Gamma|=0.574$.) In other words, at some frequency in the range 6 GHz$\leqslant f \leqslant$18 GHz, we should expect

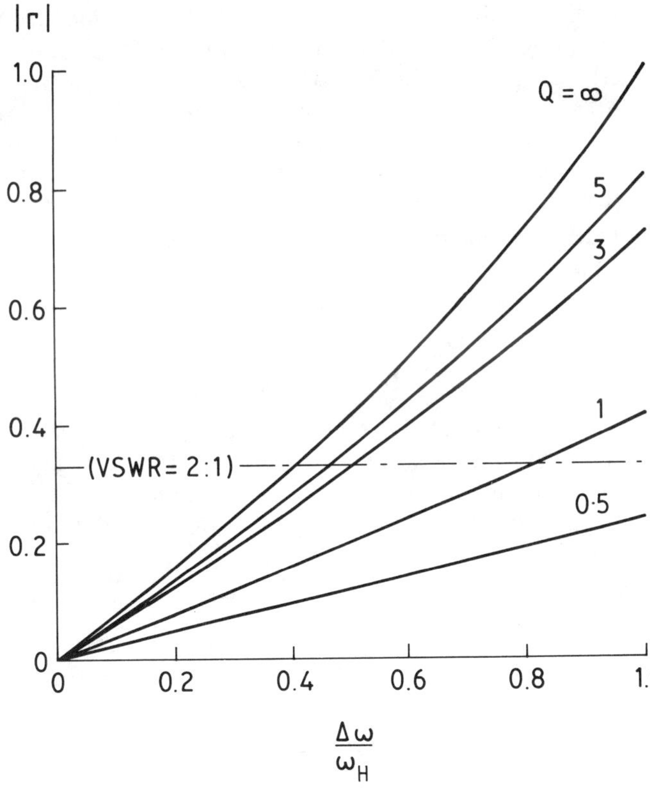

Fig. 9.17 Approximate reflection coefficient achievable depending upon the FET input Q and the fractional bandwidth ($\Delta\omega/\omega_H$).

to find the reflection coefficient has a value of at least 0.48 (or 0.574 according to whether we take the approximate value or the exact value).

Although more general treatments of the matching question are possible (see, for example, Fano's Theorem in [9.2]), the development leading to Fig. 9.17 gives tangible and useful estimates for the purpose of design limit calculations.

9.2.4 Maximum tunable gain (MTG), MAG, MSG and MUG

Generally speaking, the estimate of power gain given by eqn (9.4) is valuable only as a first, rough, approximation. A closer estimate of FET gain is almost always required as a postscript to such first calculations. In particular, a close estimate of the gain realizable *in a circuit* is valuable.

The appendix to this chapter is a summary of *S*-parameters, maximum available gain (MAG), maximum stable gain (MSG), and maximum unilateral transducer power gain (MUG), all with their respective definitions. Experience shows that none of these figures of merit gives a close enough estimate of the power gain which can be achieved when a FET of given *S*-parameters (or equivalent circuit values) is integrated in an MMIC. A better criterion has been found to be the maximum tunable gain (MTG) [3]. Because MTG is not widely appreciated as a measure of realizable circuit gain, we will provide a résumé here.

With reference to Fig. 9.18, the operating power gain of the FET, G_w, is given by

$$G_W = P_L / P_{IN}$$

where P_L is the power dissipated in the load, and P_{IN} is the power entering the input terminals of the FET (which, for a lossless input matching network, also equals the power entering the input terminals of the input matching network, i.e. of the amplifier). Following Abrie [9.3], we shall deal with the derivation of

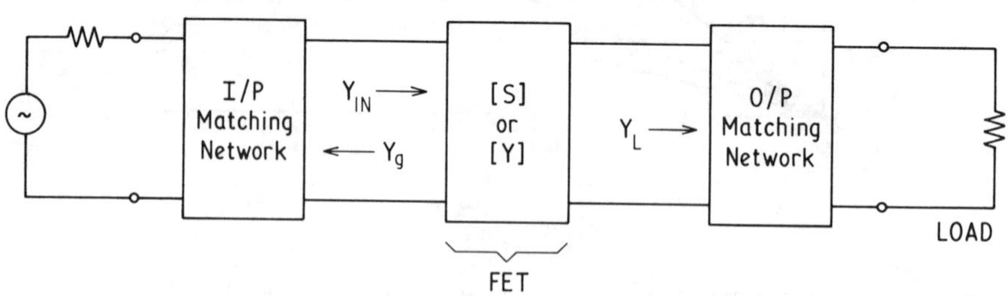

Fig. 9.18 Block diagram of FET with input and output matching networks.

MTG from the operating power gain in terms of Y-parameters. Interrelationships between S- and Y-parameters are given in the appendix to this chapter.

With $Y_L = G_L + jB_L$ in Fig. 9.19, we have (see also Section 9.A.3):

$$G_W = \left|\frac{y_{21}}{y_{22}+Y_L}\right|^2 \frac{G_L}{\mathrm{Re}\left(y_{11} - \dfrac{y_{12}y_{21}}{y_{22}+Y_L}\right)} \tag{9.20}$$

By manipulating this expression, it may be shown that the loci of admittances giving any specified constant value of power gain are circles in the $G_L - B_L$ plane [9.3]; for an inherently stable transistor, the admittance circles lie entirely in the right half plane (Fig. 9.20). Assuming the FET to be inherently stable, the maximum realizable power gain is found when the radius R_{LW} of the operating power gain circles tends to zero, where

$$R_{LW}{}^2 = G_{LW}{}^2 + \left|\frac{y_{12}y_{21}}{2g_{11}}\right|^2 (1 - 1/C^2) \tag{9.21}$$

Here $g_{11} = \mathrm{Re}(y_{11})$, and all such Y-parameters belong to the FET only (Fig. 9.18). G_{LW} is the distance of the center of the gain circle from the imaginary (B_L) axis:

$$G_{LW} = G'_L + \frac{|y_{21}|^2}{2g_{11}G_W}$$

(See eqn (9.26) for G'_L.)
C is the Linville stability factor defined by

$$C = -\frac{|y_{12}y_{21}|}{P - 2g_{11}g_{22}} \tag{9.22}$$

and P is defined by

$$P + jQ = y_{12}y_{21} \tag{9.23}$$

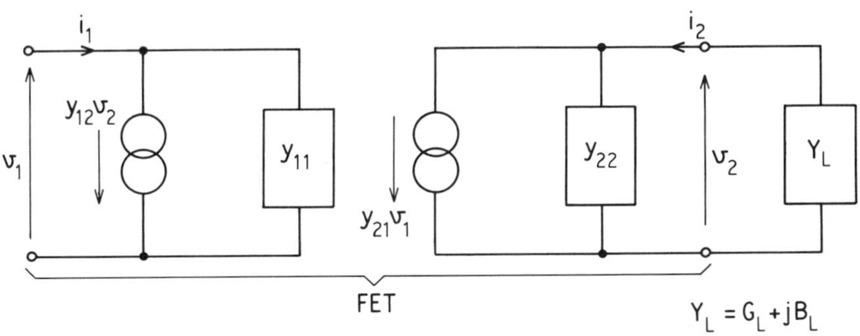

Fig. 9.19 Y-parameter equivalent network of FET and load.

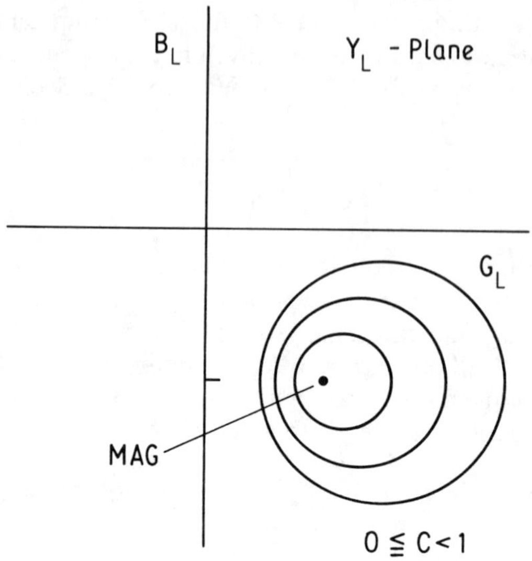

Fig. 9.20 Loci of load admittance values giving constant power gain; each circle has a constant power gain associated with it. The power gain is larger for circles of smaller radius.

When $0 \leqslant C \leqslant 1$, the device is inherently stable [9.4].

The maximum realizable power gain is obtained as R_{LW} tends to zero (Fig. 9.20):

$$\mathrm{MAG} \equiv G_{\text{W-MAX}} = \frac{|y_{21}|^2}{2g_{11}(G_{\text{LW-OPT}} - G'_{\mathrm{L}})} \tag{9.24}$$

(See eqn (9.26) for G'_{L}.) Here, $G_{\text{LW-OPT}}$ is the real part of the load termination corresponding to the maximum realizable gain:

$$Y_{\text{L-OPT}} = G_{\text{LW-OPT}} + \mathrm{j}B_{\text{LW-OPT}} = \left|\frac{y_{12}y_{21}}{2g_{11}}\right| \left(\frac{1}{C^2} - 1\right)^{1/2} + \mathrm{j}B'_{\mathrm{L}} \tag{9.25}$$

with

$$G'_{\mathrm{L}} + \mathrm{j}B'_{\mathrm{L}} = \left(\frac{P}{2g_{11}} - g_{22}\right) + \mathrm{j}\left(\frac{Q}{2g_{11}} - b_{22}\right) \tag{9.26}$$

Alternatively, $G_{\text{W-MAX}}$ may be put into the more familiar form:

$$\mathrm{MAG} \equiv G_{\text{W-MAX}} = \left|\frac{y_{21}}{y_{12}}\right| \left[\frac{1}{C} - \left(\frac{1}{C^2} - 1\right)^{1/2}\right] \tag{9.27}$$

Away from the MAG point, the operating power gain decreases the larger the radius R_{LW} becomes.

The MTG is found, by a systematic procedure, to lie on one of these (larger) circles. The systematic procedure involves assuring a prescribed value of the tunability factor, δ [9.3]. With reference to Fig. 9.18,

$$\delta = \left| \frac{\partial Y_{IN}/Y_{IN}}{\partial Y_L/Y_L} \right| \tag{9.28}$$

δ is a measure of the independence of the input admittance of the FET (Fig. 9.18) from the load admittance. The circuit or device is maximally tunable when δ is as small as possible, because then the load admittance can be adjusted without affecting the input admittance. A value of $\delta \leqslant 0.3$ is generally desirable, and is adopted as the basis for determining the MTG. It may be shown [9.3] that

$$\begin{aligned} \delta &= \frac{|y_{12}y_{21}Y_L|}{|y_{22}+Y_L|^2|Y_{IN}|} \\ &= \left|\frac{y_{12}}{y_{21}}\right| G_W \frac{\cos\theta_{IN}}{\cos\theta_L} \end{aligned} \tag{9.29}$$

where

$$\theta_{IN} = \tan^{-1}(B_{IN}/G_{IN})$$

$$\theta_L = \tan^{-1}(B_L/G_L)$$

and

$$Y_{IN} = G_{IN} + jB_{IN} = y_{11} - \frac{y_{12}y_{21}}{y_{22}+Y_L} \tag{9.30}$$

Equation (9.29) shows that tunability δ varies around each of the circles in Fig. 9.20, and that δ is directly proportional to the operating power gain, G_w. Consequently, good tunability (i.e. small values of δ) can be obtained at the expense of power gain. To find the MTG, we are concerned to find the highest value of G_W which provides a tunability factor of $\delta \leqslant 0.3$. Thus, by definition, MTG is the highest value of operating power gain at which the device is tunable (i.e. $\delta \leqslant 0.3$).

To calculate the MTG, use is made of the fact that the load admittance with the lowest tunability factor on any given gain circle is almost without exception the one furthest from the imaginary axis (Fig. 9.21). If the device is not tunable at the MAG point in Fig. 9.21 (i.e. if $\delta > 0.3$ at the MAG point where $G_W = G_{W\text{-}MAX}$), the tunability δ and the operating power gain G_W are calculated at increasing values of load conductance G_L along the line $-jB'_L$ in

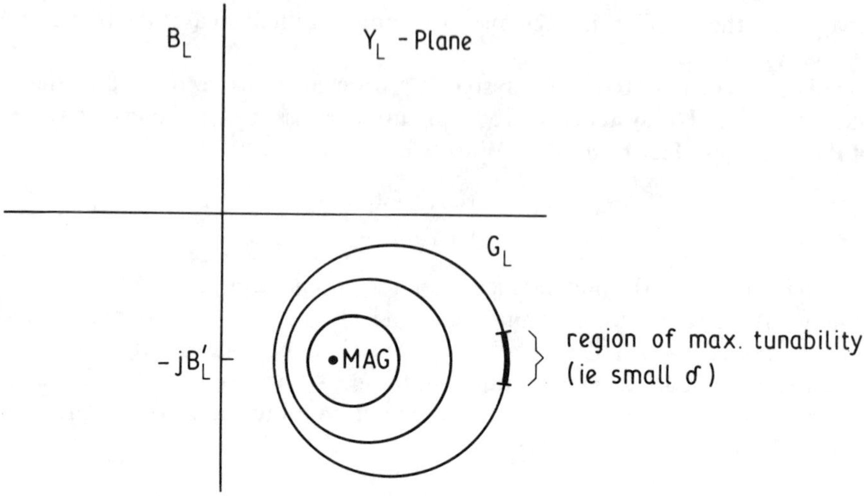

Fig. 9.21 Illustrating the concept of tunability.

Fig. 9.21. The MTG is the value of G_W at which δ falls to $\delta \leqslant 0.3$ (Fig. 9.22).

When the device is not inherently stable, it must be stabilized before attempting to find the MTG. We assume that voltage shunt feedback as in Fig. 9.23 is used to stabilize the device, in which event the Y-matrix of the new two-port network (the dotted box in Fig. 9.23) becomes

$$[Y]=\begin{bmatrix} y_{11}+g_f & y_{12}-g_f \\ y_{21}-g_f & y_{22}+g_f \end{bmatrix} \tag{9.31}$$

with $g_f=(1/R_F)$. This is now treated as a new device for which the MTG may be calculated.

In summary, to calculate the MTG:

(i) The Linville stability factor, C, is calculated from eqn (9.22). If C does not satisfy $0 \leqslant C \leqslant 1$, the device is stabilized (conceptually) in the manner of Fig. 9.23 and the device is now treated as having a new Y-matrix as in eqn (9.31). The Y-parameters of this new matrix are then used throughout all following steps.

(ii) B'_L is calculated from eqns (9.26) and (9.23), and $Y_{L\text{-OPT}}$ is computed from eqn (9.25).

(iii) $G_{W\text{-MAX}}$ ($\equiv$MAG) is calculated from eqns (9.24), (9.25) and (9.26).

(iv) The tunability factor, δ, is calculated from eqns (9.29), (9.30) at the MAG point. If $\delta \leqslant 0.3$ at the MAG point, then MTG$=G_{W\text{-MAX}}$.

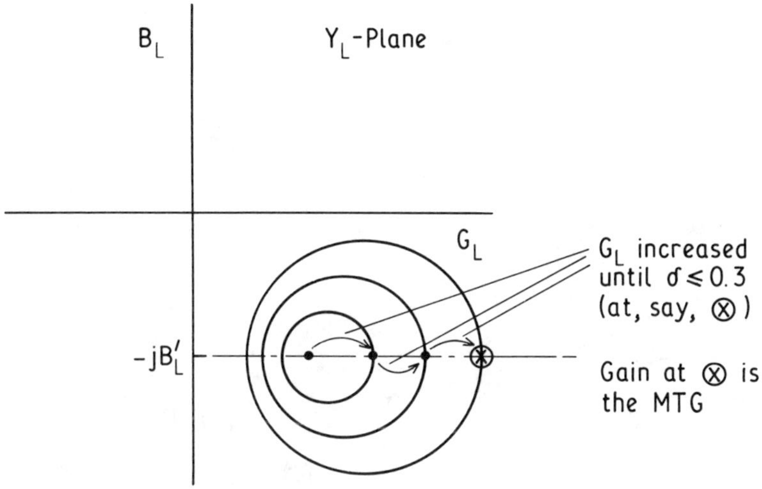

Fig. 9.22 Iteration to find the MTG.

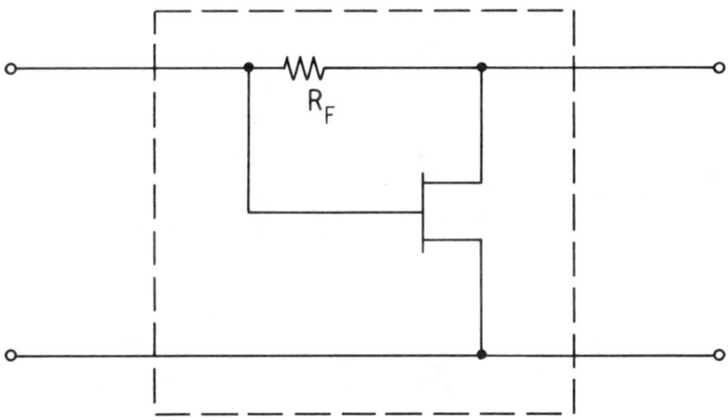

Fig. 9.23 Stabilization using a shunt feedback resistor.

(v) If $\delta > 0.3$ at the MAG point, G_L is increased progressively beyond the value $G_{LW\text{-}OPT}$ in eqn (9.25) so that the load admittance becomes

$$Y_L = G_L + jB'_L$$

At each new value of G_L, δ is calculated from eqn (9.29) and G_W is calculated from eqn (9.20). The value of G_W at which $\delta \leqslant 0.3$ is first attained is the MTG.

Figure 9.24 compares the MTG with the MUG, and the MSG or MAG, for a HEMT with a 0.5 μm by 120 μm gate. The curves relate to a device with

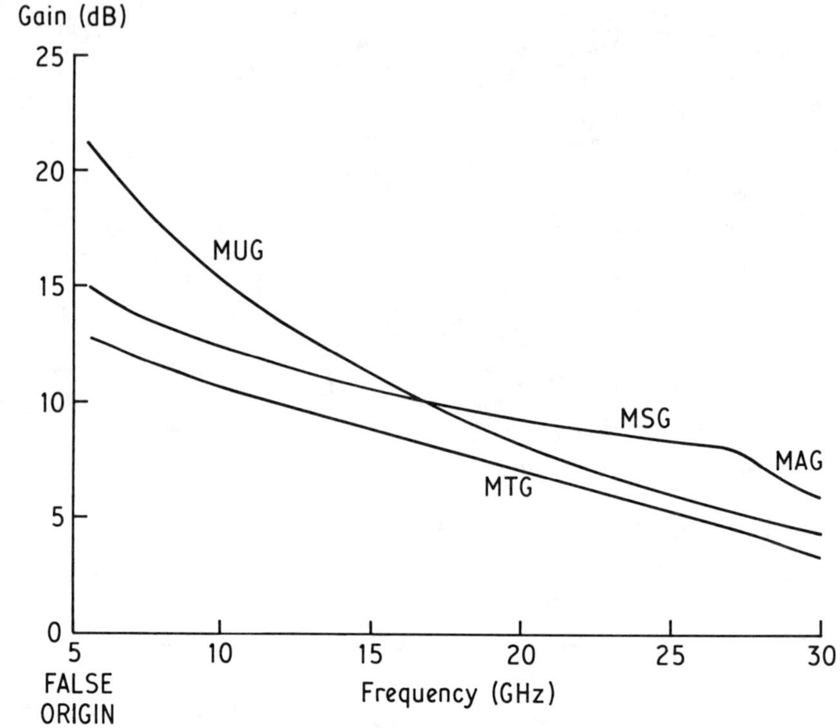

Fig. 9.24 Comparison of MTG with more commonly quoted power gains for a discrete HEMT with $L_G=0.5$ μm and $Z_G=120$ μm.

bonding pads (i.e. a discrete device). Figure 9.25 shows the anticipated effect on the various power gains if the device is integrated into an MMIC, where the geometric capacitances associated with the pads would be eliminated. In all instances the MTG is the lowest of the gain estimates, but experience shows it to be a good forecast of the gain which can be realized in a practical circuit.

9.3 Limits to large-signal (power) behavior

From eqns (6.54) to (6.56), we may write the output power of a GaAs FET per unit gatewidth as

$$\bar{P}_O=\frac{1}{16}\left[\frac{E_B L_G}{qNW}(\epsilon E_B-qNW)-\frac{qNW^2}{\epsilon}\right]qNv_{sat}(W-d_{min})$$

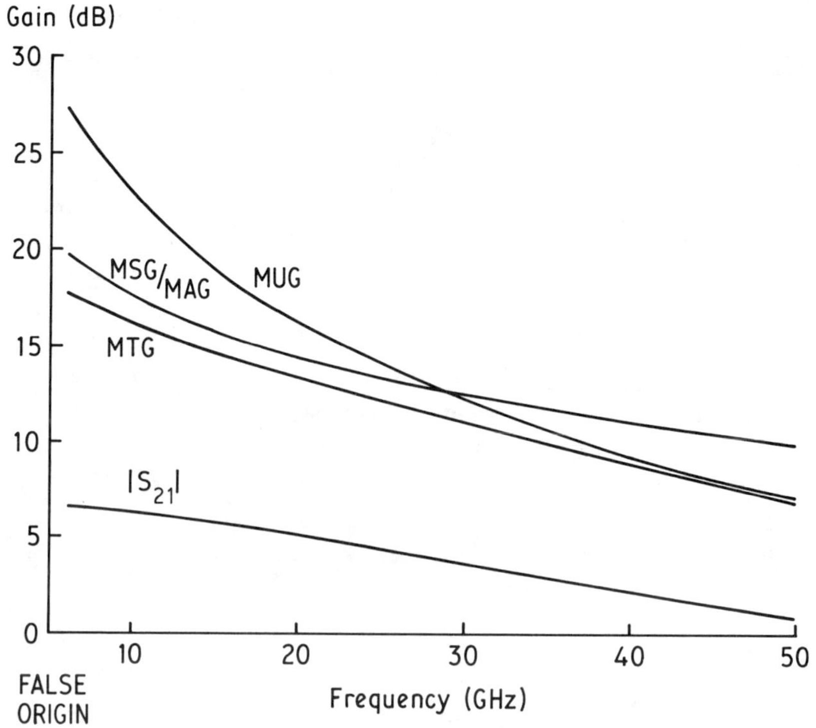

Fig. 9.25 MTG, MSG or MAG, and MUG for the HEMT in Fig. 9.24 with the pad parasitic capacitances removed.

Hence

$$\bar{P}_O < \frac{v_{sat}}{16}\left[E_B L_G(\epsilon E_B - qNW) - \frac{(qNW)^2 W}{\epsilon}\right] \quad (9.32)$$

This is a maximum when (hypothetically) $NW \approx 0$, so

$$\bar{P}_O < \frac{\epsilon v_{sat} E_B{}^2 L_G}{16} \quad (9.33)$$

This shows that, fundamentally, output power density is approximately proportional to gate length. With $E_B \approx 640$ kV cm^{-1}, we have for a 1 μm FET

$$\bar{P}_O < 3 \text{ W mm}^{-1}$$

and for a 0.5 μm FET

$$\bar{P}_O < 1.5 \text{ W mm}^{-1}$$

Equation (9.33) is an absolute limit and applies only to FETs with a cut-off frequency of $f_T \approx 0$, as can easily be seen from eqns (6.23) and (9.7). From eqn (6.23)

$$X \approx \frac{2\epsilon(V_{D'G} + V_{BO})}{qNd}$$

If $NW \approx 0$, then $Nd \approx 0$ and $X \approx \infty$. But, from eqn (9.7),

$$f_T = \frac{v_{sat}}{2\pi(L_G + X/2)} \tag{9.7}$$

Hence $f_T \approx 0$ when $NW \approx 0$.

In practical devices, the foregoing expressions may be used in reverse order to estimate $\bar{P}_O$. For example, suppose a 0.5 μm FET is to operate at 20 GHz. Then from eqn (9.7),

$$X < 2\left(\frac{v_{sat}}{2\pi \times 20\ \text{GHz}} - L_G\right) = 0.6\ \mu\text{m}$$

Then from eqn (6.23), assuming $(V_{D'G} + V_{BO})$ must cater for a voltage swing of up to 10 V, qNd must satisfy

$$qNd > \frac{2\epsilon \times 10\ \text{V}}{0.6\ \mu\text{m}} = 0.004\ \text{C m}^{-2}$$

i.e. we must have $(qNW) \geqslant 0.004\ \text{C m}^{-2}$. Taking $L_G/W \approx 3$ we thus find $W \approx 0.5\ \mu\text{m}/3 \approx 0.17\ \mu\text{m}$ so that

$$N \approx 1.5 \times 10^{23}\ \text{m}^{-3} = 1.5 \times 10^{17}\ \text{cm}^{-3}$$

Then, from eqn (9.32),

$$\bar{P}_O < 525\ \text{mW mm}^{-1}$$

This is just over one-third of the absolute mathematical limit. Similar conclusions can be drawn for 1 μm FETs so that, for practical purposes, the power density limit consistent with adequate frequency response is, for a 1 μm FET

$$\bar{P}_O \leqslant 1\ \text{W mm}^{-1}$$

and for a 0.5 μm FET

$$\bar{P}_O \leqslant 0.5\ \text{W mm}^{-1}$$

References

9.1 Glazier, E. V. D. and Lamont, H. R. L. *The Services Textbook of Radio Vol. 5—Transmission and Propagation*, 47–52, HMSO, London, 1958.
9.2 Bode, H. W. *Network Analysis and Feedback Amplifier Design*, Ch. 16, Van Nostrand, New York, 1945.
9.3 Abrie, P. L. D. *The Design of Impedance-Matching Networks For Radio-Frequency and Microwave Amplifiers*, 286–290, Artech House, Dedham, 1985.
9.4 Ibid., 278.

9.A Appendix to Chapter 9: *S*-parameters and Gain Definitions

9.A.1 *S*-parameters

Scattering parameters, or *S*-parameters, are four parameters which completely characterize a linear network loaded at both input and output ports in a 50 Ω termination (Fig. 9.A.1), or some other similarly valued system impedance (e.g. 75 Ω). A FET operated under small-signal conditions is the example of interest.

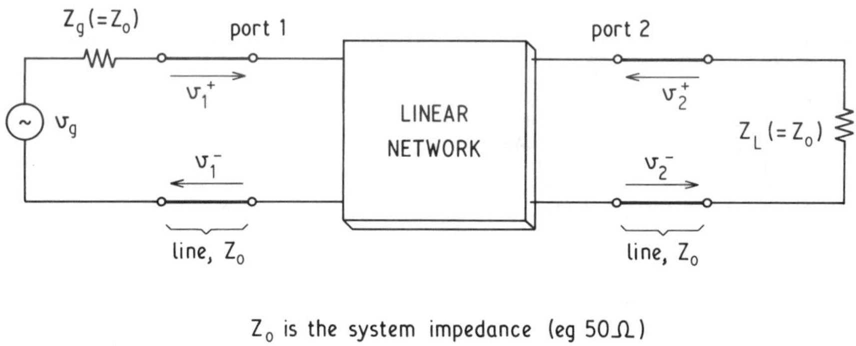

Fig. 9.A.1 Diagram to assist with the definition of *S*-parameters for a linear two-port network.

To attempt to measure the *H*-, *Y*- or *Z*-parameters of the device directly at microwave frequencies would require either a short circuit or open circuit to be presented to the network at both port 1 and port 2 in turn while measurements are made at the other ports. Such procedures are inconvenient at microwave frequencies where the provision of short circuits, and especially open circuits, is difficult. Even worse, short- or open-circuit terminations may lead to instability in the device. *S*-parameter measurements require terminations of 50 Ω (or similar), which are comparatively easy to realize; the device (suitably biased) is merely connected into a 50 Ω measurement system (Fig. 9.A.1).

Generally, the impedances seen looking from the terminations into ports 1 and 2 when the FET is connected as in Fig. 9.A.1 will be different from 50 Ω: consequently, at both ports 1 and 2 there will be an incident wave (v^+) and a reflected wave (v^-). The *S*-parameters are defined in terms of these four voltage waves as follows:

$$v_1^- = s_{11}v_1^+ + s_{12}v_2^+$$

$$v_2^- = s_{21}v_1^+ s_{22}v_2^+$$

or, in matrix notation,

$$\begin{bmatrix} v_1^- \\ v_2^- \end{bmatrix} = \begin{bmatrix} s_{11} & s_{12} \\ s_{21} & s_{22} \end{bmatrix} \begin{bmatrix} v_1^+ \\ v_2^+ \end{bmatrix}$$

The physical significance of the *S*-parameters may be extracted from these equations:

$$s_{11} = \left. \frac{v_1^-}{v_1^+} \right|_{v_2^+ = 0}$$

which is equal to the input reflection coefficient with the output terminated in a matched load (so $v_2^+ = 0$),

$$s_{22} = \left. \frac{v_2^-}{v_2^+} \right|_{v_1^+ = 0}$$

which is equal to the output reflection coefficient with the input terminated in a matched load (so $v_1^+ = 0$),

$$s_{12} = \left. \frac{v_1^-}{v_2^+} \right|_{v_1^+ = 0}$$

which is equal to the reverse transfer gain with the input terminated in a matched load (so $v_1^+ = 0$),

$$s_{21} = \left. \frac{v_2^-}{v_1^+} \right|_{v_2^+ = 0}$$

which is equal to the forward transfer gain with the output terminated in a matched load (so $v_2^+ = 0$).

Note that because the S-parameters are ratios of voltage waves, their expression in decibels requires 20 $\log_{10}(s_{ij})$ to be taken, e.g.

$$s_{11}(\mathrm{dB}) = 20\log_{10}(s_{11})$$

9.A.2 *S*- to *Y*-parameter conversions

It is important to note that the Y-parameters below are normalized to $Y_0 = Z_0^{-1}$.

$$s_{11} = \frac{(1 - Y_{11})(1 + Y_{22}) + Y_{12}Y_{21}}{(1 + Y_{11})(1 + Y_{22}) - Y_{12}Y_{21}} \qquad Y_{11} = \frac{(1 + s_{22})(1 - s_{11}) + s_{12}s_{21}}{(1 + s_{11})(1 + s_{22}) - s_{12}s_{21}}$$

$$s_{12} = \frac{-2Y_{12}}{(1 + Y_{11})(1 + Y_{22}) - Y_{12}Y_{21}} \qquad Y_{12} = \frac{-2s_{12}}{(1 + s_{11})(1 + s_{22}) - s_{12}s_{21}}$$

$$s_{21} = \frac{-2Y_{21}}{(1 + Y_{11})(1 + Y_{22}) - Y_{12}Y_{21}} \qquad Y_{21} = \frac{-2s_{21}}{(1 + s_{11})(1 + s_{22}) - s_{12}s_{21}}$$

$$s_{22} = \frac{(1 + Y_{11})(1 - Y_{22}) + Y_{12}Y_{21}}{(1 + Y_{11})(1 + Y_{22}) - Y_{12}Y_{21}} \qquad Y_{22} = \frac{(1 + s_{11})(1 - s_{22}) + s_{12}s_{21}}{(1 + s_{11})(1 + s_{22}) - s_{12}s_{21}}$$

9.A.3 Relationship between maximum operating power gain and maximum available gain

With reference to Fig. 9.A.2:

$$i_1 = v_1\left(y_{11} - \frac{y_{12}y_{21}}{y_{22} + Y_L}\right)$$

The power input to the FET (where $i_1{}^*$ is the complex conjugate of i_1) is

$$P_{IN} = \mathrm{Re}(v_1 i_1{}^*) = v_1{}^2\mathrm{Re}\left(y_{11} - \frac{y_{12}y_{21}}{y_{22} + Y_L}\right) \tag{9.A.1}$$

The output power developed in the load Y_L is

$$P_O = \left|\frac{y_{21}}{y_{22} + Y_L}\right|^2 v_1{}^2 G_L \tag{9.A.2}$$

Hence the operating power gain, $G_W = P_O/P_{IN}$, is given by

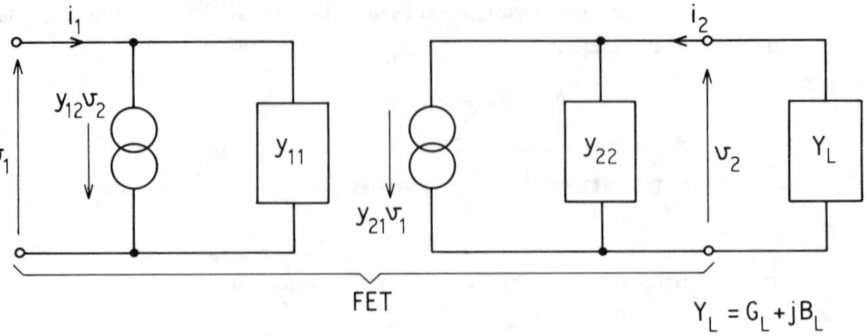

Fig. 9.A.2 *Y*-parameter equivalent network of a FET (or HEMT) and load.

$$G_W = \left|\frac{y_{21}}{y_{22}+Y_L}\right|^2 \frac{G_L}{\mathrm{Re}\left(y_{11} - \dfrac{y_{12}y_{21}}{y_{22}+Y_L}\right)} \tag{9.A.3}$$

which is eqn (9.20).

It is important to note that all results up to and including eqn (9.A.3) are completely independent of the generator and matching network that precede the FET for the simple reason that those parts of the overall circuit were ignored in the definition of G_W; G_W is the power gain between the FET input and the load.

Alternatively, the operating power gain may be expressed in terms of the scattering parameters of the FET (or HEMT) and matching networks considered together. Equation (10.13) gives such an expression based on Fig. 10.9. In that figure, when the generator and load are matched to the connecting transmission lines, $\Gamma_g = 0$ and $\Gamma_L = 0$. The operating power gain then reduces to eqn (10.14)

$$G_W = \frac{|s_{21}|^2}{1-|s_{11}|^2}$$

To emphasize the point, s_{21} and s_{11} are the overall scattering parameters of the FET (or HEMT] and matching networks viewed as a single, linear, two port. In this form, G_W has the added physical significance that, regardless of the values of s_{11} and s_{21}, the numerical value it gives for the power gain is equal to the power gain that would result if the input were conjugately matched.

A legitimate concern is the power transfer to a load, P_O, relative to the power delivered by the generator, where the generator has a fixed internal impedance of Z_g, i.e. P_O relative to P_g in Fig. 9.A.3, $G_{g\text{-}o} = P_O/P_g$, is of interest. For a lossless input matching network, $P_{IN} = P_{IM}$, and so

$$G_{g\text{-}o} = \frac{P_O}{P_g} = \frac{P_O}{P_{IM}(P_g/P_{IM})} = \frac{P_O}{P_{IN}(P_g/P_{IM})} = \frac{P_{IM}}{P_g} G_W$$

By adjusting the output matching network, such that Z_L in Fig. 9.A.3 is the complex conjugate of the FET output impedance, a maximum value of G_W can be attained, $G_{W\text{-MAX}}$ (see eqn (9.24) of the main text). Under this condition,

$$G_{g\text{-}o} = \frac{P_{IM} G_{W\text{-MAX}}}{P_g}$$

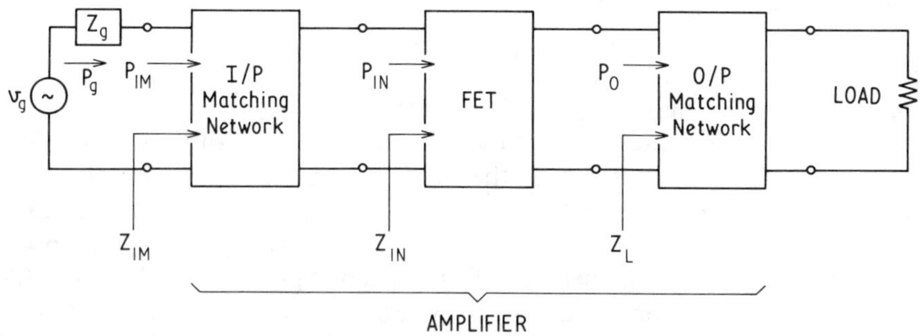

Fig. 9.A.3 Block diagram defining symbols for various power and impedance levels.

Now $G_{g\text{-}o}$ will be a maximum when the ratio (P_{IN}/P_g) is a maximum. For Z_g fixed, this condition occurs when Z_{IM} is adjusted to a value equal to the complex conjugate of Z_g, i.e.

$$\text{If } Z_{IM} = Z_g^*, \text{ then } P_{IM}/P_g = \max = 0.5$$

Thus

$$G_{g\text{-}o\ MAX} = G_{W\text{-MAX}}/2 \tag{9.A.4}$$

When $Z_{IM} = Z_g^*$, P_{IM} is termed "the power available from the generator".

In arriving at eqn (9.A.4), we have thus enforced the conditions that

(a) Z_L is conjugately matched to the FET output impedance and
(b) the amplifier is conjugately matched to the generator.

Under these conditions,

(c) the power delivered to Z_L (which is equal to the power delivered to the load for a lossless output matching network) is a maximum, and is said to be "the power available from the FET" and

(d) the power delivered to the amplifier input from the generator is a maximum, and is said to be "the power available from the generator".

If we now define the maximum available gain (MAG) as

$$\text{MAG} = \frac{\text{power available from FET}}{\text{power available from generator}} \tag{9.A.5}$$

then we can write (9.A.4) as

$$G_{\text{g-o MAX}} = G_{\text{W-MAX}}/2 = \text{MAG}/2 \tag{9.A.6}$$

since the MAG is identical to $G_{\text{W-MAX}}$.

The curiosity here is that $G_{\text{W-MAX}}$ was defined completely independently of any input matching considerations, whereas the definition of MAG apparently made reference to an input generator matching condition. It is important to realize, however, that the denominator in eqn (9.A.5) is still in the nature of a power input to the FET. Thus, viewed in isolation, the definition of MAG in eqn (9.A.5) is superfluous in the sense that it is already embodied in G_W a special case of the operating power gain (G_W) when the output is conjugately matched. The insistence on "power available from the generator" in the denominator of eqn (9.A.5) is an over-complication since "power input to the FET" would have served just as well, and would have given the same numerical result.

The real importance of the MAG is in the context of the power transfer from the generator to the output, eqn (9.A.6), for it is in seeking a maximum in $G_{\text{g-o}}$ that the conjugately-matched input condition implied by the denominator of eqn (9.A.5) becomes important.

9.A.4 Maximum available gain (MAG)

MAG has the physical significance discussed towards the end of Section 9.A.3, and is defined only for a FET which is inherently stable, i.e. for a FET for which the Linville stability factor C falls in the range $0 \leqslant C \leqslant 1$. Despite the fact that MAG is principally important in determining the overall power gain of which an amplifier might be capable (eqn (9.A.6)), it is nevertheless a quantity which can be regarded as belonging to a FET. MAG is given by

$$\text{MAG} = \left|\frac{y_{21}}{y_{12}}\right| \left[\frac{1}{C} - \left(\frac{1}{C^2} - 1\right)^{1/2}\right] \tag{9.A.7}$$

(see eqn (9.22) of the text for an expression for C).

In terms of equivalent circuit elements, the MAG is given by (where $\mathscr{L}_s$ is the source inductance):

$$\text{MAG} = \frac{(f_T/f)^2}{4g_o(R_g + R_i + R_s + \pi f_T \mathscr{L}_s) + 4\pi f_T C_{gd}(2R_g + R_i + R_s + 2\pi f_T \mathscr{L}_s)}$$

9.A.5 Maximum stable gain (MSG)

When the device is not inherently stable, i.e. C does not lie in the range $0 \leqslant C \leqslant 1$, the MAG is theoretically infinite (the FET can be made to deliver output power for no input power, given certain critical values for the input and output terminating impedances, i.e. the device will oscillate). The bracketed term in C in eqn (9.A.7) increases as C increases, and has a value of unity as C approaches unity. Thus the maximum gain which is possible with unconditional stability is

$$\mathrm{MSG} = \left|\frac{y_{21}}{y_{12}}\right| = \left|\frac{s_{21}}{s_{12}}\right| \tag{9.A.8}$$

9.A.6 Transducer power gain (G_T)

The transducer power gain is defined as

$$G_T = \frac{\text{Power delivered to load}}{\text{Power available from source}}$$

$$G_T = \frac{|s_{21}|^2(1-|\Gamma_g|^2)(1-|\Gamma_L|^2)}{|(1-s_{11}\Gamma_g)(1-s_{22}\Gamma_L)-s_{12}s_{21}\Gamma_L\Gamma_g|^2} \tag{9.A.9}$$

where (Fig. 9.A.1):

$$\Gamma_g = \frac{Z_g - Z_0}{Z_g + Z_0}$$

$$\Gamma_L = \frac{Z_L - Z_0}{Z_L + Z_0}$$

Apart from the MTG, the transducer power gain is probably the gain of most interest. G_T is of relevance where the input is not matched, so the input reflection coefficient is not small and power in the incident wave is reflected from the input. In such a circumstance, what we want from a circuit is typically a constant level of (amplified) output power for a constant level of power incident upon the network, regardless of how much of that incident power is reflected from (or otherwise not absorbed by) the network and transistor.

G_T is also of use where the device input is conjugately matched to the generator, but the load impedance varies (as, for instance, where the antenna (load) element in one channel of a phased array is exposed to fields from surrounding elements but the input termination remains a conjugate match (or approximately so). This last condition is only met for small tunability factors, δ (see Section 9.2.4)).

G_T is a property of the FET and circuit viewed as an overall network.

9.A.7 Maximum unilateral transducer power gain (MUG)

MUG is derived from G_T in two distinct and separate steps:

(i) s_{12} is set equal to zero (i.e. $s_{12} \approx 0$) without altering the other three S-parameters and

(ii) the generator and load reflection coefficients are so adjusted that

$$\Gamma_g = s_{11}^*; \; \Gamma_L = s_{22}^*$$

Inserting these conditions in eqn (9.A.9), we obtain

$$\mathrm{MUG} = \frac{|s_{21}|^2}{(1-|s_{11}|^2)(1-|s_{22}|^2)} \tag{9.A.10}$$

As eqn (9.A.10) involves the S-parameters of a FET only, it may be viewed as a property which can be ascribed to that FET.

Condition (i) is strictly not a valid approximation for the band of frequencies covered by present-day MMICs that are FET or HEMT based (i.e. s_{12} cannot be regarded as negligible), and so MUG must be seen as an overly optimistic estimate of the in-circuit gain which can be expected from a device.

Chapter 10

FETs in amplifiers

10.1 Introduction

The task of designing a monolithic microwave integrated circuit amplifier consists in part of applying the results of Chapters 6 and 7 to ensure the FETs and HEMTs around which design is done are well understood and technologically repeatable, of applying the results of Chapter 8 to synthesize such devices to meet particular MMIC specifications, partly of selecting an amplifier topology capable of providing the performance required and, lastly, of calculating all the circuit element values. Subsequent amendments to the first overall design may need to be made when the circuit is laid out in the general form shown in Fig. 3.1, because at that stage parasitic reactances and inter-component coupling can be quantified and allowed for†. The purpose of this chapter is to give the MMIC designer sufficient information to allow a selection of an amplifier topology, and also to allow a first estimate of circuit values in a form that could then be finalized by a computer-aided scheme (such as TOUCHSTONE© or SUPERCOMPACT©). The FET or HEMT is regarded as a designable part of the circuit in these procedures.

The FET and HEMT have an operating power gain versus frequency characteristic of the following fundamental form:

$$G_p = \left(\frac{f_T}{f}\right)^2 \frac{R}{(R_g + R_i + R_s)} \tag{10.1}$$

Equation (10.1) is a slight expansion of eqn (9.4) in that it allows for the source resistance R_s which exists in real FETs and HEMTs, but which was

†A three-dimensional computer solution of the electromagnetic field in MMICs is now available from advanced software packages. Such a tool is important to chip condensation and all the economic benefits that should follow from it.

omitted from eqn (9.4)†. Here, as before, R is the load on the output of the device, and strictly should include the output conductance of the transistor itself, g_o. To realize an overall flat gain from an amplifier therefore requires networks in series with the device to provide a complementary gain slope, i.e. a gain which rises with frequency or, more correctly, an attenuation which decreases with frequency.

The design of microwave broadband amplifiers is a challenging theoretical problem involving the threefold requirements of gain slope compensation, broad-band matching of the transistor's reactance and broad-band impedance transformation with large impedance ratios.

The amplifier is a more complicated broad-band matching problem than that considered in Section 9.2.3, because that problem involved a one-port. The amplifier, however, is a two port in which the input (port 1) is coupled to the output (port 2) by the gate-drain capacitance, C_{gd}. The match at port 1 is dependent upon the network connected to port 2 (and vice versa). A device to which the principles of tunability have been applied (Chapter 9) eases the design task. Further, the matching problem treated in Section 9.2.3 involved an ideal transformer to effect broad-band impedance transformation. The ideal transformer is a reciprocal, passive and lossless two-port network. In practice, ideal transformers have to be approximated by spiral transformers [10.1], which may eventually prove to be viable at tens of gigahertz. More commonly, matching networks have consisted of lumped inductors, capacitors, and transmission lines. Considerations of chip size dictate that the numbers of matching network components should be limited.

The concurrent tasks in amplifier design, therefore, are to equalize the frequency-dependent response of the transistor input and output networks, and to transform the impedance levels at the input and output to effect a conjugately matched condition for maximum power transfer from the generator to the transistor, and from the transistor to the load‡. At the same time it is necessary to achieve some prescribed overall power gain versus frequency characteristic, often a constant gain approximately independent of

†In terms of equivalent circuit elements, the MAG is given by

$$\mathrm{MAG} = \frac{(f_T/f)^2}{4g_o(R_g + R_i + R_s + \pi f_T \mathscr{L}_s) + 4\pi f_T C_{gd}(2R_g + R_i + R_s + 2\pi f_T \mathscr{L}_s)}$$

where $\mathscr{L}_s$ is the source inductance. This has the same dependence on frequency f as the simplified form, eqn (10.1).

‡Conjugate matching for maximum power transfer may not be feasible at frequencies in the low gigahertz range, especially if spiral transformers are not available, by virtue of the excessive chip area that would be occupied by the elements of the matching network. In such cases, design principles akin to low-frequency voltage-gain design can be employed, and any power gain deficiency made good by inclusion of an extra FET or two in the MMIC.

frequency. For low-noise amplifiers, the objective of the input network is not to provide a conjugate match but a pre-determined mismatch (eqn (6.53)). Regarding the FET or HEMT as a designable part of an MMIC while attempting to satisfy the foregoing requirements can:

(i) meet performance specifications that might otherwise be unattainable,

(ii) afford improved margins of process tolerance so that yield increases and

(iii) potentially save on the number of elements or transistors required in matching networks, thus leading to small-area chips and lower chip cost.

10.2 Amplifier topologies and design principles

In this section we consider the following amplifier types:

(a) reactively matched,
(b) lossy matched,
(c) feedback,
(d) distributed and
(e) balanced.

We shall use the terms "wideband amplifiers" and "broadband amplifiers" to refer to amplifiers with a bandwidth of typically one or two octaves, i.e. amplifiers which essentially have a passband characteristic. Such amplifiers are often of the reactively matched type and find application in phased-array radars, electronic counter-measure (ECM) systems, civil radar and satellite communication systems (SATCOMS). Amplifiers having bandwidths of a few tens of percent mostly employ reactive matching, and find application in communications such as direct broadcast from satellite (DBS) systems.

Ultra-broadband amplifiers, as a class, have a bandwidth of several octaves, and often have a low-frequency response down to dc, or at most a few hundred megahertz. Those circuits which respond at dc can rightly be regarded as low-pass. Ultra-broadband (or ultra-wideband) MMICs are required for electronic warfare (EW) systems, guided weapons (GW), instrumentation and pulse amplifiers.

As a guide, reactively-matched amplifiers allow the most gain to be derived from a given FET, and they also offer the lowest noise. Their bandwidth tends to be the most restricted, however. For all that, it is often worth exploring the gain and reflection coefficients which can be achieved using reactive matching even over quite wide bandwidths. Lossy matching is a broad-banding scheme, and can be viewed as an interleaving of two amplifiers in one: a reactively-matched circuit to handle the higher frequencies, and a resistively-coupled circuit to cater for the lower frequencies. Reasonable gain is available, but the noise properties of a purely reactively-matched amplifier are to some extent sacrificed. Feedback amplifiers allow broadband amplification, and can be designed for a measure of robustness against process variations. They are generally more noisy than reactively-matched amplifiers, and give less gain. Distributed amplifiers are the prime example of "additive" amplifiers, and offer ultra-broadband amplification with moderate gain and rather poor noise figure. They also tend to require large amounts of dc power per decibel of power gain produced. Balanced amplifiers are amplifiers that use two nominally identical MMIC chips on a hybrid MIC to achieve an excellent match over about an octave bandwidth with reasonable gain.

It is worth remarking that some of the applications mentioned earlier require good phase tracking characteristics of MMICs, in addition to specified gain (and perhaps power). In turn, such a requirement transfers to a requirement of phase characteristic repeatability from transistor-to-transistor, and from chip-to-chip, which places the most severe demands on technology. One way phase repeatability might be assured is to design the FETs (or HEMTs) in the MMIC to have only a small output phase shift (maybe 20° or less) at the highest operating frequency in the band of interest. Unfortunately, it can be shown that small output phase shift and high output power are directly conflicting requirements†, so that the realization of 6–18 GHz, 1 W power MMICs, for example, to work into a load of 50 Ω (or thereabouts) is a fundamentally difficult problem, as well as being very demanding of materials and process technology.

As reference to Fig. 9.17 will suggest, wideband-to-ultra-wideband amplifiers with reactive matching networks will have a fairly high reflection coefficient. To prevent the reflected energy re-entering the source (or generator), a balanced amplifier configuration is employed in which two nominally identical MMICs are fed, in parallel branches, from a quadrature coupler. The outputs of the chips are similarly combined in a second quadrature coupler. Any power reflected from the chips is then dumped in a resistor connected to the isolation port of the coupler. Balanced amplifiers will be discussed in section 10.2.7.

Difficulties of ultra-wideband matching can be overcome with a

†It is reasonable to suggest that much remains to be done to improve active layer uniformity, to improve gate length control, and to reduce the density of electrically active bulk- and surface-deep levels (or at least to improve their repeatability), in the interests of manufacturing MMICs with good phase tracking to deliver high power levels into loads of typically 50 Ω.

distributed amplifier topology, provided the artificial transmission lines can be designed to have a high enough cut-off frequency f_C (Fig. 9.14) to accommodate the band of interest. In this case in particular, the facility of being able to tailor the active devices is of potentially great benefit, as the example of Section 10.5 will illustrate. A high cut-off frequency implies, however, a small value for the FET input capacitance, most of which is comprised of space-charge capacitance in the shadow of the gate (eqn (6.10) for C_I). For a FET of a given gatelength, comparison of eqn (6.10) with eqn (6.18) will then show that a device having a relatively low input capacitance will also have a relatively low transconductance, so we arrive at the idea that the distributed amplifier is a topology requiring low gain, broadband active devices.

Somewhat in contrast to the distributed amplifier, feedback amplifiers generally require devices having high transconductance, as will be seen in Section 10.2.4. The implication is that FETs, or HEMTs, can be designed to render one or another of the available topologies viable, and to explore yield among the various options consistent with performance objectives.

10.2.1 Reactively matched amplifiers

To begin with, we examine some of the constraints which govern the gain and bandwidth, as well as reflection coefficient, when designing lossless matching networks as shown in Fig. 10.1. Replacing the transistor by the simple high pass C_2–R_2 series combination of Fig. 9.11, the input may be represented as in Fig. 10.2, in which the device capacitance is regarded as a

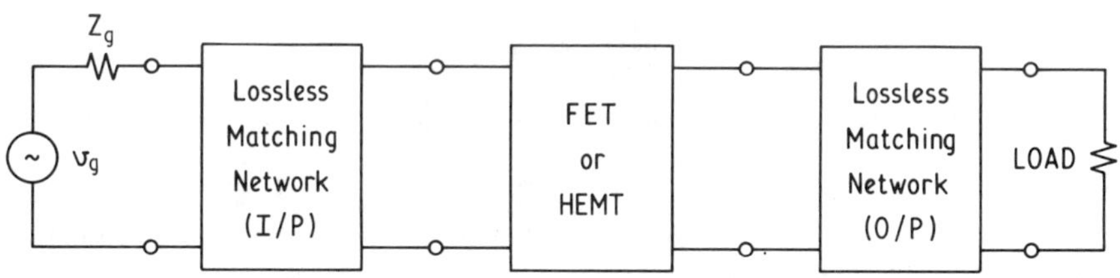

Fig. 10.1 Block diagram of FET (or HEMT) with matching networks.

part of the lossless input matching network. The scattering matrix $[S_M]$ is normalized to Z_g at port 1, and to R_2 at port 2. It may be shown that such a lossless, passive network has the following properties [10.2 p. 30], all of which are basically expressions of the requirement that the power reflected from a network which has no gain or loss should equal the power incident upon that network, minus the power transmitted:

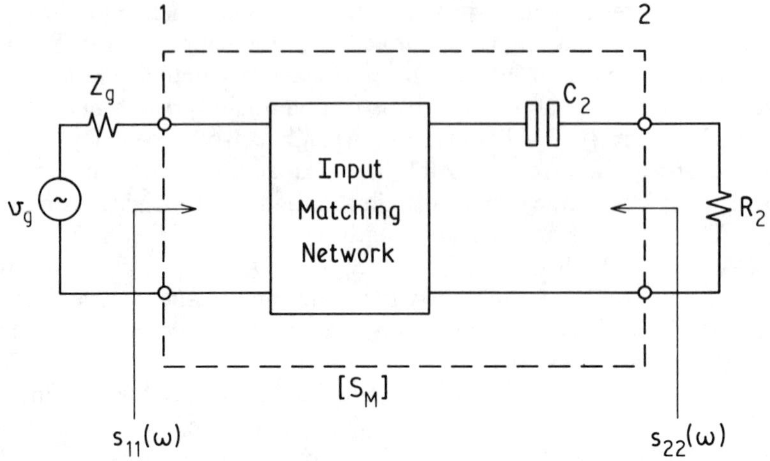

Fig. 10.2 Input matching network loaded with simplified equivalent input circuit of a FET.

$$|s_{11}(\omega)|^2 = 1 - |s_{12}(\omega)|^2 \tag{10.1}$$

$$|s_{11}(\omega)|^2 = 1 - |s_{21}(\omega)|^2 \tag{10.2}$$

$$|s_{22}(\omega)|^2 = 1 - |s_{21}(\omega)|^2 \tag{10.3}$$

$$|s_{22}(\omega)|^2 = 1 - |s_{12}(\omega)|^2 \tag{10.4}$$

Furthermore, all the elements of the matrix $[S_M]$ obey the condition:

$$|s_{ij}(\omega)| \leqslant 1 \tag{10.5}$$

Writing $|\Gamma|$ in place of $|s_{11}|$ and $G_M(\omega)$ in place of $|s_{21}(\omega)|^2$, eqn (10.2) becomes

$$|\Gamma|^2 = 1 - G_M(\omega) \tag{10.6}$$

i.e. the power transmission (or gain) of the matching network, G_M, and the reflection coefficient at the input to the matching network, are related.

The power gain characteristic of both FETs and HEMTs is given approximately by (eqns (9.5) and (9.6)):

$$G_p \approx (f_{max}/f)^2 = (\omega_{max}/\omega)^2 \tag{10.7}$$

Hence the operating power gain of the device loaded resistively has a gain roll-off of 6 dB per octave, or 20 dB per decade, as shown in Fig. 10.3. If the overall amplifier is to have a gain that is independent of frequency, then the gain slope of the transistor must be compensated by a complementary gain

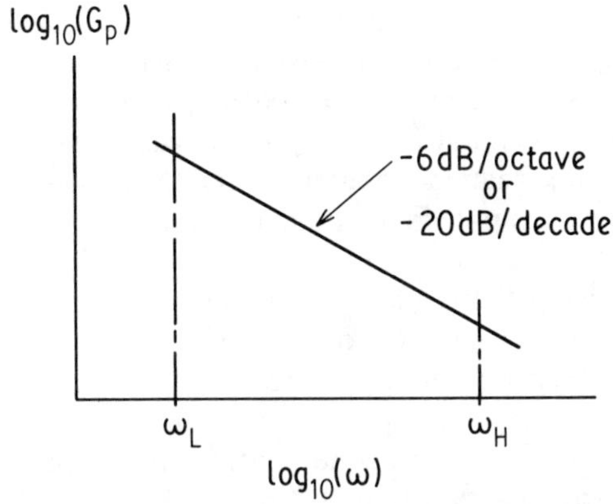

Fig. 10.3 Basic power gain versus frequency characteristic of unilateralized FET or HEMT.

slope in either or both of the matching networks. Assuming, for example, that compensation is to be achieved in the input matching network of Fig. 10.1, for flat overall gain over the band $\omega_L \leqslant \omega \leqslant \omega_H$ we must have

$$G_M(\omega) = (\omega/\omega_M)^2 \tag{10.8}$$

Since $G_M(\omega)$ must satisfy $G_M(\omega) \leqslant 1$ by virtue of eqn (10.5), it is necessary that

$$\omega_M \geqslant \omega_H$$

Writing $\Delta\omega = (\omega_H - \omega)$, eqn (10.8) becomes

$$G_M(\omega) = \left(\frac{\omega_H}{\omega_M}\right)^2 \left(1 - \frac{\Delta\omega}{\omega_H}\right)^2$$

or

$$G_M(\omega) = K\left(1 - \frac{\Delta\omega}{\omega_H}\right)^2 \qquad K \leqslant 1 \tag{10.9}$$

Substituting eqn (10.9) into eqn (10.6) then yields

$$|\Gamma| = \left[1 - K\left(1 - \frac{\Delta\omega}{\omega_H}\right)^2\right]^{1/2} \tag{10.10}$$

Equation (10.10) is plotted in Fig. 10.4 for the case of $K = 1$, which is the case that gives the minimum reflection coefficient. The overall gain is found from eqns (10.8) and (10.7) as

$$G(\omega) = G_M(\omega) G_p(\omega) = (\omega_{max}/\omega_M)^2$$
$$= (\omega_{max}/\omega_H)^2 \text{ for } K = 1$$

Figure 10.4 shows that gain leveling can be achieved only at the price of mismatch. As $(\Delta\omega/\omega_H)$ increases, the mismatch increases, i.e. as $\omega=(\omega_H-\Delta\omega)$ decreases, an increasing mismatch must be provided if flat overall gain is to be realized. In simple terms, flattening of the increasing power level evident in Fig. 10.3 at decreasing frequencies requires that the lossless matching network should reflect the excess power, as indicated in Fig. 10.5.

Curves of $|\Gamma|$ versus $(\Delta\omega/\omega_H)$ are also given in Fig. 10.4 for $K=0.4$, 0.6 and 0.8. For decreasing K, the gain of the lossless matching network decreases (eqn (10.9)). This reduction in gain of the lossless matching network can only be realized by increased reflection of power from the network, i.e. by $|\Gamma|$ increasing.

Figure 10.6 contrasts the reflection coefficient required for flat amplifier gain with the approximate worst reflection coefficient in the band when the amplifier is matched at a particular frequency with the type of matching network shown in Fig. 9.16. The two sets of curves do not intersect, meaning that the type of network in Fig. 9.16 cannot provide gain slope compensation at the same time as minimum reflection coefficient. For this reason it is

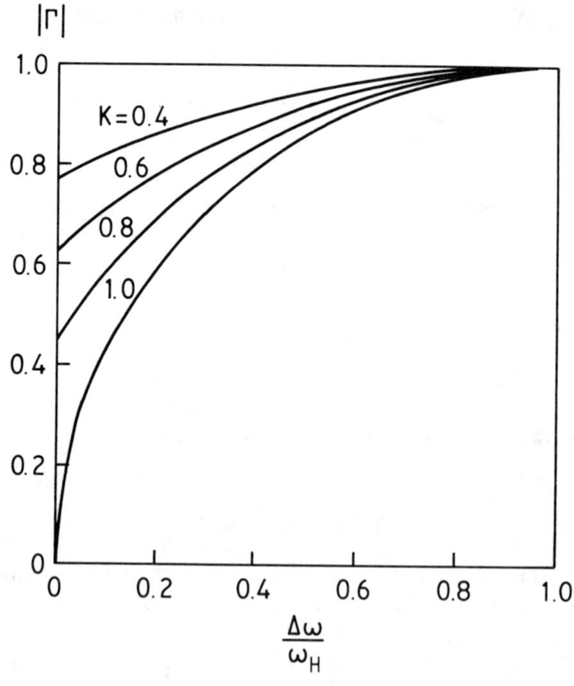

Fig. 10.4 Necessary reflection coefficient using a lossless equalizer for gain leveling as a function of equalizer gain (K) and fractional bandwidth $(\Delta\omega/\omega_H)$.

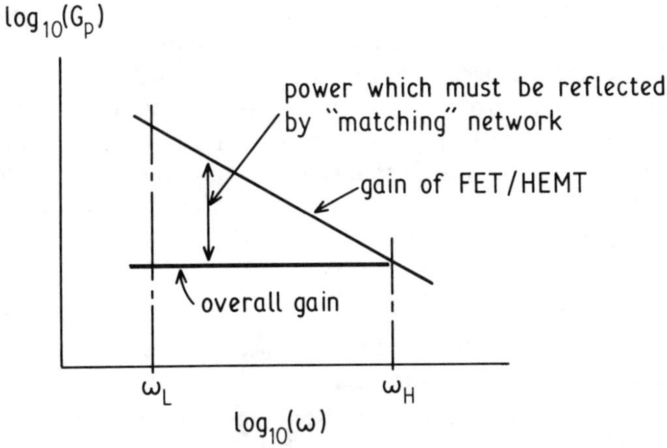

Fig. 10.5 Mechanism of achieving level gain using a lossless equalizer or matching network.

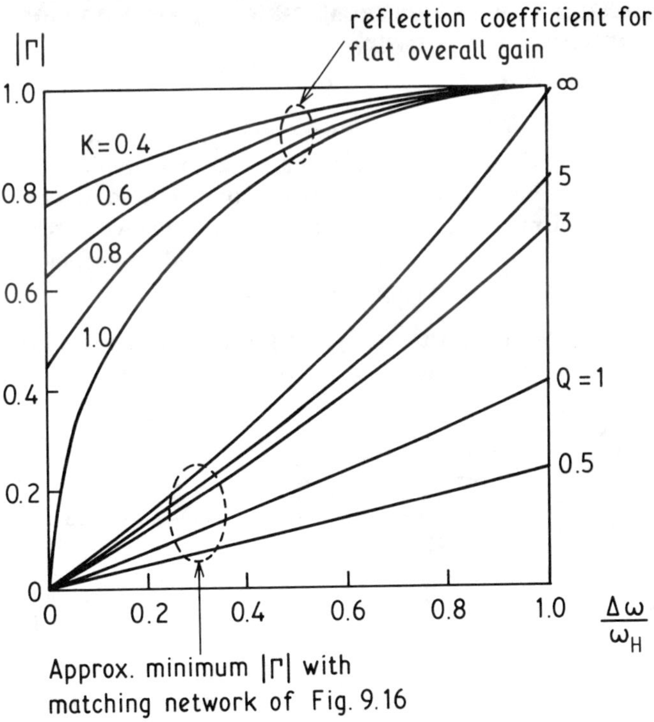

Fig. 10.6 Contrast of reflection coefficient for flat overall gain with case for matching using an idealized transformer.

necessary to achieve gain slope compensation and impedance matching in separate parts of the amplifier, for example by designing an inter-stage network to provide the necessary gain slope, eqn (10.8).

On fundamental grounds the curves of Fig. 10.4 show that, if an acceptable match (generally implying a VSWR of 2:1 or less) is to be achieved over a wide bandwidth, a means of absorbing the power not absorbed by the active part of the network must be provided. An example is shown in Fig. 10.7. The widely practised methods are:

(i) use of a circulator (as in Fig. 10.7),

(ii) a balanced amplifier configuration, with 3 dB quadrature couplers at the input and output (see Section 10.2.7),

(iii) absorption on-chip using a distributed transmission line matching scheme (as covered in Section 9.2.3) and

(iv) lossy matching.

10.2.2 Design of a two-element matching or gain slope compensation network

Although sophisticated procedures exist for synthesizing matching networks [10.2–10.4] such procedures may lead to matching networks that contain many elements. Requirements of chip compactness, however, render it relevant to examine matching networks containing only two or three elements. In this section we shall examine the capability of a simple two-element inductive network (Fig. 10.8) to provide:

(a) a conjugate match (naturally, over a limited bandwidth) and

(b) gain leveling over a wide bandwidth.

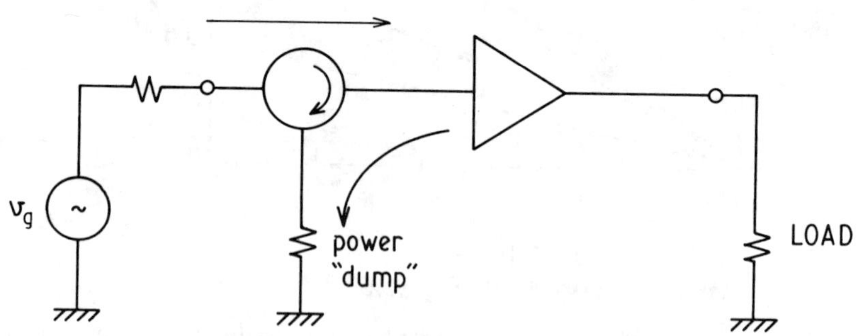

Fig. 10.7 Pseudo-matching by power dumping.

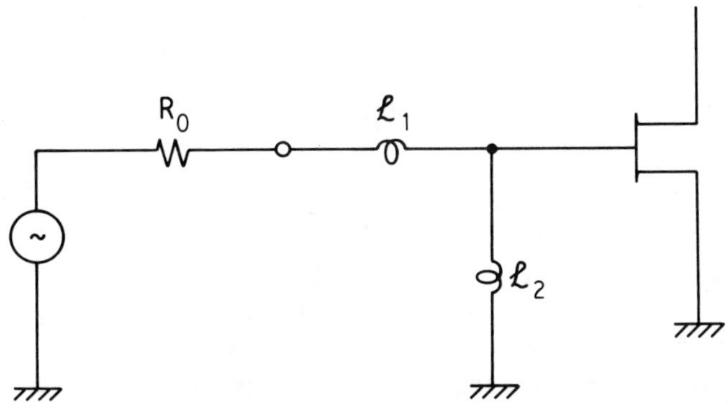

Fig. 10.8 Simple two-inductor matching network suitable for MMICs.

In the gain leveling application, we shall consider an example covering the frequency range 6–18 GHz, for which $\Delta\omega/\omega_H = 0.67$. From Fig. 10.4, the reflection coefficient can be expected to be $|\Gamma| \geqslant 0.95$ at some point in the band. For illustrative purposes, gain leveling will be achieved in the input network; it could equally well be achieved in an inter-stage network [10.4].

In real-world applications we are principally interested in achieving constant power out for a given constant power available from the source†. For the general connection shown in Fig. 10.9, the unilateral transducer power gain is, from eqn (9.A.9) with $s_{12} = 0$,

$$G_{TU} = \frac{\text{power delivered to load}}{\text{power available from generator}}$$

$$= \frac{|s_{21}|^2 (1 - |\Gamma_g|^2)(1 - |\Gamma_L|^2)}{|1 - s_{11}\Gamma_g|^2 |1 - s_{22}\Gamma_L|^2} \tag{10.11}$$

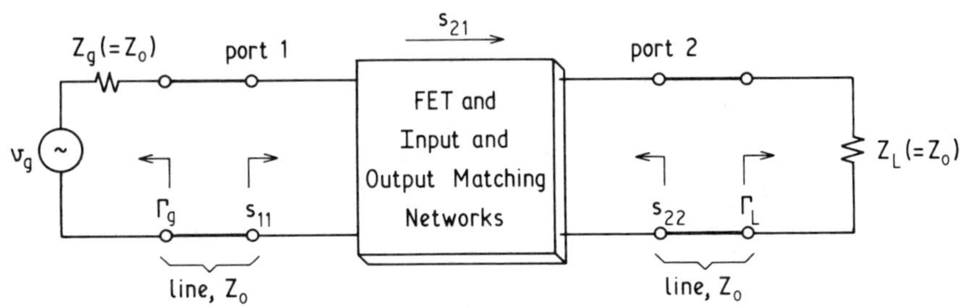

Fig. 10.9 FET and matching networks considered as a single, linear, two-port network.

†The word "constant" here means independent of frequency.

When Z_g and Z_L are matched to the interconnecting transmission line sections,

$$\Gamma_g = 0 \quad \text{and} \quad \Gamma_L = 0$$

and then,

$$G_{TU} = |s_{21}|^2 \tag{10.12}$$

It is important to emphasize, as Fig. 10.9 stresses, that s_{21} here is the forward transfer scattering parameter of the FET and any input and output matching networks considered all together as a linear two-port.

In this circumstance, the requirement for constant power to be developed in the load for a given constant power available from the source can be satisfied by designing the circuit to have $|s_{21}|^2$ independent of frequency, or approximately so. To tackle the gain leveling task, therefore, we shall have to design the $|s_{21}|$ of the FET and matching circuit to be independent of frequency. Note, however, that the power in the reverse wave reflected from port 1 of the amplifier is ignored by $|s_{21}|^2$, so $|s_{21}|^2$ is not a good measure of the operating power gain.

The operating power gain may be written [10.5] as

$$G_W = \frac{\text{power delivered to load}}{\text{power input to amplifier}}$$

$$G_W = \frac{|s_{21}|^2\,(1-|\Gamma_L|^2)}{(1-|s_{11}|^2)+|\Gamma_L|^2(|s_{22}|^2-|D|^2)-2\mathrm{Re}(\Gamma_L N)} \tag{10.13}$$

where

$$D = s_{11}s_{22} - s_{12}s_{21} \quad \text{and} \quad N = s_{22} - Ds_{11}{}^*$$

Hence, when $\Gamma_g = 0$ and $\Gamma_L = 0$, G_W reduces to

$$G_W = \frac{|s_{21}|^2}{1-|s_{11}|^2} \tag{10.14}$$

Again, s_{21} and s_{11} are scattering parameters not just of the FET but of the FET and matching networks all considered together (refer to Fig. (10.9)).

Taking the ratio of the definitions of G_{TU} and G_W in eqns (10.11) and (10.14) in the case $\Gamma_g = 0$ and $\Gamma_L = 0$,

$$P_{IN} = P_{AV}(1-|s_{11}|^2) \tag{10.15}$$

where P_{AV} is the power available from the generator, and P_{IN} is the power input to the network. Equation (10.15) is a useful basis for calculating $|s_{11}|$, as we shall shortly see. It also serves as a reminder that, when the input is conjugately matched, $|s_{11}|=0$ and the operating power gain (eqn (10.14)) is equal to the unilateral transducer power gain (eqn (10.12)). Viewed another way, if the device is not conjugately matched to the generator (so $|s_{11}|\neq 0$), eqn (10.14) for the operating power gain, G_W, gives a numerical value equal to the power gain that could be achieved if the input were now, in fact, conjugately matched with a lossless matching network.

Lastly, since the interest is principally in achieving constant amplified power out, P_O, for a constant power available from the generator P_{AV}, using a power dump (e.g. Fig. 10.7) is the only way of achieving a reasonable reflection coefficient at the input to the amplifier (presuming the gain leveling is achieved in the input network).

We consider the network shown in Fig. 10.10, and aim to deduce analytic expressions for the values of $\mathscr{L}_1$ and $\mathscr{L}_2$ required to match the device conjugately to the generator, i.e. Z_{IN} in Fig. 10.11, must equal $R_g=R_0$. The impedance Z may be written from straightforward network analysis as

$$Z=\frac{(\omega^2C_2\mathscr{L}_2)^2+j\omega\mathscr{L}_2[1-\omega^2C_2\mathscr{L}_2+(\omega C_2R_2)^2]}{(1-\omega^2C_2\mathscr{L}_2)^2+(\omega C_2R_2)^2} \tag{10.16}$$

Since $Z_{IN}=j\omega\mathscr{L}_1+Z$ and we require $\mathrm{Re}(Z_{IN})=R_0$, we must have

$$\frac{x^2R_2}{(1-x)^2+D^2}=R_0$$

where $x=\omega^2C_2\mathscr{L}_2$, and $D=(1/Q)=\omega C_2R_2$. Solving for x,

$$x=\frac{(R_0/R_2)+\{(R_0/R_2)^2-(R_0/R_2)[(R_0/R_2)-1](D^2+1)\}^{1/2}}{(R_0/R_2)-1} \tag{10.17}$$

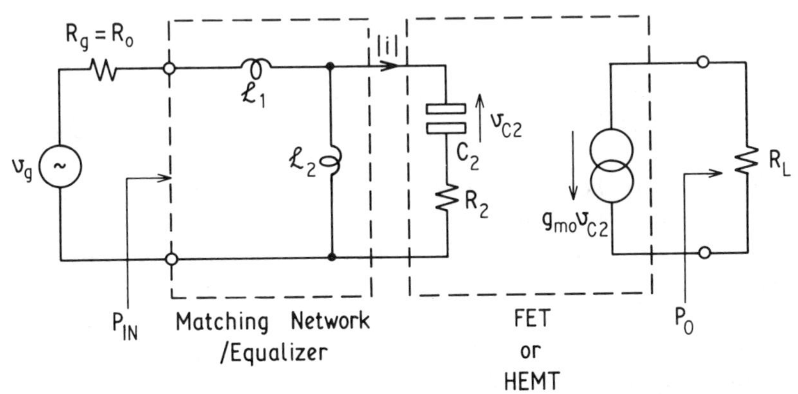

Fig. 10.10 Basic FET unilateral equivalent circuit and two-element matching network for analysis.

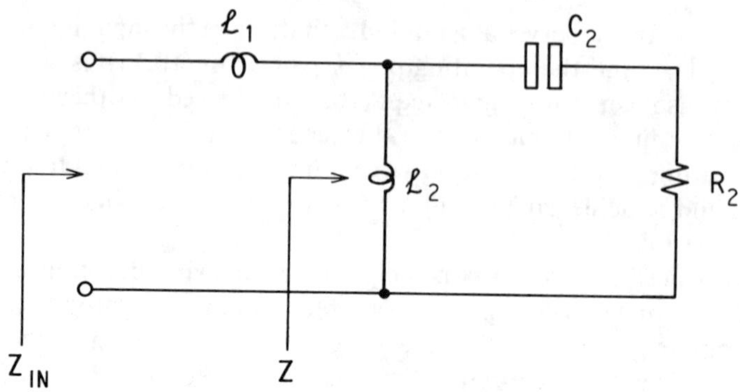

Fig. 10.11 Definition of Z_{IN} and Z for analysis.

Note that for the assumed network to be capable of providing a conjugate match, x must be positive and real, requiring that (from eqn (10.17)),

$$R_0 > R_2$$

$$\text{and} \qquad Q > [(R_0/R_2) - 1]^{1/2} \qquad \text{where} \qquad Q = (1/D)$$

Subject to these conditions being satisfied, the solution for x from eqn (10.17) gives $\mathscr{L}_2$ as

$$\mathscr{L}_2 = x/\omega^2 C_2 \tag{10.18}$$

The second major condition required for a conjugate match is that the imaginary part of Z_{IN} in Fig. 10.11 should equal zero. Recalling that $Z_{IN} = j\omega\mathscr{L}_1 + Z$, with Z given by eqn (10.16), we must therefore have

$$\mathscr{L}_1 = \mathscr{L}_2 \left[\frac{x - 1 - D^2}{(1-x)^2 + D^2} \right] \tag{10.19}$$

Equations (10.18) and (10.19) give the values of the matching network elements required to match the FET or HEMT conjugately to the generator; x is found from eqn (10.17).

For many applications this suffices; the overall network can now be simulated using a proprietary network analysis program such as TOUCHSTONE© or SUPERCOMPACT©, and the overall behavior checked for sensitivity to element tolerances and so on. For completeness, however, we shall continue to develop equations to calculate the unilateral transducer power gain, G_{TU}, and the input reflection coefficient magnitude, $|s_{11}|$, thus making available an equation set from which a program for a Hewlett Packard HP-41C, for example, can be readily written. The first step is to transform the circuit of Fig. 10.10 into the form of Fig. 10.12 using

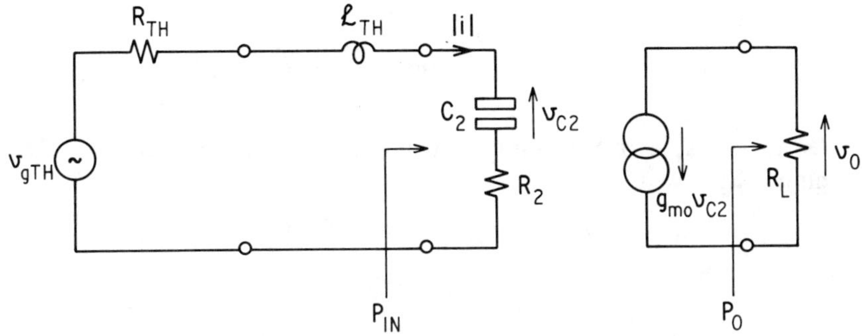

Fig. 10.12 Thévenin resolved network.

Thévenin's theorem. Then, for Fig. 10.12, the following set of equations applies:

$$v_{gTH} = v_g \frac{\omega \mathcal{L}_2}{[R_0{}^2 + \omega^2(\mathcal{L}_1 + \mathcal{L}_2)^2]^{1/2}}$$

$$R_{TH} = \frac{\omega^2 \mathcal{L}_2{}^2 R_0}{R_0{}^2 + \omega^2(\mathcal{L}_1 + \mathcal{L}_2)^2}$$

$$\mathcal{L}_{TH} = \frac{\mathcal{L}_2[R_0{}^2 + \omega^2 \mathcal{L}_1(\mathcal{L}_1 + \mathcal{L}_2)]}{R_0{}^2 + \omega^2(\mathcal{L}_1 + \mathcal{L}_2)^2}$$

$$X_{\mathcal{L}TH} = j\omega \mathcal{L}_{TH} \qquad X_{C_2} = \frac{1}{\omega C_2}$$

$$|i| = \frac{v_{gTH}}{[(R_{TH} + R_2)^2 + (X_{\mathcal{L}TH} - X_{C_2})^2]^{1/2}}$$

$P_{IN} = |i|^2 R_2$ (power input to network)

$P_{AV} = \dfrac{|v_g|^2}{4R_0}$ (power available from generator)

$|v_{C2}| = |i| X_{C2}$ (signal across C_2)

$|v_o| = g_{m0}|v_{C2}| R_L$ (signal across load)

$P_L = \dfrac{|v_o|^2}{R_L}$ (power in load)

$G_{TU} = 10 \log_{10}\left[\dfrac{P_L}{P_{AV}}\right]$ (unilateral transducer power gain) and, from eqn (10.15),

$$|s_{11}| = [1 - (P_{IN}/P_{AV})]^{1/2} \qquad \text{(input reflection coefficient of FET and matching network)}$$

As an example of matching network design we consider the task of matching a FET with

$$C_2 = 0.2 \text{ pF} \qquad R_2 = 10\,\Omega$$

$$g_{m0} = 30 \text{ mS}$$

at 18 GHz to a generator with $R_g = 50\,\Omega$. The load is $R_L = 50\,\Omega$. The response will be calculated using the foregoing equation set (and checked using TOUCHSTONE©) over the frequency range 6–18 GHz for later comparison with the case where $\mathscr{L}_1$ and $\mathscr{L}_2$ are re-designed to achieve gain leveling. Applying eqns (10.17–10.19) gives

$$\mathscr{L}_1 = 0.78 \text{ nH}$$

$$\mathscr{L}_2 = 0.684 \text{ nH}$$

$|s_{11}|$ and G_{TU} are plotted in Fig. 10.13. The plots confirm that the FET is conjugately matched at 18 GHz, with a gain of 9.44 dB. A point of interest is that the maximum gain occurs just below the frequency at which the device is conjugately matched to the source.

We turn next to the use of the $\mathscr{L}_1$–$\mathscr{L}_2$ input network, Fig. 10.8, for gain slope compensation or gain leveling. To achieve gain leveling, the values of $\mathscr{L}_1$ and $\mathscr{L}_2$ must be changed. Compared with the gain response of Fig. 10.13, the gain needs to be lifted at the lower end of the band, and possibly depressed at the higher end of the band. Such a change requires $\mathscr{L}_1$ to be reduced in value, and $\mathscr{L}_2$ increased in value. Simple iterative schemes can be developed to determine new values. With

$$\mathscr{L}_1 = 0.4 \text{ nH} \qquad \text{and} \qquad \mathscr{L}_2 = 1.8 \text{ nH}$$

the response becomes that shown in Fig. 10.14. Compared with the previous case, $|s_{11}|$ is reduced in value below about 15 GHz (i.e. the match is improved), whereas above 15 GHz $|s_{11}|$ is increased in value (worse match). In Fig. 10.14 is also shown the effect on the gain of capacitance errors in the FET (as might arise from gate length errors). Very roughly, such response variations could be expected for gate length variations in the range 0.4 μm to 0.6 μm if the nominal (design) FET is 0.5 μm.

From these results it can be seen that a reasonably flat gain can be realized over a wide bandwidth, even with a simple two-element matching network. The gain ripple is about 1.5 dB as designed. The consequences of FET

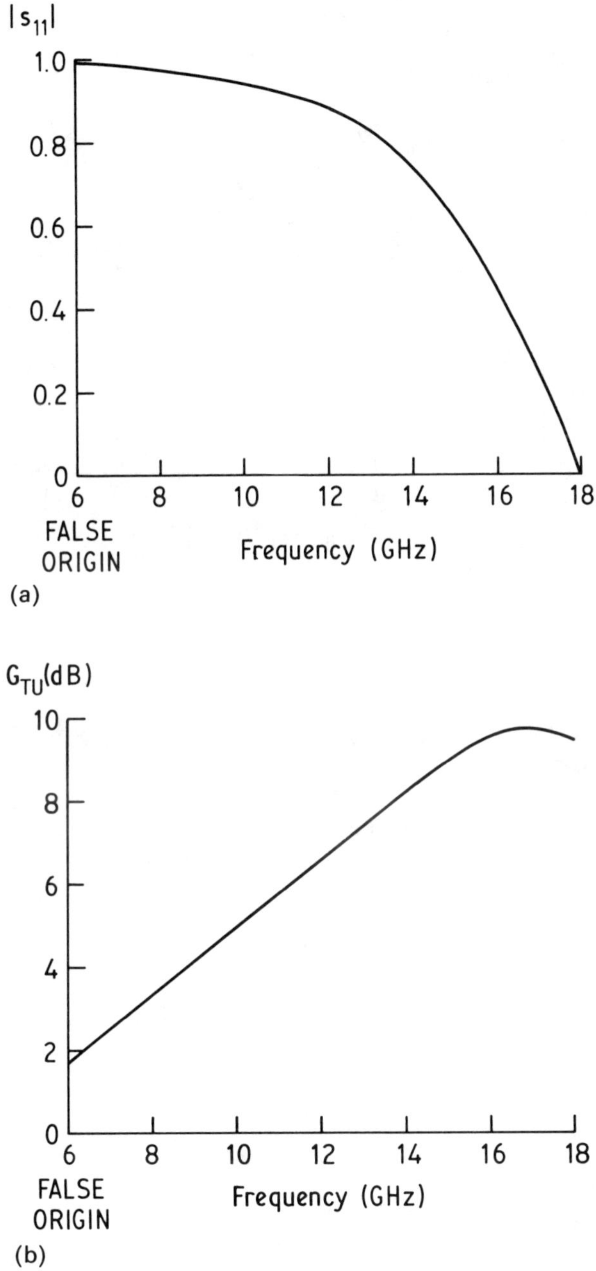

Fig. 10.13 Input reflection coefficient (a) and unilateral transducer power gain (b) for example amplifier matched at 18 GHz.

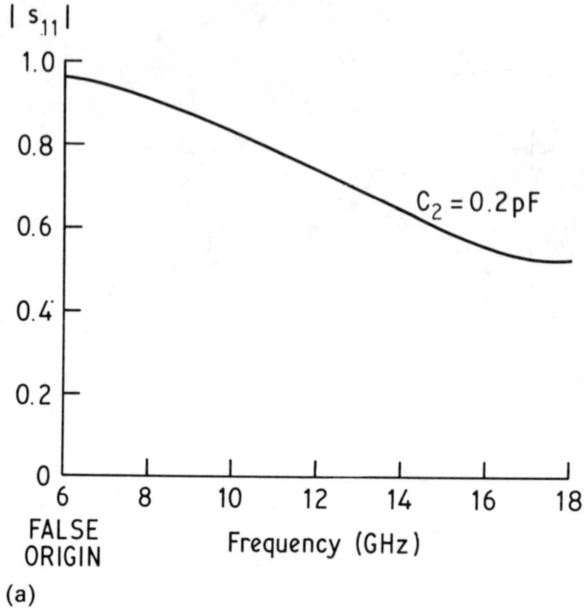

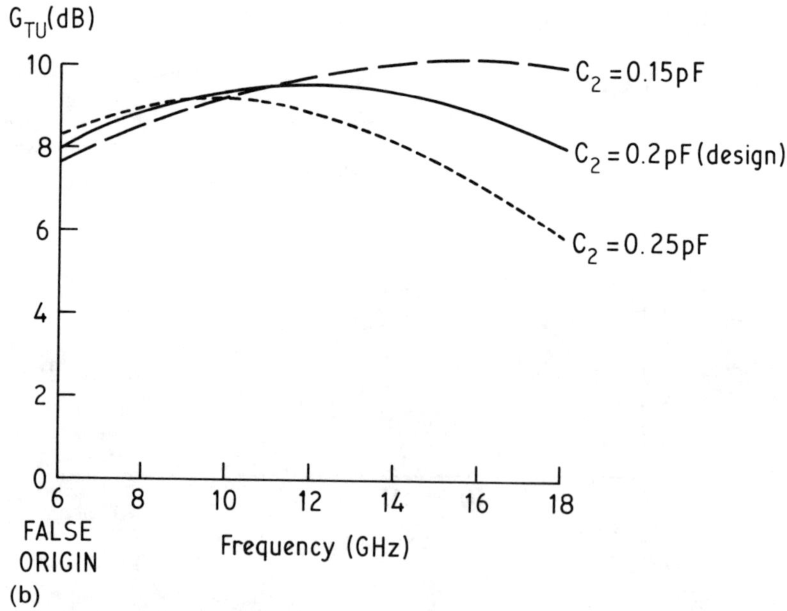

Fig. 10.14 Input reflection coefficient (a) and unilateral transducer power gain (b) for example amplifier with 8 dB gain over 6–18 GHz.

variations are reasonably severe, however: the in-band gain variation can become as large as 3 dB (c.f. the case with $C_2 = 0.25$ pF), and the response peak shifts up and down in the band. Nevertheless, when allied to good FET process control and local feedback design techniques, satisfactory chip response can be achieved. Either spiral inductors or transmission-line inductive elements, commonly microstrip, can be used to realize $\mathscr{L}_1$ and $\mathscr{L}_2$, depending upon the frequency at which the circuit must operate.

Figure 10.15 shows examples of other lossless matching networks, each of which might find application in hybrid MIC realizations, but which might be less extensively employed in MMICs because they consume too much chip area. In MMIC engineering, small chip size is paramount. An extra degree of flexibility is available to the MMIC designer by designing the active device concurrently with the passive network—an option which is rarely open to the MIC engineer. Even so, there is limited room for maneuver in some MMIC design tasks because only the values of the FET equivalent circuit can be adjusted by device design, not the configuration of the circuit.

10.2.3 Lossy matched amplifiers

Because the gain of a FET or HEMT decreases roughly at 6 dB per octave

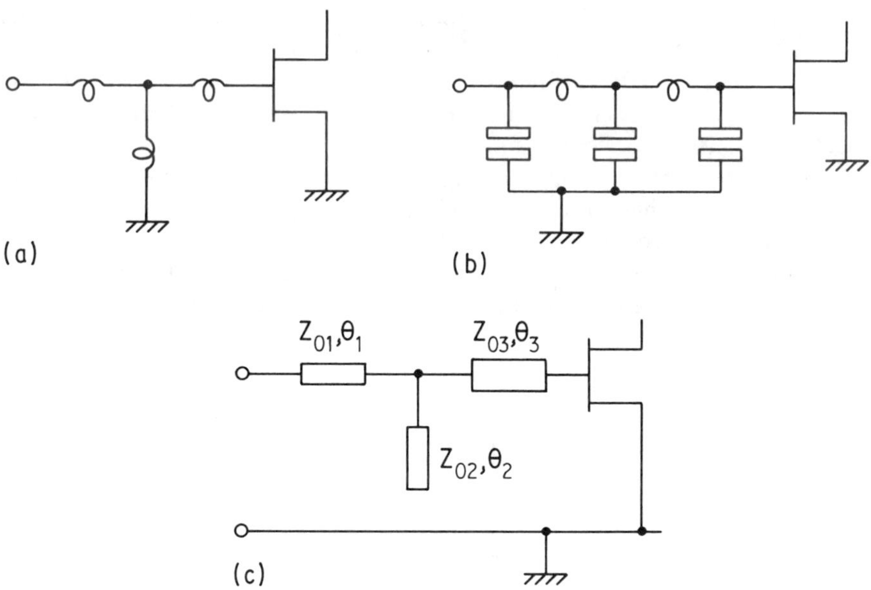

Fig. 10.15 Various reactive matching network configurations.

as shown in Fig. 10.3, an amplifier gain characteristic which is approximately independent of frequency requires power to be reflected at the lower frequencies, generally as in Fig. 10.5. In other words, the input match or interstage match, as appropriate, must become worse at low frequencies. In the example of Fig. 10.14, $|s_{11}|$ illustrates this behavior.

One means of improving the match at low frequencies is to provide resistive elements which become more tightly coupled to the main signal path through the amplifier at low frequencies (Fig. 10.16). M_1, M_2 and M_3 are lossless impedance matching networks, and F_1, F_2 and F_3 are low-pass lossless impedance transformers. At high frequencies in the band specified, the transformers F_1, F_2 and F_3 transform the values of R_1, R_2 and R_3 into high impedances that do not load the signal path through the amplifier. In this limit the circuit behaves as a reactively-matched amplifier. At low frequencies in the band of interest, the transformers F_1, F_2 and F_3 are designed to be electrically short, thus coupling R_1, R_2 and R_3 across the main signal path. The gain of the amplifier, and the matches at the input, at the inter-stage, and at the drain of the second FET, are all then dependent upon the resistor values. In a sense, therefore, a lossy matched amplifier consists of two interleaved circuits both having the same gain: a resistively-coupled amplifier providing gain and matching at the lower frequencies in the band to be covered, and a reactively-matched amplifier providing gain and matching at the higher frequencies. The circuit is thus applicable to flat-gain, ultra-broadband amplification [10.6]. An MMIC of this type is the TGA8014 GaAs power MMIC described by Nelson *et al.* [10.7]. A later MMIC also using lossy-reactive matching is the EG8017 chip [10.8]. Both circuits are power MMICs covering the 6–18 GHz band; both were designed and fabricated by Texas Instruments, with electronic warfare applications in mind.

At the lowest frequencies in the range of operation, a single-stage lossy matched amplifier has an equivalent circuit as in Fig. 10.17. A resistor of value Z_0 has been shown connected across the FET input terminals to provide input matching, and a resistor also of value Z_0 is connected across the drain-source terminals to provide output matching. The power delivered to the external

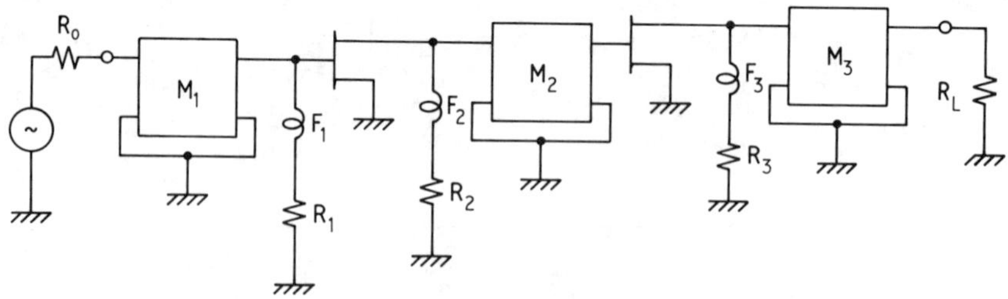

Fig. 10.16 Configuration of lossy matched amplifier.

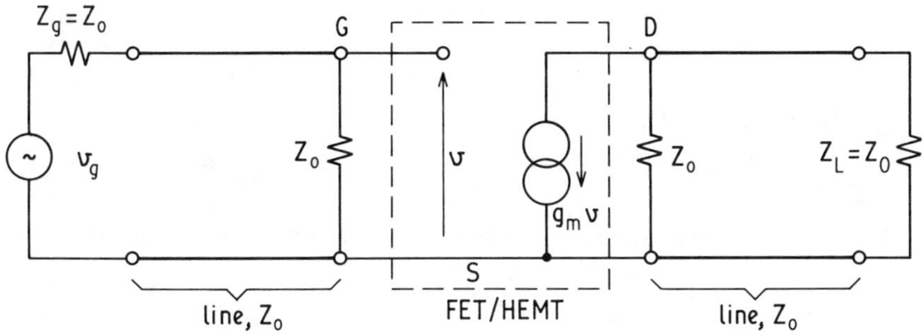

Fig. 10.17 Single-stage lossy matched amplifier at dc.

load, Z_L, is then $(g_m v/2)^2 Z_0$, and the power available from the generator is (v^2/Z_0). Hence the transducer power gain comes to

$$G_{TU} = (g_m Z_0/2)^2 \tag{10.20}$$

For $Z_0 = 50\ \Omega$, a transconductance of 80 mS is required to give a gain of 6 dB. Typically, this in turn implies a FET with a total gatewidth of 0.5 mm to 1 mm. Lossy matching, therefore, is not appropriate for small-signal ultra-broadband amplifiers where the FETs in each stage may have a gatewidth of, perhaps, 150 μm to 300 μm, and a corresponding transconductance of around 15 mS to 30 mS.

10.2.4 Feedback amplifiers

In this section we shall be concerned with feedback amplifiers of the type shown in Fig. 10.18. Our purpose will be to illustrate a method of first-trial design, finding circuit values which can then be iterated by computer methods

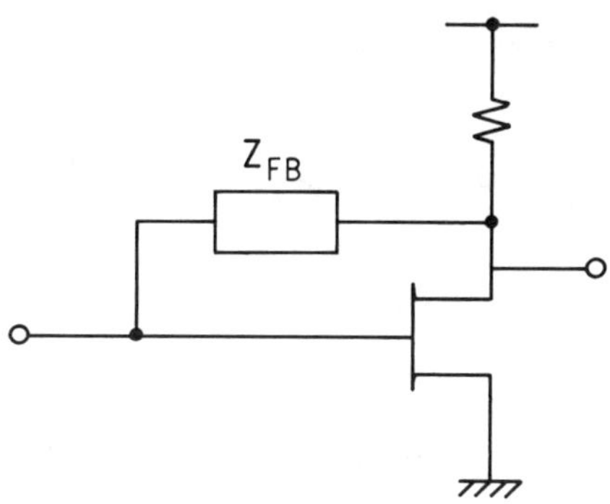

Fig. 10.18 Shunt feedback single-stage amplifier.

to achieve the performance desired. The connection has the following attributes.

(a) If offers gain equalization over a fairly wide bandwidth.

(b) It can improve the $|s_{21}|$ at the highest frequencies, compared with the bare FET.

(c) The matches can be improved concurrently with (a) and (b), thus simplifying the task of matching network design.

(d) It reduces the sensitivity of the overall circuit to device variations (i.e. the circuit offers a degree of robustness against process tolerances).

As so often is the case, these benefits are not entirely without cost. At low frequencies the gain is degenerated by feedback (we shall see how gain equalization is achieved partly by this mechanism). Secondly, devices having a high transconductance are required, especially if frequencies approaching dc are to be amplified with matched circuits. We shall deal briefly with this point first.

Figure 10.19 shows an equivalent circuit of a FET (or HEMT) at dc with a feedback resistor R_{FB} connected. The load impedance is assumed to be a general value $Z_L = 1/G_L$. Straightforward circuit analysis gives the input resistance of the device with feedback as

$$R_{in} = \frac{v}{i} = \frac{1 + (g_o + G_L)R_{FB}}{g_m + g_o + G_L} \tag{10.21}$$

If the input is to be matched to Z_0 in this limit, the requirement $R_{IN} = Z_0$ gives, for R_{FB},

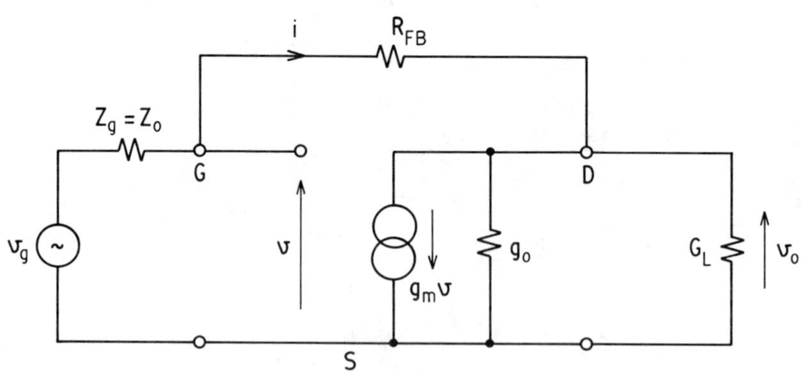

Fig. 10.19 dc analysis of shunt feedback amplifier.

$$R_{\mathrm{FB}}=\frac{g_{\mathrm{m}}+g_{\mathrm{o}}+G_{\mathrm{L}}-G_0}{G_0(g_{\mathrm{o}}+G_{\mathrm{L}})} \tag{10.22}$$

where $G_0=1/Z_0$.

The power gain of the circuit under the matched input condition eqn (10.22), is found as follows. Assuming Z_{L} to be real only, the output power is

$$P_{\mathrm{O}}=v_{\mathrm{o}}^2/Z_{\mathrm{L}}=v_{\mathrm{o}}^2 G_{\mathrm{L}} \tag{10.23}$$

where

$$v_{\mathrm{o}}=\frac{-g_{\mathrm{m}}v+i}{g_{\mathrm{o}}+G_{\mathrm{L}}}$$

From eqn (10.21),

$$i=\frac{v(g_{\mathrm{m}}+g_{\mathrm{o}}+G_{\mathrm{L}})}{1+(g_{\mathrm{o}}+G_{\mathrm{L}})R_{\mathrm{FB}}}$$

Substituting,

$$\frac{v_{\mathrm{o}}}{v}=\frac{1}{g_{\mathrm{o}}+G_{\mathrm{L}}}\left\{\frac{-g_{\mathrm{m}}(g_{\mathrm{o}}+G_{\mathrm{L}})R_{\mathrm{FB}}+g_{\mathrm{o}}+G_{\mathrm{L}}}{1+(g_{\mathrm{o}}+G_{\mathrm{L}})R_{\mathrm{FB}}}\right\}$$

Constraining the value of R_{FB} to the matching condition, eqn (10.22),

$$\frac{v_{\mathrm{o}}}{v}=\frac{G_{\mathrm{o}}-g_{\mathrm{m}}}{g_{\mathrm{o}}+G_{\mathrm{L}}} \tag{10.24}$$

The power available from the generator is $P_{\mathrm{AV}}=v^2/Z_0$. For the input matched condition the transducer power gain and the operating power gain are equal. Calling this power gain G_{p},

$$G_{\mathrm{p}}=\frac{P_{\mathrm{O}}}{P_{\mathrm{AV}}}=\left(\frac{G_{\mathrm{o}}-g_{\mathrm{m}}}{g_0+G_{\mathrm{L}}}\right)^2 Z_0 G_{\mathrm{L}} \tag{10.25}$$

When $g_{\mathrm{o}} \ll G_{\mathrm{L}}$,

$$G_{\mathrm{p}}\approx(G_{\mathrm{o}}-g_{\mathrm{m}})^2 Z_0 Z_{\mathrm{L}}$$

From these equations it is clear that to exhibit power gain we must have $g_{\mathrm{m}}>G_{\mathrm{o}}$. For example, if $Z_{\mathrm{L}}=Z_0=50\ \Omega$, a gain of 6 dB requires a FET with $g_{\mathrm{m}}=60$ mS.

Most design tasks, however, require amplification of signals over a band of microwave frequencies where the electrical parameters of the device are strong functions of frequency. To deal with this case, we need to develop approximate methods of analysis and design which offer good physical insight into the effects of feedback. An extremely useful graphical technique has been published by Perez and Ortega [10.9], whose method we shall now follow.

In Fig. 10.20 is shown a schematic of the FET represented by its $[Y]$ matrix with the feedback network represented by an admittance matrix $[Y']$ connected around it. If Z_{FB} in Fig. 10.18 is a single resistor, then

$$[Y'] = \begin{bmatrix} y'_{11} & y'_{12} \\ y'_{21} & y'_{22} \end{bmatrix} = \begin{bmatrix} 1/R_{FB} & -1/R_{FB} \\ -1/R_{FB} & 1/R_{FB} \end{bmatrix} \tag{10.26}$$

The overall admittance matrix of the FET with feedback is given by the sum of the two individual matrices:

$$[Y^T] = [Y] + [Y'] \tag{10.27}$$

The next stage in the development requires us to return to the general relationships which exist between S- and Y-parameters. In terms of Y-parameters normalized to Y_0, Section 9.A.2 of the appendix to Chapter 9 gives the S-parameters of a linear two port network as (using capital Y to distinguish the normalized form from the actual parameter y),

$$s_{11} = \frac{(1 - Y_{11})(1 + Y_{22}) + Y_{12}Y_{21}}{\Delta} \tag{10.28}$$

$$s_{12} = \frac{-2Y_{12}}{\Delta} \tag{10.29}$$

$$s_{21} = \frac{-2Y_{21}}{\Delta} \tag{10.30}$$

$$s_{22} = \frac{(1 + Y_{11})(1 - Y_{22}) + Y_{12}Y_{21}}{\Delta} \tag{10.31}$$

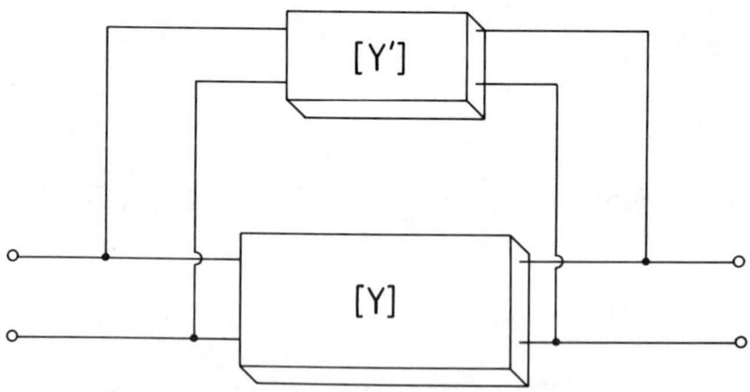

Fig. 10.20 Representation of FET and feedback network by Y-matrices for analysis.

$$\Delta = (1+Y_{11})(1+Y_{22}) - Y_{12}Y_{21} \tag{10.32}$$

Using the notation $[\mathbf{S}^{\mathrm{T}}]$ to refer to the scattering matrix of the FET with feedback, we may deduce some simple relationships for $(2s_{11}{}^{\mathrm{T}}/s_{21}{}^{\mathrm{T}})$, $(2s_{22}{}^{\mathrm{T}}/s_{21}{}^{\mathrm{T}})$ and $(2/s_{21}{}^{\mathrm{T}})$ providing certain approximations apply regarding the magnitude of the Y'-parameters relative to the Y-parameters, as follows:

$$Y'_{11} \ll Y_{11}$$

$$Y'_{22} \ll Y_{22}$$

$$Y'_{21} \ll Y_{21}$$

$$\text{Also} \qquad Y_{11} \ll Y_{21} \qquad \text{and} \qquad Y_{22} \ll Y_{21}$$

Although it is not always clear that these assumptions are valid, we shall nevertheless continue with them to gain the clear physical insight which follows from them. Dividing eqn (10.28) by eqn (10.30) results in

$$\frac{s_{11}}{s_{21}} = \frac{(1-Y_{11})(1+Y_{22}) + Y_{12}Y_{21}}{-2Y_{21}} \tag{10.33}$$

But, from eqn (10.27),

$$Y_{11}{}^{\mathrm{T}} = Y_{11} + Y'_{11}$$

$$Y_{22}{}^{\mathrm{T}} = Y_{22} + Y'_{22}$$

Hence, using the assumption that $Y'_{21} \ll Y_{21}$,

$$\frac{s_{11}{}^{\mathrm{T}}}{s_{21}{}^{\mathrm{T}}} \approx \frac{(1-Y_{11}-Y'_{11})(1+Y_{22}+Y'_{22}) + (Y_{12}+Y'_{12})Y_{21}}{-2Y_{21}}$$

Neglecting product terms in $Y'_{11}Y'_{22}$, this may be expanded into

$$\frac{s_{11}{}^{\mathrm{T}}}{s_{21}{}^{\mathrm{T}}} \approx \frac{(1-Y_{11})(1+Y_{22}) + Y_{12}Y_{21}}{-2Y_{21}} + \frac{-Y'_{11}(1+Y_{22}) + Y'_{22}(1-Y_{11}) + Y'_{12}Y_{21}}{-2Y_{21}}$$

Substituting eqn (10.33), and using the fact that $Y'_{11} = -Y'_{12}$, $Y'_{22} = -Y'_{12}$,

$$\frac{2s_{11}{}^{\mathrm{T}}}{s_{21}{}^{\mathrm{T}}} \approx \frac{2s_{11}}{s_{21}} - Y'_{12}\left(1 + \frac{Y_{11}+Y_{22}}{Y_{21}}\right)$$

So, finally,

$$\frac{2s_{11}{}^{\mathrm{T}}}{s_{21}{}^{\mathrm{T}}} \approx \frac{2s_{11}}{s_{21}} - Y'_{12} \tag{10.34}$$

This is the first of the results required. The second can be written by inspection from the symmetry of the [S] and [Y] matrix relationships and (10.34):

$$\frac{2s_{22}{}^{\mathrm{T}}}{s_{21}{}^{\mathrm{T}}} \approx \frac{2s_{22}}{s_{21}} - Y'_{12} \tag{10.35}$$

To derive the third result, for $(2/s_{21}{}^{\mathrm{T}})$, we start with eqn (10.30) and eqn (10.32) and again expand $Y_{11}{}^{\mathrm{T}}$ as $Y_{11} + Y'_{11}$ etc.:

$$s_{21}{}^{\mathrm{T}} \approx \frac{-2Y_{21}}{(1+Y_{11}+Y'_{11})(1+Y_{22}+Y'_{22}) - (Y_{12}+Y'_{12})Y_{21}}$$

Rearranging, and again neglecting product terms in $Y'_{11}Y'_{22}$,

$$(1+Y_{11})(1+Y_{22}) - Y_{12}Y_{21} + Y'_{11}(1+Y_{22}) + Y'_{22}(1+Y_{11}) - Y'_{12}Y_{21} = \frac{-2Y_{21}}{s_{21}{}^{\mathrm{T}}}$$

Reusing eqns (10.30) and (10.32), and again recognizing that $Y'_{11} = -Y'_{12} = Y'_{22}$, we have

$$\frac{-2Y_{21}}{s_{21}} - Y'_{12}(2+Y_{11}+Y_{22}+Y_{21}) = \frac{-2Y_{21}}{s_{21}{}^{\mathrm{T}}}$$

or

$$\frac{2}{s_{21}{}^{\mathrm{T}}} \approx \frac{2}{s_{21}} + Y'_{12} + \frac{2}{Y_{21}}$$

Perez and Ortega [10.9] chose to ignore the last term in this expression, which is a rather doubtful step. Bearing in mind that Y_{21} is normalized to $Y_0 = 1/Z_0$, the requirement is

$$2 \ll Y_{21} = y_{21}Z_0$$

To make a rough estimate of $y_{21}Z_0$, we might assume $y_{21} \approx g_{\mathrm{m}}$. Thus for a device with $g_{\mathrm{m}} = 60$ mS and $Z_0 = 50\ \Omega$, we find that

$$y_{21}Z_0 \approx 3$$

which hardly satisfies the inequality. Nevertheless, we shall assume the inequality applies so that finally

$$\frac{2}{s_{21}{}^{\mathrm{T}}} \approx \frac{2}{s_{21}} + Y'_{12} \tag{10.36}$$

Eqns (10.34), (10.35) and (10.36) show that, to rough order, the effect of the feedback is to translate the $(2s_{11}/s_{21})$, $(2s_{22}/s_{21})$ and $(2/s_{21})$ loci of the device by the vector $\pm Y'_{12}$ at any frequency. This observation is the basis of a simple graphical construction which, despite the approximations involved in its derivation, leads to a useful first design for subsequent computer-aided design.

As a first step in the graphical construction, the ratios $(2s_{11}/s_{21})$, $(2s_{22}/s_{21})$ and $(2/s_{21})$ of the FET S-parameters are plotted in a complex plane. For example, suppose we have a FET with

$$s_{21}=2.81 \qquad \text{angle } 108^\circ \text{ at } 6 \text{ GHz}$$

In a complex plane, we can then plot $(2/s_{21})=0.71$; angle (-108°). By carrying out this simple procedure for all frequencies in the band of interest, a locus of $(2/s_{21})$ can be drawn. Similarly, loci of $(2s_{11}/s_{21})$ and $(2s_{22}/s_{21})$ can be plotted. Figure 10.21 shows this construction carried out over the frequency range 6–18 GHz for a power FET element having nominal gate dimensions of 0.6 μm by 500 μm†.

Figure 10.21 also shows loci of constant $|s_{21}|$. For instance, if $|s_{21}|=6$ dB, then the actual value‡ of $|s_{21}|$ is 2. A locus of constant $|s_{21}|$ plotted on co-ordinate axes of $2/|s_{21}|$ is a circle of radius $2/|s_{21}|=2/2=1$. Using these constant $|s_{21}|$ circles as a guide, we can see that, for the example FET, $|s_{21}|$ varies from about 8.9 dB at 6 GHz to 0.9 dB at 18 GHz.

We next consider the consequences of introducing a feedback resistor R_{FB} so that, from eqn (10.26),

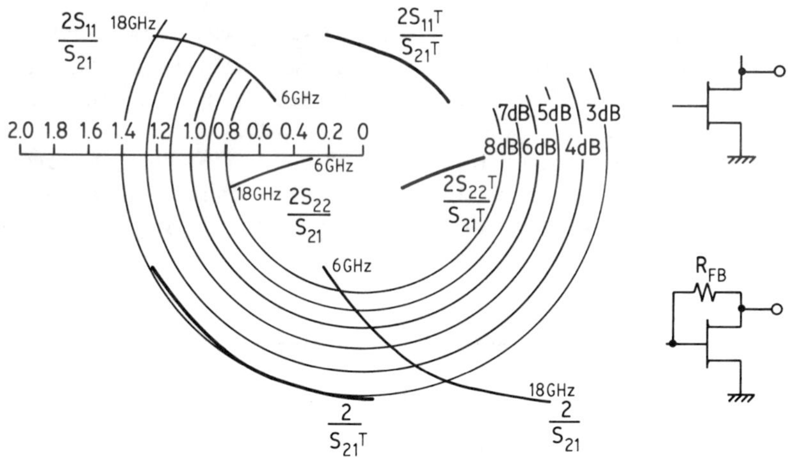

Fig. 10.21 Graphical design of single FET amplifier for flat gain with feedback resistor R_{FB}.

†Note that the FET is an MMIC element, so there are no bondwire and package parasitic contributions to the S-parameters.

‡Recall that S-parameters are ratios of voltage waves, so to express s_{21} in decibels it is necessary to take $20 \log_{10}(s_{21})$.

$$Y'_{12} = -1/R_{FB}Y_0 = -Z_0/R_{FB}$$

From eqn (10.36), the effect is to translate the locus of $2/s_{21}$ to the left; from eqns (10.34) and (10.35), the loci of $(2s_{11}/s_{21})$ and $(2s_{22}/s_{21})$ are translated to the right. This simple graphical construction has been carried out for $Y'_{12}=1$ in Fig. 10.21, resulting in an overall $s_{21}{}^{T}$ of 3 dB uniformly across the entire 6–18 GHz band. Such gain equalization has been achieved by reducing the gain at low frequencies in the band, and increasing it at high frequencies. Physically, this has arisen because the feedback is negative in the third quadrant of the $2/s_{21}$ complex plane but positive in the fourth quadrant; the positive feedback boosts the gain. From eqn (10.12) we are reminded that a constant value of $|s_{21}|$ gives a transducer power gain that is also constant.

The effect of the feedback on s_{11} can be judged from the shifted $(2s_{11}{}^{T}/s_{21}{}^{T})$ locus. The locus is now nearer the origin than it was. Had the feedback caused the locus to pass through the origin at some frequency, then at that point the amplifier would have been exactly matched. As it is, the input match is merely improved. So far as the output match is concerned, there is a degradation in the match at the lower frequencies but an improvement at the higher frequencies, as may be readily checked by reading $(2s_{22}{}^{T}/s_{21}{}^{T})$ and $(2/s_{21}{}^{T})$ from the shifted loci and forming $s_{22}{}^{T}=(2s_{22}{}^{T}/s_{21}{}^{T})/(2/s_{21}{}^{T})$. Thus, the feedback has achieved gain equalization, with qualified improvement in the input and output matches.

In the example of Fig. 10.21, the feedback was chosen with gain equalization in mind. As an alternative, it could have been chosen to achieve as closely as possible a matched condition at either the input or the output over a specified range of frequencies, if that were a priority.

More gain can be realized from the same FET with feedback by the simple expedient of including an inductance $\mathscr{L}_D$ in series with the drain as shown in Fig. 10.22. The important first-order consequence of including $\mathscr{L}_D$ is to rotate

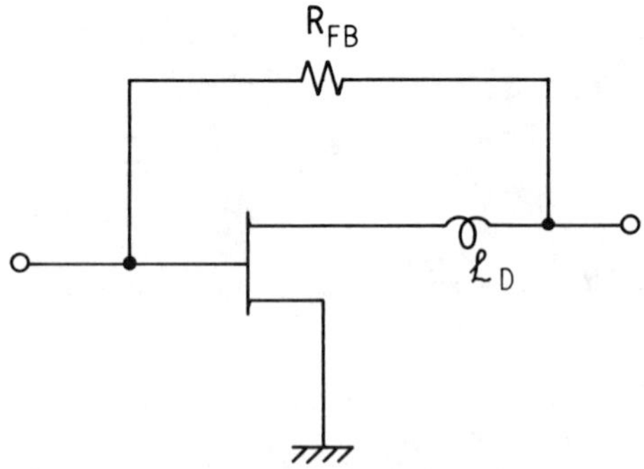

Fig. 10.22 Inclusion of a drain series inductor.

the $2/s_{21}$ plot in an anti-clockwise direction, approximately pivoting about the low frequency end of the plot. This rotation may be seen by comparing the $2/s_{21}$ locus in Fig. 10.23 for $\mathscr{L}_D = 300$ pH with the original in Fig. 10.21. If this new locus is now translated to the left by the application of a feedback resistor, a value of $|s_{21}|$ varying between 5 dB to 5.7 dB and back to 5 dB can be achieved over the band 6–18 GHz. In this case the translation vector has a magnitude $Y'_{12} = -0.73$, implying $R_{FB} \approx 68\ \Omega$. As before, the input reflection coefficient $|s_{11}|$ is reduced in magnitude.

As previously argued, graphical solutions of this kind offer useful physical insight into the effects of feedback upon the magnitude and phase of the principal scattering parameters of an amplifier stage, and they provide a good starting point for computer iteration.

10.2.5 Distributed amplifiers

With a conventional amplifier, any attempt to increase the gain by increasing the FET's transconductance usually results in an increase in input capacitance as well. Reference to Table 6.1 will show that, regardless of whether increases in gatewidth or channel doping density are contemplated as a means of increasing the transconductance, the outcome is always the same. The distributed amplifier, sometimes also called a traveling wave amplifier for reasons that will become obvious, offers a means of combining the transconductance of several FETs without combining their input capacitances.

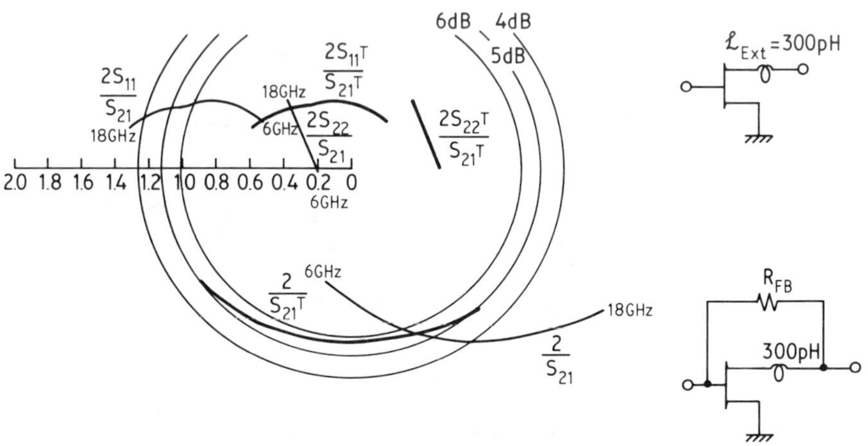

Fig. 10.23 Graphical construction for reasonably level gain with gain increase by inclusion of drain series inductor.

The physical basis of the distributed amplifier is shown in Fig. 10.24. Considering the FET gate electrodes first, the impedance looking into a FET gate terminal is capacitive, with some series resistance. By connecting the gates across a series of inductances $\mathscr{L}_G$, as shown in Fig. 10.24, an artificial transmission line is formed of the type analyzed in Fig. 9.12. If the artificial line is terminated at its right end by an on-chip resistor of value Z_{OG} equal to the characteristic impedance of the line, then a signal input at the left end will travel along the line exciting each FET as it passes. Upon reaching the right end of the artificial transmission line, the signal is absorbed by the on-chip matched load, Z_{OG}.

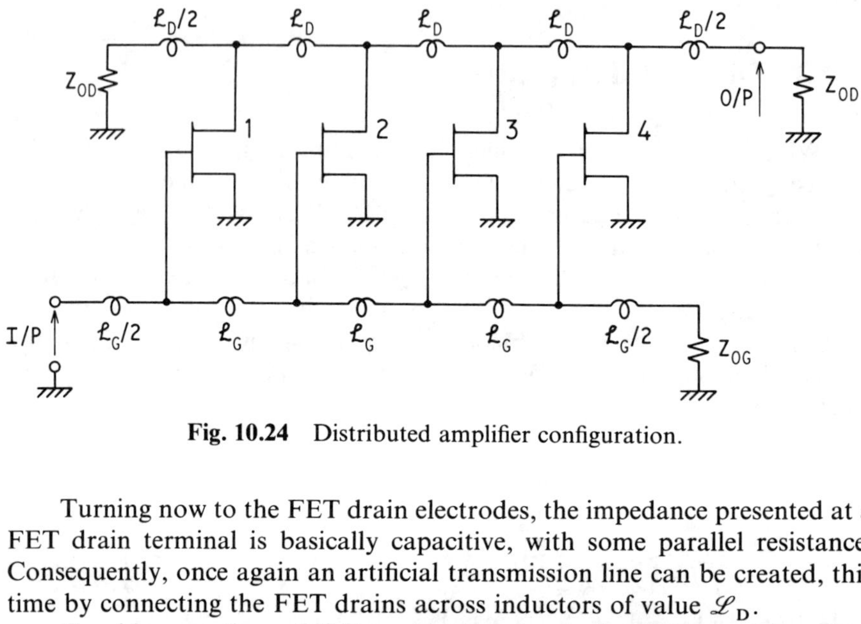

Fig. 10.24 Distributed amplifier configuration.

Turning now to the FET drain electrodes, the impedance presented at a FET drain terminal is basically capacitive, with some parallel resistance. Consequently, once again an artificial transmission line can be created, this time by connecting the FET drains across inductors of value $\mathscr{L}_D$.

Consider now the path followed by the signal when FET no. 1 is excited by the wave traveling down the gate artificial transmission line. The signal gives rise to a current in the drain, roughly of value $g_{m0}v_g$, which flows from the FET into the node where the drain is connected to the drain artificial transmission line. We will assume for the time being that the impedance seen looking to the left of the node is the same as that looking to the right, so that the current splits into two: half flows to the left towards the on-chip dummy load, Z_{OD}, and half flows to the right as a wave which travels along the artificial line until it is dissipated in the external matched load, also of value Z_{OD}. By "external", we mean that the resistor is off-chip.

The signal dissipated in the external load lags in phase with respect to the signal input at the left end of the gate line; it has had to travel part-way along the gate line, undergo phase delay as well as amplification in passing through FET no. 1, and travel along the drain line to the load. The same is true of paths through FETs 2, 3 and 4; half the drain current from each FET gives rise to a wave traveling to the right, eventually to be dissipated in the external load. By so designing the amplifier that the total phase delay along the signal paths through FETs 1 to 4 is the same, the signals all arrive at the external load in phase with one another, thereby developing the maximum power. Conceptually, equal phase delay can be assured by making the phase shift per section of the gate line equal to the phase shift per section in the drain line. All four FETs are assumed to be identical also.

We have yet to answer for the waves that travel to the left along the drain line, towards the on-chip dummy load. As we shall shortly show, waves propagating towards the dummy load do not combine in the same way as the forward-propagating waves at microwave frequencies; in fact, they partly or totally cancel one another. The signal current at the left end of the drain line is thus generally smaller than the signal current delivered to the external load. Such net current as remains is absorbed by the matched dummy load.

In principle, the distributed amplifier may have any number of FETs, say N, not just four. Fig. 11.5, for example, shows a five-stage MMIC traveling wave amplifier. Practical considerations, mainly having to do with gate line attenuation arising from FET input resistance, normally limit the number of FETs to, typically, seven.

Our analysis of the distributed amplifier will be conducted in four parts. First, a forward- and reverse-wave analysis will be carried out to determine the ratio of currents in the drain-line dummy load and the external load. Secondly, an approximate expression will be derived for the complex propagation constants of the artificial transmission lines. Thirdly, detailed expressions will be derived for the attenuation coefficient and the phase coefficient for the gate and drain lines separately. Finally, the overall power gain expression for the amplifier will be developed. We start with the forward- and reverse-wave analysis.

Figure 10.25 shows an idealized schematic in which the gate- and drain artificial transmission lines are represented as continuous, lossless structures. The phase shift per unit distance is taken to be the same for both lines, and equal to β. A transconductance of $2\overline{g_{m0}}$ per unit distance along x is assumed to couple the gate line unidirectionally and uniformly to the drain line along the length of the lines ($0 \leqslant x \leqslant L$). The term "unidirectionally" refers to the assumption that there is no coupling from the drain line to the gate line, only from the gate line to the drain line. The factor of 2 in $2\overline{g_{m0}}$ is merely introduced to avoid having to carry the awkward factor of $\frac{1}{2}$ throughout the analysis to account for the 50–50 split of the drain current into forward- and reverse-

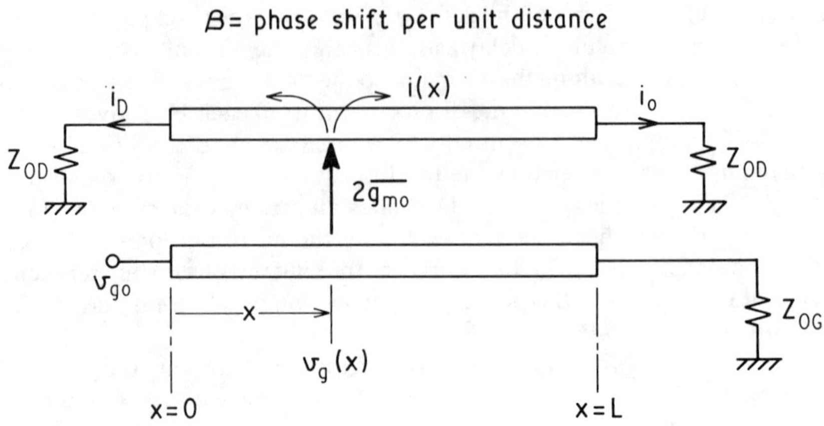

Fig. 10.25 Idealized representation of distributed amplifier, mainly for analysis of reverse-propagating waves on drain line.

propagating waves. With such a model the total transconductance integrated over the length of the lines is $2\overline{g_{m0}}L$, whereas in a real amplifier it is Ng_{m0}, g_{m0} being the transconductance per FET. Therefore, to relate the present model to the practical case, it is merely necessary to have

$$\overline{g_{m0}} = \frac{Ng_{m0}}{2L}$$

With the foregoing assumptions, the current injected onto the drain line at any point x splits into two equal components $i(x)$, where

$$i(x) = \overline{g_{m0}} v_g(x) \tag{10.37}$$

and $v_g(x)$ is the signal voltage on the gate line at position x,

$$v_g(x) = v_{g0} \exp(-j\beta x) \tag{10.38}$$

The forward wave propagating to the right along the drain line results in a current i_o in the external load of

$$i_o = i(x)\exp[-j\beta(L-x)] = \overline{g_{m0}} v_{g0} \exp(-j\beta L)$$

By integrating over 0 to L, we find the total signal current in the external load:

$$i_{oTOT} = \int_0^L i_o \, dx = g_{m0} v_{g0} \exp(-j\beta L) \tag{10.39}$$

where $g_{m0} = \int_0^L \overline{g_{m0}} \, dx = \overline{g_{m0}} L$.

We now consider reverse propagating waves, i.e. waves on the drain line propagating to the left towards the dummy load. The current $i(x)$ at x gives rise to a current i_D in the dummy load of

$$i_D = i(x)\exp(-j\beta x)$$

The phase factor βx accounts for the phase delay as the wave propagates backwards over distance x along the drain line until it reaches the dummy load. Using eqns (10.37) and (10.38) we obtain

$$i_D = \overline{g_{m0}} v_{g0} \exp(-j2\beta x)$$

The total current in the dummy load, i_{DTOT}, is obtained by integrating the signal components i_D over the range 0 to L:

$$i_{DTOT} = \int_0^L i_D \, dx = \overline{g_{m0}} v_{g0} \int_0^L \exp(-j2\beta x)\, dx$$
$$= (\overline{g_{m0}} v_{g0}/2\beta)\{\sin(2\beta L) - j[1-\cos(2\beta L)]\}$$

Multiplying top and bottom by L, writing $\theta_{TOT} = \beta L$, and recalling that $g_{m0} = \overline{g_{m0}} L$,

$$i_{DTOT} = (g_{m0} v_{g0}/2\theta_{TOT})\{\sin(2\theta_{TOT}) - j[1-\cos(2\theta_{TOT})]\} \qquad (10.40)$$

We are concerned to find the magnitude of the total signal currents in the dummy load and external drain-line load; the phase factors have no bearing upon the power developed in these loads. After a little manipulation with trigonometric functions we find, from eqn (10.40),

$$|i_{DTOT}| = g_{m0} v_{g0}[\sin(\theta_{TOT})/\theta_{TOT}] = g_{m0} v_{g0} \operatorname{sinc}(\theta_{TOT}) \qquad (10.41)$$

whereas, from eqn (10.39),

$$|i_{oTOT}| = g_{m0} v_{g0} \qquad (10.42)$$

Taking the ratio, we obtain

$$|i_{DTOT}/i_{oTOT}| = \operatorname{sinc}(\theta_{TOT}) \qquad (10.43)$$

The sinc function is drawn in Fig. 10.26. If the inductance per unit length of the lines is $\bar{\mathscr{L}}$, and the capacitance per unit length is $\bar{C}$, then the phase factor is determined by the propagation coefficient $\beta = \omega(\bar{\mathscr{L}}\bar{C})^{1/2}$ in accordance with the expression

$$\theta_{TOT} = \beta L = \omega(\bar{\mathscr{L}}\bar{C})^{1/2} L$$

Consequently, the horizontal axis in Fig. 10.26 could equally well be read as frequency. We can draw the following conclusions.

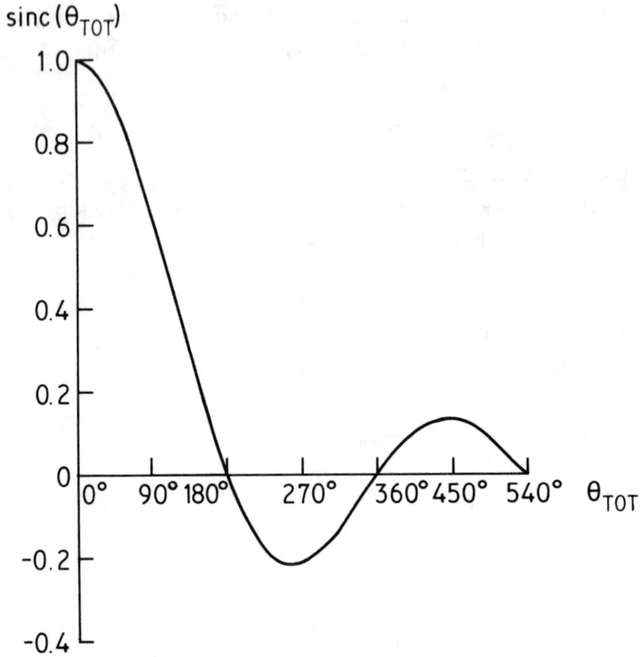

Fig. 10.26 The sinc function.

(a) The signal current in the external load is independent of frequency (eqn (10.42)). The distributed amplifier thus produces an output power and a power gain that is constant.

(b) At dc the signal power dissipated in the dummy drain line load is equal to the power developed in the external load (eqn (10.43) with $\theta_{TOT}=0$ at $\omega=0$), but as the frequency increases the reverse-propagating waves tend to combine destructively, thereby causing the power dissipated in the dummy load to collapse.

In a practical amplifier, conclusion (a) above does not apply without limit. The lumped nature of the artificial lines in a real amplifier means that the lines have a cut-off frequency given by eqn (9.13). Suppose, for example, we wish to build a distributed amplifier to work in a 50-ohm system using FETs having an input capacitance of $C_2=0.2$ pF. Equation (9.13) then gives the line cut-off frequency as

$$f_C=\frac{1}{\pi\times 50\times 0.2\times 10^{-12}}=32\text{ GHz}$$

If the lines are to be matched to the generator and load with a VSWR of better than 2:1, Fig. 9.14 tells us that the maximum frequency of operation will be $0.86 f_C$, i.e.

$$\text{Maximum frequency of operation} = 27 \text{ GHz}$$

The conclusion regarding the power dissipated in the dummy load is substantially unchanged in a real MMIC, as may be verified either by direct probe measurement on an experimental chip, or by computer simulation.

The fact that the artificial lines in a real MMIC are lossy due to the FET resistances loading them has no important consequence for the reverse-propagating waves, but it does affect the forward waves, and hence power output, in a way which cannot be ignored. To deal with this question we must first develop an expression for the complex propagation coefficient of a lossy line of the type shown in Fig. 10.27. Generally,

$$\gamma = [(R + j\omega\mathscr{L})(G + j\omega C)]^{1/2}$$

In this expression, and in Fig. 10.27, the elements R, $\mathscr{L}$, G and C may be expressed in per section terms for a lumped component artificial line, or in per unit distance terms for a uniform waveguide or transmission line. We shall denote the second by the use of a bar above the symbols, e.g. $\bar{R}$.

R, or $\bar{R}$, is the element of resistance in series with the inductors $\mathscr{L}$; in Fig. 10.27, R has been assumed negligible. The characteristic impedance of the line, Z_0, is

$$Z_0 = \left(\frac{R + j\omega\mathscr{L}}{G + j\omega C}\right)^{1/2}$$

Combining these two equations gives

$$\gamma = (G + j\omega C)Z_0$$

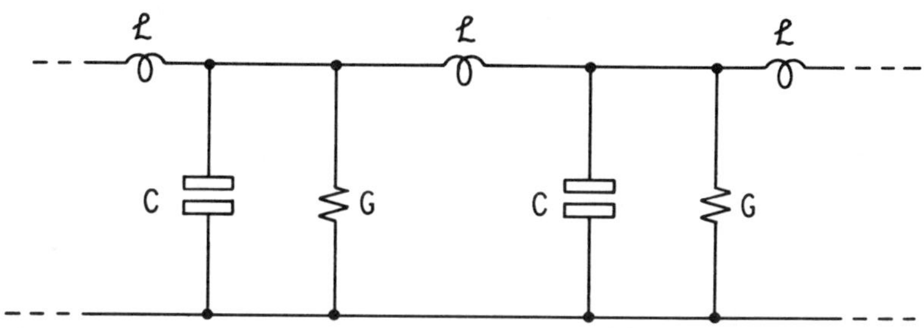

Fig. 10.27 Lumped transmission line with shunt loss.

For the case of $R \approx 0$ and G small, γ may be approximated as follows:

$$\gamma = (G + j\omega C)\left(\frac{j\omega \mathscr{L}}{G + j\omega C}\right)^{1/2}$$

i.e.

$$\alpha + j\beta = \gamma = (G + j\omega C)\left(\frac{\mathscr{L}}{C}\right)^{1/2}\left(1 + \frac{G}{j\omega C}\right)^{-1/2}$$

$$\approx (G + j\omega C)\left(\frac{\mathscr{L}}{C}\right)^{1/2}\left(1 - \frac{G}{j2\omega C}\right)$$

$$\approx \left(\frac{G}{2} + j\omega C\right)\left(\frac{\mathscr{L}}{C}\right)^{1/2}$$

i.e.

$$\alpha \approx \frac{G}{2}\left(\frac{\mathscr{L}}{C}\right)^{1/2} \tag{10.44}$$

$$\beta \approx \omega(\mathscr{L}C)^{1/2} \tag{10.45}$$

For low loss, we can also approximate Z_0 with,

$$Z_0 \approx \left(\frac{\mathscr{L}}{C}\right)^{1/2} \tag{10.46}$$

The foregoing results may be applied to the drain and gate lines in turn.

Figure 10.28 shows a schematic of a practical amplifier in which the inductors are realized with lengths of microstripline. To prevent the chip from becoming too large, the microstrip sections may be meandered (as in the

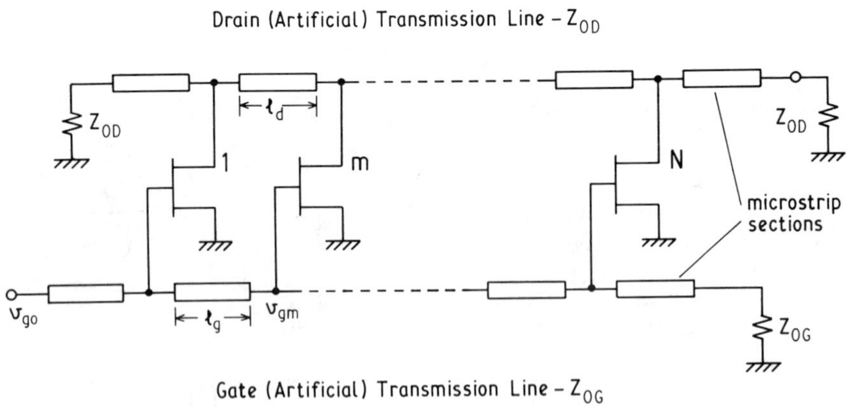

Fig. 10.28 Use of microstrip sections to realize the inductors in a distributed amplifier.

photograph of Fig. 11.5). Alternatively, the microstriplines may be coiled into single-turn inductors [10.10]. Regardless, the inductors not only have inductance $\bar{\mathscr{L}}$ per unit length, they also have capacitance. This capacitance must be included together with the device capacitance in the analysis.

For the drain line, the microstrip sections between the FETs are taken to be of electrical length l_d. The total effective inductance per unit length of microstripline will be termed $\bar{\mathscr{L}}_d$. We use the symbol $\bar{\mathscr{L}}_D$ to refer to the inductance per unit length of the drain artificial transmission line. The act of meandering or coiling the microstrip sections gives rise to mutual coupling which causes $\bar{\mathscr{L}}_D$ to differ from $\bar{\mathscr{L}}_d$. We shall ignore such effects for present purposes, so that

$$\bar{\mathscr{L}}_D \approx \frac{\bar{\mathscr{L}}_d l_d}{l_d} = \bar{\mathscr{L}}_d$$

If the capacitance per unit length of the microstripline is $\bar{C}_d$, the total capacitance per unit length of the artificial drain line is

$$\bar{C}_D = (C_o + \bar{C}_d l_d)/l_d = \bar{C}_d + C_o/l_d$$

where C_o is the output capacitance of the FET. Figure 10.29 shows one section of the drain line, from which we also have the shunt conductance per unit length as $G_D = (1/r_o) \times 1/l_d$.

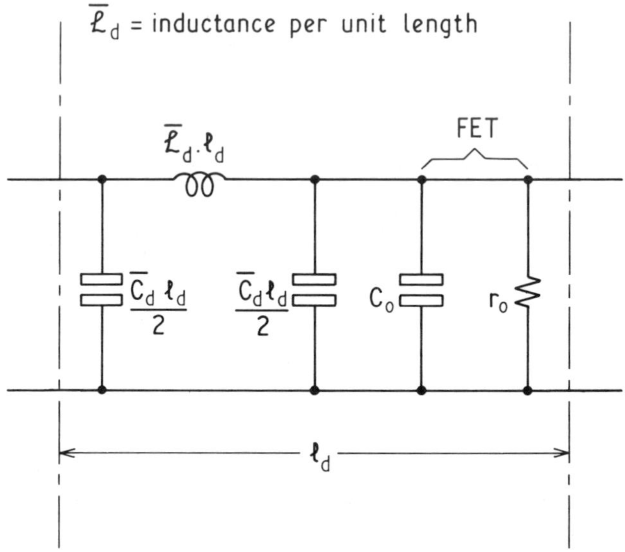

Fig. 10.29 One section of the drain artificial line.

Using the results of eqns (10.44), (10.45) and (10.46), we find for the attenuation per unit length along the drain line,

$$\alpha_D \approx \frac{1}{2r_o l_d}\left[\frac{\bar{\mathscr{L}}_d}{(\bar{C}_d + C_o/l_d)}\right]^{1/2} \tag{10.47}$$

The phase constant in radians per unit distance along the drain line is

$$\beta_D \approx \omega[\bar{\mathscr{L}}_d(\bar{C}_d + C_o/l_d)]^{1/2} \tag{10.48}$$

The characteristic impedance is

$$Z_{OD} \approx \left(\frac{\bar{\mathscr{L}}_d}{\bar{C}_d + C_o/l_d}\right)^{1/2} \tag{10.49}$$

We turn next to the gate line, which is similar to the drain line except for the fact that the equivalent circuit of the FET is now a series C–R network (Fig. 10.30). The first task is to transform this series network into an equivalent parallel C–R network at any frequency ω. The impedance of the series network is

$$Z = R_2 + \frac{1}{j\omega C_2} = \frac{1 + j\omega C_2 R_2}{j\omega C_2}$$

The admittance is

$$Y = Z^{-1} = \frac{j\omega C_2 + \omega^2 C_2{}^2 R_2}{1 + \omega^2 C_2{}^2 R_2{}^2}$$

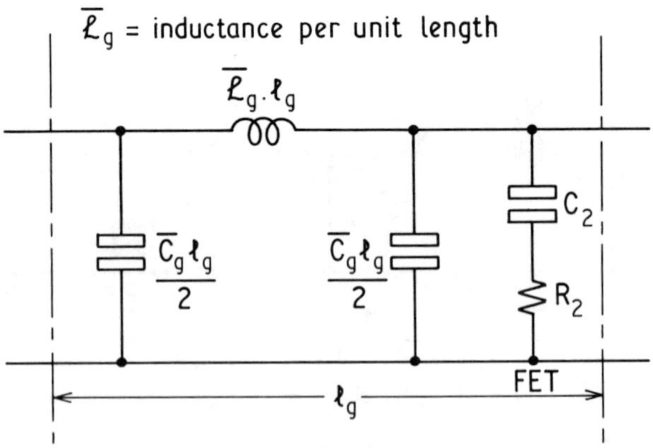

Fig. 10.30 One section of the gate line.

Typical values are $f=20$ GHz, $C_2=0.2$ pF and $R_2=10\ \Omega$. Then $\omega^2 C_2{}^2 R_2{}^2 \approx 0.06$, so that we may take the denominator to roughly equal unity. Hence

$$Y \approx \omega^2 C_2{}^2 R_2 + \mathrm{j}\omega C_2$$

The first term on the right hand side represents the conductance branch of an equivalent parallel G–C network, and the second represents the equivalent parallel capacitance. Following a procedure identical to that employed for the drain line, we can now write, for the gate line, the inductance per unit length:

$$\bar{\mathscr{L}}_{\mathrm{G}} = \frac{\bar{\mathscr{L}}_{\mathrm{g}} l_{\mathrm{g}}}{l_{\mathrm{g}}} = \bar{\mathscr{L}}_{\mathrm{g}}$$

the capacitance per unit length:

$$\bar{C}_{\mathrm{G}} = (\bar{C}_{\mathrm{g}} + C_2/l_{\mathrm{g}})$$

and the shunt conductance per unit length:

$$\bar{G}_{\mathrm{G}} = \frac{\omega^2 C_2{}^2 R_2}{l_{\mathrm{g}}}$$

Hence, again making use of eqns (10.44–10.46),

$$\alpha_{\mathrm{G}} \approx \frac{\omega^2 C_2{}^2 R_2}{2 l_{\mathrm{g}}} \left(\frac{\bar{\mathscr{L}}_{\mathrm{g}}}{\bar{C}_{\mathrm{g}} + C_2/l_{\mathrm{g}}} \right)^{1/2} \tag{10.50}$$

$$\beta_{\mathrm{G}} \approx \omega[\bar{\mathscr{L}}_{\mathrm{g}}(\bar{C}_{\mathrm{g}} + C_2/l_{\mathrm{g}})]^{1/2} \tag{10.51}$$

$$Z_{0\mathrm{G}} \approx \left(\frac{\bar{\mathscr{L}}_{\mathrm{g}}}{\bar{C}_{\mathrm{g}} + C_2/l_{\mathrm{g}}} \right)^{1/2} \tag{10.52}$$

Equations (10.47) and (10.50) respectively give the attenuation coefficients of signals propagating along the drain and gate lines. Similarly, eqns (10.48) and (10.51) give the phase delay per unit length along the two lines, and eqns (10.49) and (10.52) give their characteristic impedances. All these quantities appear in the expression for the power gain of the amplifier, which we shall now derive.

Consider the path from the input, via FET m, to the output, as shown in Fig. 10.31. The signal wave propagating along the gate line has an amplitude as it passes FET m of

$$v_{\mathrm{gm}} = v_{\mathrm{go}} \exp(-\alpha_{\mathrm{G}} m l_{\mathrm{g}})$$

FET m generates a current i_{dm} in response. Assuming half of i_{dm} goes into the wave propagating towards the external load, the current actually entering the load has an amplitude of

$$i_{\mathrm{om}} = (i_{\mathrm{dm}}/2) \exp[-\alpha_{\mathrm{D}}(N-m) l_{\mathrm{d}}]$$

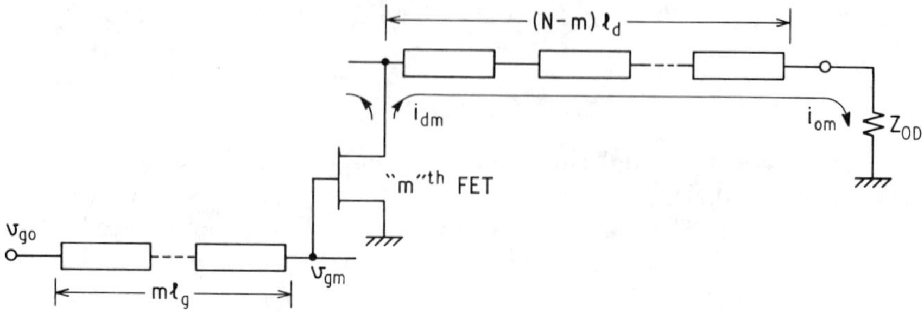

Fig. 10.31 Signal path from the input to the output via the "*m*"th FET.

But

$$i_{dm} = g_{m0}v_{gm} = g_{m0}v_{go}\exp(-\alpha_G l_g m)$$

Therefore

$$i_{om} = (g_{m0}/2)v_{go}\exp(-\alpha_D N l_d)\exp[m(\alpha_D l_d - \alpha_G l_g)] \qquad (10.53)$$

If we now assume that the phase delay per section of the gate line is equal to the phase delay per section of the drain line, all components i_{om} arising from the individual FETs will add in-phase so that the total current in the external load, i_{oTOT}, can be found by summing the magnitudes i_{om} given by eqn (10.53). The sum is taken over $1 \leqslant m \leqslant N$. We shall replace this summation by an integral. The question then arises as to what the limits of integration should be. Taking limits 1 to N would be appropriate to the number of FETs. Taking limits 0 to N is more appropriate given the total length of the lines involved, and hence for determining the total attenuation. Since the integration affects the last term in eqn (10.53), a term which expresses attenuation, we shall assume limits of 0 to N, thus,

$$i_{oTOT} = \frac{g_{m0}}{2} v_{go}\exp(-\alpha_D N l_d) \int_0^N \exp[m(\alpha_D l_d - \alpha_G l_g)]\, dm$$

$$= \frac{g_{m0}}{2} v_{go} \frac{\exp(-\alpha_G N l_g) - \exp(-\alpha_D N l_d)}{\alpha_D l_d - \alpha_G l_g}$$

The output power is $i_{oTOT}^2 Z_{OD}$; the input power to the amplifier is v_{go}^2/Z_{OG}. Consequently, the operating power gain is

$$G_p = \frac{g_{m0}^2 Z_{OG} Z_{OD}}{4} \frac{[\exp(-\alpha_G N l_g) - \exp(-\alpha_D N l_d)]^2}{(\alpha_D l_d - \alpha_G l_g)^2} \qquad (10.54)$$

In the limit $(\alpha_D N l_d) \to 0$ and $(\alpha_G N l_g) \ll 1$,

$$\exp(-\alpha_G N l_g) \approx (1 - \alpha_G N l_g) \quad \text{and} \quad \exp(-\alpha_D N l_d) \approx 1$$

Then,

$$G_p \approx \frac{g_{m0}{}^2 Z_{0G} Z_{0D}}{4} \left[\frac{1 - \alpha_G N l_g - 1}{-\alpha_G l_g} \right]^2$$

i.e.

$$G_p \approx (N g_{m0}/2)^2 Z_{0G} Z_{0D} \tag{10.55}$$

Equations (10.54) and (10.55), together with eqns (10.47) to (10.52), are the results quoted by Ayasli *et al.* [10.11].

There are many practical implications which follow from this analysis. If we contemplate constructing a distributed amplifier with a large number of FETs, N, the exponential terms in eqn (10.54) could reduce the gain to almost zero. At the other extreme, when N is a small number we might expect eqn (10.55) to apply, in which case the gain increases as N^2. Depending upon the detailed values of $\alpha_G l_g$ and $\alpha_D l_d$ there is some limit to N beyond which the gain does not improve for the addition of extra FETs, and indeed falls. Typically, distributed amplifiers are constructed from between four and seven FET elements. Extra gain is obtained by cascading chips.

Examination of the attenuation coefficients, eqns (10.47) and (10.50), shows that α_D is frequency independent, whereas α_G is proportional to ω^2. A processing error which affects the FET output network, C_o–r_o, thus tends to shift the gain up or down in value, uniformly across the band. An error which affects the input network, C_2–R_2, is likely to give rise to a gain slope error. For instance, oversized gates give rise to C_2 being too large via the space-charge component C_{gc}, and so a negative gain slope could be anticipated. (Such effects will be apparent in Figs 11.9 and 11.10.)

From eqns (10.45) and (10.46) we may write

$$\beta \approx \omega C Z_0$$

If we wish to satisfy the requirement for equal phase delay per section along the gate and drain lines, and also have the same chaacteristic impedance for both lines, we must ensure that

$$\beta_G \approx \omega(\bar{C}_g + C_2/l_g) l_g Z_0 = \beta_D \approx \omega(\bar{C}_d + C_o/l_d) l_d Z_0$$

i.e. we require

$$(C_2 + \bar{C}_g l_g) \approx (C_o + \bar{C}_d l_d) \tag{10.56}$$

There is a difficulty here. Typically, C_2 might be a factor of four or five larger than C_o. Thus, C_o needs to be increased somehow. This may be done either by adding parallel metal–insulator–metal overlay capacitors at the drains on the

MMIC (as in the photograph† of Fig. 11.5), or by transforming the FET output admittance using a series transforming section of microstripline (as in the Raytheon chip of Fig. 11.25).

Monolithic integration is ideal for the practical realization of distributed amplifiers because there are no package and mounting parasitics to contend with. Such imponderables could render the amplifier response very variable from circuit-to-circuit in a hybrid MIC solution (see [10.12] for an account of a hybrid version). At worst, it may not prove feasible to design an amplifier to meet a given specification in the first place. Careful consideration of the equations developed here will show that meeting the requirements of line cut-off frequency, impedance level, line attenuation coefficients and gain, may mean that a fairly special FET must be available. Methods of deducing the material and structural specifications of a FET from its equivalent circuit can be employed to synthesize FET elements needed for distributed amplifier MMICs.

Further details on distributed amplifiers can be found in [10.10, 10.11, 10.13 and 10.14]. Section 10.5 will deal with the synthesis of FET elements for distributed power amplifiers: a computer-aided design approach to the same problem is discussed in Section 11.6. Section 11.2 considers the sensitivity of a typical distributed amplifier MMIC to process variations.

10.2.6 The matrix amplifier

The matrix amplifier has been proposed fairly recently [10.15] as a variant of the distributed amplifier principle. The paper by Niclas and Pereira [10.15] is of interest for the clear way in which it distinguishes between multiplicative and additive amplification, and the way it argues the matrix topology embodies the benefits of both.

Multiplicative amplification results in an overall gain that is proportional to the product of the individual gains of two or more stages of amplification. Multiplicative amplification is achieved through cascading individual stages. In contrast, additive amplification results from the sum, or addition, of the signals from individual stages into a common load; the distributed amplifier is the dominant example. Their individual merits are that multiplicative amplification generally gives higher overall gain from a given number of devices (e.g. FETs), whereas the additive principle offers larger bandwidths.

Suppose a systems application calls for a single MMIC chip to provide at least 15 dB gain over the band 0.5–20 GHz. A first design of a distributed

†The metal connect to the MIM capacitors can just be seen as the diagonal bridges passing across the top left hand corner of each FET.

amplifier may suggest that four stages are possible (i.e. $N=4$), each stage having a FET with $g_{m0}=30$ mS. Assuming the chip is to work in a 50 Ω system, eqn (10.55) gives a rough estimate for the power gain as

$$G_p=(4\times 0.03/2)^2\times 50\times 50\equiv 9.5 \text{ dB}$$

To achieve the gain specified, the multiplicative gain principle could now be applied by cascading two such four-stage circuits as shown in Fig. 10.32. A length of microstripline of characteristic impedance $Z_0=Z_{0G}=Z_{0D}=50\ \Omega$ connects the input of the second amplifier to the output of the first, and the whole circuit could easily be designed and laid-out as a single chip.

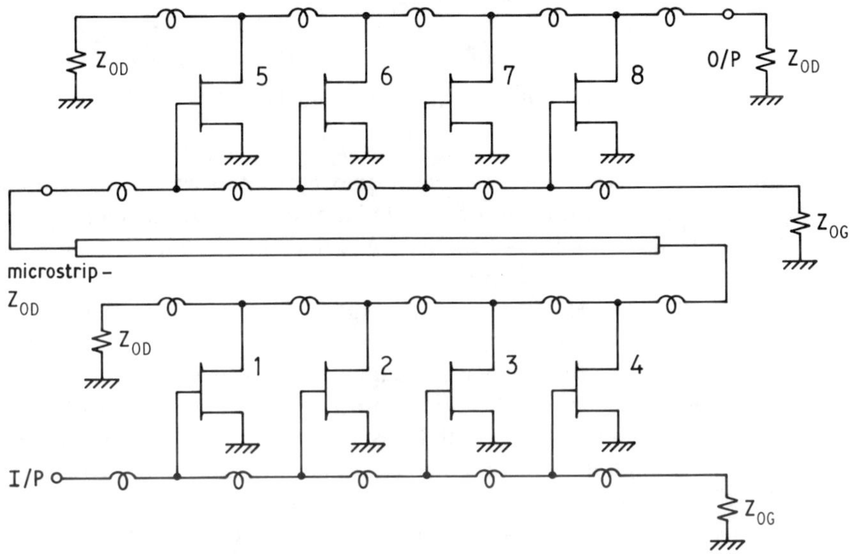

Fig. 10.32 Cascade connection or integration of two distributed amplifiers.

The matrix amplifier is an alternative scheme to a cascade of distributed amplifiers. Instead of the second group of FETs (FETs 5 to 8) being fed in series from the first group (FETs 1 to 4), the matrix amplifier excites the gate of FET 5 from the drain of FET 1, the gate of FET 6 from the drain of FET 2, and so on, i.e. there is an artificial transmission line between the two groups of FETs that is common to both groups (Fig. 10.33). The total device capacitance per section of this central line now consists of input capacitance C_2 from the upper FET, and output capacitance C_o from the lower FET. Bearing in mind that $C_o \ll C_2$, in order to have all three artificial lines in Fig. 10.33 the same, capacitance C_2 would have to be added to each section of the drain line,

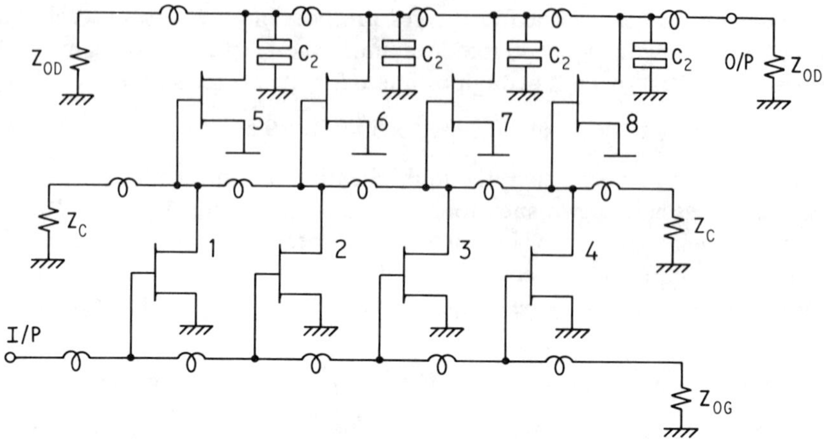

Fig. 10.33 The matrix amplifier.

as shown explicitly in the figure. The addition of capacitance to the drain line is similar to the requirement of the distributed amplifier, as discussed following eqn (10.56). Niclas and Pereira [10.15] argue that the regular array of FETs suggests the name "matrix" amplifier.

In the limit of an idealized FET consisting of input capacitance, output capacitance and transconductance only, with no resistances or inductances anywhere in the equivalent circuit†, the gain of the matrix arrangement is not independent of frequency, i.e. the gain is not flat. It has been shown [10.15] that the absolute gain and the gain flatness can be improved by doubling the inductance per section in the center line. Analysis using a complete equivalent circuit for the FET results in degradation of both the gain and bandwidth. Simple adjustment of the values of the dummy loads, the line inductances, and the loading capacitors allow the degradations to be partly offset. By abandoning the idea of uniformly distributed identical lines and subjecting all the elements to individual adjustment, and also by adding input and output matching networks, further useful performance enhancements are possible.

While admitting that the bandwidth of the matrix amplifier, for a given FET, is slightly smaller than a cascade of two distributed amplifiers based on the same FET, Niclas and Pereira suggest that:

(a) the gain of the two circuits is comparable,
(b) the matrix amplifier may have better input and output matches,
(c) lower noise figure should result, and

†Thus the FET equivalent circuit appears in Fig. 1.1(b).

(d) especially in monolithic form, the matrix amplifier should occupy less chip area, and hence cost less to produce.

For the time being, at least, these speculated advantages have yet to be demonstrated in practice. As with many new ideas, the true worth of the matrix amplifier will not become apparent until it has been developed and thoroughly evaluated by a number of organizations. In the end, issues such as how designable the circuit is, and how robust it is to fabrication process variations, could be deciding factors. One comparison between the amplifiers of Figs 10.32 and 10.33 not made by Niclas and Pereira is the phase delay between input and output for a given gain and bandwidth. In the absence of a detailed study it is only possible to surmise that the matrix amplifier should have roughly half the phase delay, which could be important in applications where good phase tracking between channels is required in systems such as phased-array radars.

10.2.7 Balanced amplifiers

From Fig. 10.4, the reflection coefficient for a wideband reactively matched amplifier can easily be 0.8 or greater, i.e. a return loss as high as -2 dB. To prevent the reflected energy giving rise to standing waves on any transmission line between the generator (or source) and the amplifier, and to prevent the energy being dissipated in the source, a balanced amplifier configuration may be employed. Two quadrature couplers are used: one at the input, to split the signal from the generator into two equal parts to be fed to two amplifiers operating in parallel, the other to combine the outputs of the amplifiers into a single load. The principle of the amplifier becomes clearer if the properties of a 3 dB quadrature coupler are reviewed first.

Figure 10.34 is a schematic of a 3 dB quadrature coupler. The number (i.e. 3 dB) refers to a division of the input power into two equal signal powers, each equal to one half of, or -3 dB relative to, the input power. Assuming the split signals propagate on lines of the same characteristic impedance, the voltage amplitude of the two waves is $v/\sqrt{2}$, where v is the amplitude of the input signal.

The coupler is a four-port, linear, passive network. The port at the same end of the coupler as the input port is termed the coupled port; the signal leaving this port has amplitude and phase $(v/\sqrt{2}, 0°)$ relative to the input. The diametrically opposed port at the other end of the coupler is called the direct port, and supports a signal $(v/\sqrt{2}, 90°)$ or $\mathrm{j}(v/\sqrt{2})$. Finally, at the other end of the coupler from the input port, but on the same side, is the isolated port: ideally the strength of the signal coupled out of this port is zero. Note that,

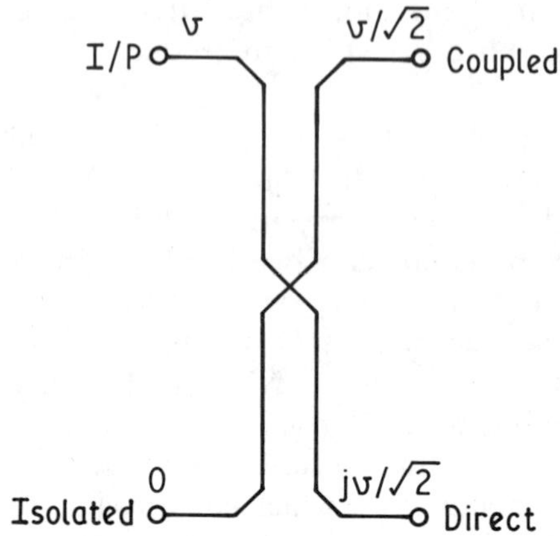

Fig. 10.34 Schematic of a -3 dB coupler.

depending upon the detailed type and layout of the coupler, the various named ports may appear in different positions relative to the input port. A useful diagrammatic summary can be found in [10.16].

Looking again at Fig. 10.34, it is perhaps worth emphasizing that, whichever port of the four ports the signal is incident upon, the coupled, direct and isolated ports all bear the same geometric relationship to that input port. For example, for a signal incident on the bottom right port, the bottom left port is the coupled port (signal $v/\sqrt{2}$), the top left port is the direct port (signal $jv/\sqrt{2}$), and the top right port becomes the isolated port.

The balanced amplifier uses two 3 dB quadrature couplers and two nominally identical MMIC amplifiers connected as shown in Fig. 10.35. Normally the amplifiers would consist of input and output matching networks, as well as the active devices and interstage matching or equalizing networks. Figure 10.35 traces the signal paths, amplitudes and phases. Regarding the whole arrangement now as a two-port, with the input port at the top left and the output port at the bottom right, the total amplitude of the signals backscattered into the input port is

$$(s_{11}{}^{(a)} - s_{11}{}^{(b)})\, v/2$$

Similarly, the total signal amplitude transmitted to the output port is

$$j(s_{21}{}^{(a)} + s_{21}{}^{(b)})\, v/2$$

Hence for the input and forward transmission S-parameters of the balanced amplifier, we have

$$s_{11}{}^{T} = (s_{11}{}^{(a)} - s_{11}{}^{(b)})/2 \tag{10.57}$$

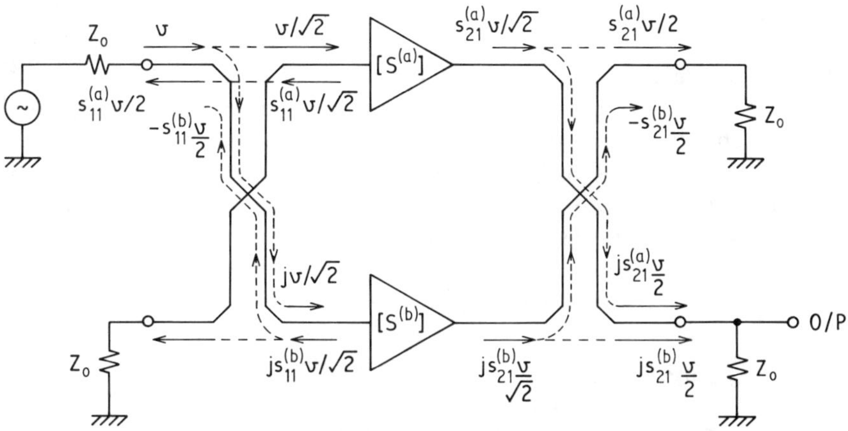

Fig. 10.35 Schematic of a balanced amplifier showing signal flow paths.

$$s_{21}{}^{\mathrm{T}}=\mathrm{j}(s_{21}{}^{(\mathrm{a})}+s_{21}{}^{(\mathrm{b})})/2 \tag{10.58}$$

From symmetry considerations, we can also immediately write

$$s_{22}{}^{\mathrm{T}}=(s_{22}{}^{(\mathrm{b})}-s_{22}{}^{(\mathrm{a})})/2 \tag{10.59}$$

$$s_{12}{}^{\mathrm{T}}=\mathrm{j}(s_{12}{}^{(\mathrm{b})}+s_{12}{}^{(\mathrm{a})})/2 \tag{10.60}$$

If the amplifier is truly balanced, $s_{11}{}^{(\mathrm{a})}=s_{11}{}^{(\mathrm{b})}$, and $s_{22}{}^{(\mathrm{b})}=s_{22}{}^{(\mathrm{a})}$. Regardless of the values of these reflection coefficients, the reflection coefficients at the input and the output of the balanced amplifier are zero. Furthermore, eqn (10.58) shows that, in the case of perfect balance, the magnitude of $s_{21}{}^{\mathrm{T}}$ is unchanged from the s_{21} of the original amplifiers.

In conclusion, despite the use of two nominally identical chips, there is no gain increase. Two chips instead of just one, and two 3 dB couplers, is the price we have to pay to achieve good matches at both input and output. The balanced amplifier is another example of additive amplification, however, and it will deliver a maximum power equal to twice the maximum power available from one of the MMICs alone. An excellent example of a balanced amplifier is the circuit using two power MMIC chips for the 6–18 GHz band described by Texas Instruments [10.8].

10.3 First trial device synthesis

In this section we attempt to make a preliminary identification of the material and device structural requirements of FETs designed to give a desired amplifier gain at a specified frequency. A case in point would be the FET required to give a gain of about 10 dB at 18 GHz, which is the gain achieved in the conjugate matching example of Fig. 10.13 in Section 10.2.2. We shall continue with this example to illustrate the procedures involved. The methods are an extension of those developed in Section 9.2.1.

When the input of the FET is conjugately matched to the generator, the unilateral transducer power gain is equal to the operating power gain. For a load of $R = R_L$, the power gain is given roughly by

$$G_p \approx (f_T/f)^2 \left(\frac{R_L}{R_g + R_i + R_s} \right)^{1/2} \tag{10.1}$$

Re-arranging for f_T,

$$f_T = f(G_p(R_g + R_i + R_s)/R_L)^{1/2} \tag{10.61}$$

If, at some frequency f, we require a prescribed gain G_p from a FET operating into a load R_L, we can estimate the transition frequency (f_T) needed from the device provided we know (or can estimate, or guess) the value of the resistance sum $(R_g + R_i + R_s)$, which we shall call R_Σ. Having estimated f_T, eqn (9.8) can be rearranged to find the gate length of the FET required:

$$L_G = \frac{v_{sat}}{2\pi f_T} - \frac{\epsilon(V_{D'G} + V_{BO})}{qNd} \tag{10.62}$$

In this equation it is necessary to assume reasonable values for the sheet charge density in the depletion layer, Nd. The drain-gate voltage $V_{D'G}$ is normally known from the bias condition at which it is required to operate the FET in the amplifier.

To pursue the example, we require $G_p = 10$ dB at $f = 18$ GHz into $R_L = 50\ \Omega$. We then find the alternatives given in Table 10.1.

Values of Nd in the range $1.5 \times 10^{12}\ \text{cm}^{-2} < Nd < 3 \times 10^{12}\ \text{cm}^{-2}$ might be considered to be the practical limits, yielding the alternative solutions for the gate length given in Fig. 10.36. The figure shows that the total sum resistance or, loosely, the total input resistance of the FET has a marked effect on the gate length the FET must have in order to realize the gain required; the higher the input resistance, the shorter the gate length must be. The sheet charge in the depletion layer has a less pronounced effect; increasing Nd allows larger gate lengths to be used. The limit on Nd is set by such constraints as gate-drain breakdown, gate-drain capacitance and the open channel current. One implication of these findings is that MMIC reproducibility may suffer if the input resistance of the FETs is poorly controlled.

To continue with the example, the FET in the matching network design calculation of Section 10.2.2 had $R_2 \approx 10\ \Omega$. Taking $R_2 \approx R_\Sigma$ then requires our

Table 10.1 Transition frequency f_T required for example FET, depending upon R_Σ

$R_\Sigma = (R_g + R_i + R_s)$ (Ω)	f_T (GHz)
5	18
10	25.5
15	31.2

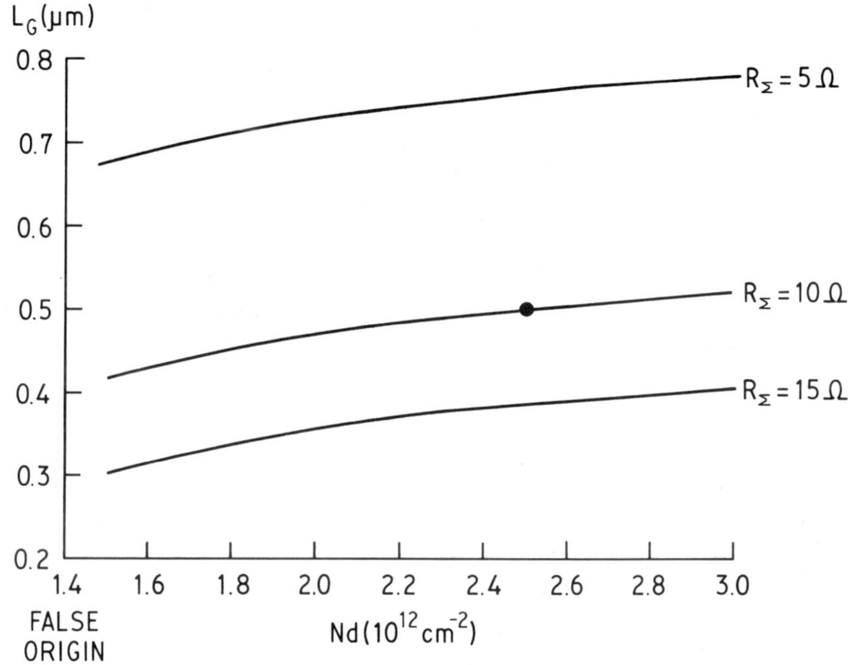

Fig. 10.36 Design chart for first trial FET design example.

FET to have technological parameters lying somewhere on the central curve in Fig. 10.36. We shall select the clinical value of $L_G = 0.5$ μm, which then implies an Nd product of 2.5×10^{12} cm^{-2} (the point marked on the figure). From Section 10.2.2, the FET also had

$$C_2 = 0.2 \text{ pF} \qquad g_{m0} = 30 \text{ mS}$$

The major contribution to C_2 will, of course, be the space-charge gate-channel capacitance, C_{gc}. The remainder will be made up of geometric capacitance, and Miller capacitance.

We may find the ratio of the depletion depth d to gatewidth Z_G by rearranging eqn (6.18):

$$(d/Z_G) = (\epsilon v_{sat}/g_{m0}) = 3.83 \times 10^{-4}$$

The choices are then as listed in Table 10.2.

Table 10.2 Choices for depletion depth, d, and associated channel doping, N, for example FET design

Z_G (μm)	d (μm)	N (10^{17} cm^{-3})
150	0.058	4.35
200	0.077	3.26
250	0.096	2.61

We shall select $Z_G = 200$ μm as offering a reasonable compromise between minimizing gate propagation delay and having a doping density in the channel that is not too high. From eqn (6.8) we can determine the source-gate bias at which the FET must be operated:

$$V_{S'G} = (qNd^2/2\epsilon) - V_{BO} = 0.534 \text{ V}$$

(i.e. the gate bias must be $V_{GS'} = -0.534$ V).

From eqn (6.24) we next find the space-charge layer extension into the gate-drain space, X,

$$X \leqslant \left[\frac{2\epsilon}{qN(V_{S'G} + V_{BO})}\right]^{1/2} (V_{D'G} + V_{BO}) = 0.25 \text{ μm}$$

Then, from eqn (6.15),

$$C_{gd} = \frac{2\epsilon Z_G}{1 + 2X/L_G} = 0.023 \text{ pF}$$

Hence, our design FET would have a space-charge component of gate-drain capacitance of 0.023 pF. A full simulation of the amplifier consisting of the FET and the matching networks would, of course, include this value of C_{gd} to check for the effect of including all the elements omitted from the first-order synthesis procedure being used here.

It remains to find the channel layer thickness, W. If we assume a profile with a sharp tail, the drain current will mainly consist of channel current, I_{CH}, because the substrate current will be small for the high interfacial potential barrier associated with a sharp profile tail (refer to Fig. 6.30). From eqn (6.11),

$$I_D \approx I_{CH} = qNv_{sat}(W-d)Z_G$$

A value for I_{CH} can be found from the requirements that $R_\Sigma = 10\ \Omega$ and that R_i must (obviously) be less than R_Σ by the amount $(R_g + R_s)$. Allowing $R_g \approx 2\ \Omega$ and $R_s \approx 2\ \Omega$ also, gives $R_i \approx 6\ \Omega$. From eqn (6.7), we must then have

$$I_{CH} \approx (v_{sat}L_G/\mu_0 R_i) = 27.8\ \text{mA}$$

From eqn (6.11), we then find W,

$$W \approx 0.104\ \mu\text{m}$$

Finally, the synthesized device must be checked for adequate breakdown voltage, which entails first finding $\overline{I_{DSS}}$:

$$\overline{I_{DSS}} = qNv_{sat}(W-d_0)\ \text{mA mm}^{-1}$$

where d_0 is the depth of depletion with $V_{S'G} = 0$ V applied. Equation (6.8) gives $d_0 \approx 0.06\ \mu$m; then $\overline{I_{DSS}} \approx 210\ \text{mA mm}^{-1}$. From Fig. 6.60 we find that the gate-drain breakdown voltage V_{DGB} should be about 17 V, which is easily adequate (in fact, we could afford to use a slightly thicker channel layer in the interests of keeping R_Σ within bounds).

To complete the results checking, the equations of Table 6.1 can be used to back-calculate the equivalent circuit element values of the FET from the synthesized technological data. A summary is given in Table 10.3 in which practical, rounded, values have been used.

After rigorous checking for satisfactory microwave behavior of the

Table 10.3 Back-calculation of equivalent circuit element values from synthesized technological data for FET

Input data	*Output data*
$L_G = 0.5\ \mu$m	$\mathscr{L}_g = 10$ pH
$Z_G = 200\ \mu$m	$C_{gc} = 0.17$ pF
$W = 0.1\ \mu$m	$C_{gd} = 0.023$ pF
$N = 3.5 \times 10^{17}\ \text{cm}^{-3}$	$I_{CH} = 30$ mA
$\mu_0 = 3000\ \text{cm}^2\ \text{Vs}^{-1}$	$R_i = 5.5\ \Omega$
$m = 2$ (gate fingers)	$g_{m0} = 31$ mS
$V_{D'S'} = 3$ V	$\tau_{gm} = 0.5$ ps
$V_{S'G} = 0.5$ V	

overall amplifier using a network simulation program, the circuit could be constructed with the FET specification given in the left column in Table 10.3.

10.4 FET synthesis by reverse modeling

Sometimes, initial attempts at the design of a circuit result in a FET equivalent circuit that the designer would like to have available to him. Situations of this kind arise when the equivalent circuits of the FETs ordinarily available, perhaps from a foundry, are inadequate in some respect. A case in point is the distributed amplifier†; not only must the input capacitance and input resistance be kept within bounds to achieve acceptable gate-line attenuation (see eqn (10.50)), and not only must the transconductance be large enough to give the overall power gain desired (eqn (10.54)), but the gate-drain capacitance must be kept small to prevent unwanted effects arising from coupling between the drain line and the gate line. A full computer simulation which incorporates feedback through the FET (or finite s_{12}) makes the risks apparent. In such cases, the reverse modeling principle of Chapter 8 can be used to provide the circuit designer and FET technologist with an idea of the FET structure they require.

Suppose the standard 0.5 μm by 250 μm FET offered by a foundry or other process line has the following circuit element values at $V_{S'G}=0$ V:

$$C_{gc}=0.2\text{ pF} \qquad C_{gd}=0.025\text{ pF} \qquad g_{m0}=30\text{ mS}$$

By simulation, let us suppose it has become obvious that the transconductance offers little margin for error in the gain, and that the gate-drain capacitance is too large. By substituting different values it has been ascertained that the FET we would really like has (at $V_{S'G}=0$ V bias)

$$C_{gc}=0.2\text{ pF} \qquad C_{gd}=0.015\text{ pF} \qquad g_{m0}=35\text{ mS}$$

The question is, what structure is needed to give these values?

Reverse modeling provides an answer‡. Assuming various gatewidths,

†We are here assuming that the FET used is a single gate device, not a dual gate FET or a cascode connection.

‡We should bear in mind that the equations in this book assume that the gate recess depth equals the depth of depletion at the surface in the gate-source and gate-drain spaces. In practical FETs, this condition is rarely if ever met, if only because the surface-state occupancy changes during FET operation. Gate recess depth and the presence of N^+ cap layers affect C_{gd} and, to some extent, C_{gc} and g_{m0}. The full set of equations which abandon the simple assumption give slightly different results for L_G from those in Table 10.4. Nevertheless, the conclusion that L_G must be reduced by approaching 30% is correct. Section 8.3 gives a discussion of likely errors.

eqns (8.6) and (8.8) give for L_G and N_u respectively the values listed in Table 10.4.

Table 10.4 FET structure synthesis by reverse modeling

Z_G (μm)	L_G (μm)	N_u (10^{17} cm^{-2})
250	0.33	1.7
200	0.38	2.7
150	0.43	4.7

As a matter of practical choice, the 200 μm gatewidth option might be the best compromise; 0.33 μm is more difficult to hold to tolerance than 0.38 μm, even with electron-beam direct writing and, at the other extreme, 4.7×10^{17} cm^{-3} doping is getting a little large, particularly if activation of an ion implant is part of the process.

The channel layer thickness and the desired source-gate spacing are obtained from the device resistances. Suppose a value of $R_\Sigma = R_g + R_i + R_s = 7.5\ \Omega$ is required, and that $R_g = 1.5\ \Omega$. The requirement is thus for $(R_i + R_s) = 6\ \Omega$. Assuming an ohmic contact specific resistance of $\rho_c \approx 0.2\ \Omega$ mm then gives $R_{contact} = 1\ \Omega$ for a 200 μm gatewidth FET (see eqn (6.31)). That leaves 5 Ω to be accounted for by R_{SG} (eqn (6.32)) and R_i (eqn (6.7)). Because one FET is to be operated at $I_D = I_{DSS}$ (i.e. at $V_{S'G} = 0$ V), substituting from eqn (6.11) for $qNWZ_G$ into eqn (6.32) while assuming $d \ll W$ allows us to write

$$R_{SG} \approx L_{SG} v_{sat} / \mu_0 I_{DSS} \tag{10.63}$$

whereas eqn (6.7) gives, for R_i,

$$R_i \approx L_G v_{sat} / \mu_0 I_{DSS} \tag{6.7}$$

i.e. roughly, $(R_{SG}/R_i) \approx L_{SG}/L_G$. Taking $L_{SG} = 0.5$ μm and $L_G = 0.38$ μm means we must have

$$R_{SG} \approx (0.5/0.88) \times 5\ \Omega \approx 2.8\ \Omega$$

$$\text{and} \qquad R_i \approx 2.2\ \Omega$$

Taking $\mu_0 \approx 3000$ cm^2 Vs^{-1} ($= 0.3$ m^2 Vs^{-1}) then gives, from eqn (6.7) above,

$$I_{DSS} \approx 58 \text{ mA}$$

$$\text{or} \qquad \overline{I_{DSS}} \approx 290 \text{ mA mm}^{-1}$$

Once again, this number can be used with Fig. 6.60 to check that the gate-drain breakdown voltage is adequate for the intended application. To calculate the channel layer thickness under the gate, W, use is made of eqn (6.11) once more. First, the depletion depth at $V_{S'G}=0$ V must be found. Using eqn (6.8) with $N=N_u=2.7\times10^{17}$ cm^{-3} and $V_{BO}=0.8$ V,

$$d_0=0.065\ \mu\text{m}$$

Finally, from eqn (6.11),

$$W\approx\frac{I_{DSS}}{qNv_{sat}Z_G}+d_0\approx0.13\ \mu\text{m}$$

The structural specification of the synthesized FET is summarized in Table 10.5.

Table 10.5 Summary specification of synthesized FET structure

L_G (μm)	Z_G (μm)	N (10^{17} cm^{-3})	W (μm)
0.38	200	2.7	0.13

Methods such as these can be very valuable in helping MMIC designers and process technologists understand one another's needs. In Section 10.5, an extension of the foregoing techniques to the synthesis of power FET elements will be discussed. Note that the FET specification in Table 10.5 is superficially similar to that in Table 10.3. The sheet concentration of the layer after gate recess etching, NW, is the same: 3.5×10^{12} cm^{-2}. Consequently, a process engineer etching to the time-honoured criterion of a pre-decided value of ungated channel current would fail to distinguish between the two layers. Moreover, 0.5 μm sounds similar enough to 0.4 μm for the gatelength. Despite these apparent similarities, it is highly unlikely these two FETs could be substituted for one another.

10.5 FET synthesis for distributed power amplifiers

If many small FETs are integrated in parallel in an effort to combine their individual output powers, the overall dimensions of the device may become a significant fraction of the wavelength of electromagnetic waves in GaAs at the frequency of interest. In that circumstance the FET elements will be driven with signals of progressive phase lag the further away from the main feed point the element is. Also the number of FETs in the parallel array may become so large that the impedance level is lowered to values making matching difficult over anything but a narrow band ($\Delta\omega/\omega \approx 5\%$). The traveling-wave amplifier, or distributed amplifier, offers a means of realizing relatively high power levels over a broad band and, as such, is an alternative to using two lossy-matched MMICs in a balanced amplifier [10.7–10.8].

As usual, we start with a question: how are we to know whether a desirable specification for an amplifier with certain gain, output power, frequency response and load impedance, is fundamentally realizable? To answer this question, we first of all develop some large-signal equations for these quantities by analysis of the distributed amplifier structure.

10.5.1 Power-impedance considerations

Referring to Fig. 10.37, each of the N GaAs FETs has finite input resistance as well as capacitance, so that the gate line attenuates signals propagating along it. For simplicity we shall assume that topology amendments can be made to compensate for the progressive loss of signal amplitude so that all elements experience equal gate drive [10.17].

At dc, half the total drain current flows into the external load, and half flows into the dummy load (assuming the loads to be of equal value). Under RF conditions, partial phase cancelation of the drain current signals propagating towards the dummy load occurs but, for a well-designed amplifier, those propagating toward the external load still combine to produce the same power in the load as at dc. Thus, as a first approximation,

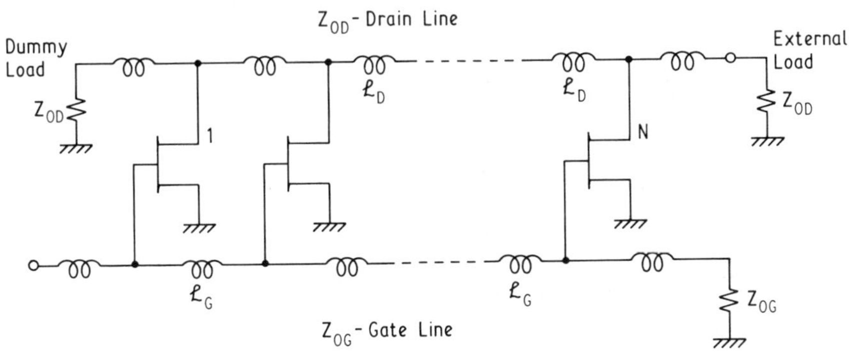

Fig. 10.37 Distributed amplifier configuration for power traveling-wave amplifier analysis.

expressions for the RF power and gain can be written by treating the amplifier at dc [10.17].

Assuming uniform gate- and drain-line voltage distributions and an overall response which is sensibly independent of frequency, the useful power delivered to the external load in Fig. 10.37 may be written as

$$P_O = n\bar{P}_O/2 \tag{10.64}$$

where n is the total gatewidth of all N FETs added together, and $\bar{P}_O$ is the power per millimetre of gate width of any one FET (all FET elements are assumed to be identical).

Referring to Fig. 10.38, to extract optimum power from the FET elements the total load at the drain of each FET must equal the value

$$R_L = (V_{DSB} - V_K)/n\bar{I}_F \tag{10.65}$$

which is the load that will maximize the output power at the value $n\bar{P}_O$ from the FETs having a total gate periphery of n millimeters, all FETs operating at a drain current of $\bar{I}_F$ mA mm^{-1} under forward gate drive of typically $V_{GS} = V_F = +0.5$ V. Thus we must have

$$\frac{Z_{OD}}{2} = \frac{V_{DSB} - V_{K)}}{n\bar{I}_F} \tag{10.66}$$

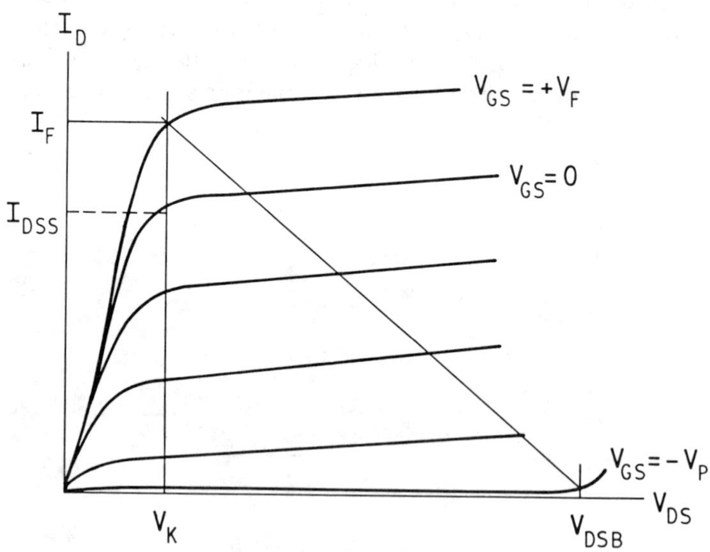

Fig. 10.38 FET limits defining available current and voltage peak-to-peak amplitudes.

Combining eqns (10.64) and (10.66) we obtain

$$Z_{0D} = \frac{(V_{DSB} - V_K)}{\bar{I}_F} \frac{\bar{P}_O}{P_O} \tag{10.67}$$

Independently of all else, this expression fixes the external drain line load in terms of the FET limiting parameters $\bar{P}_O$, V_{DSB}, and $\bar{I}_F$, and the desired total output power, P_O.

Assuming that full output power is achieved when the FET channels are fully modulated between the open-channel condition ($V_{GS} = +V_F$) and pinch off ($V_{GS} = -V_p$), the RF power input to the amplifier is

$$P_{IN} = \frac{(V_p + V_F)^2}{8Z_{0G}} \tag{10.68}$$

where the factor of 8 converts the peak-to-peak voltage swing $(V_p + V_F)$ to rms The gain G is simply obtained as the output power P_O divided by the input power, P_{IN},

$$G = \frac{8P_O Z_{0G}}{(V_p + V_F)^2} \tag{10.69}$$

Rearranging, we have

$$Z_{0G} = \frac{G(V_p + V_F)^2}{8P_O} \tag{10.70}$$

Hence the gate-line impedance is constrained by the gain required, G, the FET pinch-off V_p, and the desired power output, P_O.

10.5.2 Frequency considerations

If the capacitance per FET drain is C_D (Fig. 10.39(a)), the drain-line cut-off frequency is, in analogy with eqn (9.13),

$$f_{CD} = 1/\pi(\mathscr{L}_D C_D)^{1/2} = 1/\pi C_D Z_{0D} \tag{10.71}$$

Similarly, with reference to Fig. 10.39(b), the gate-line cut-off frequency is

$$f_{CG} = 1/\pi(\mathscr{L}_G C_G)^{1/2} = 1/\pi C_G Z_{0G} \tag{10.72}$$

Assuming f_{CG} must exceed the maximum operating frequency (or bandwidth) f_B by some factor, say 2, then

$$f_{CG} \geqslant 2f_B$$

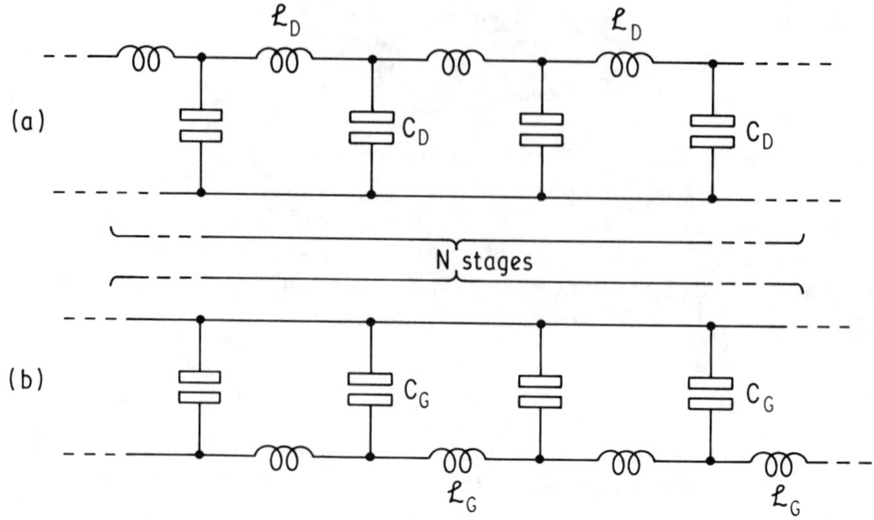

Fig. 10.39 Simplified representation of drain and gate lines.

or

$$C_G \leqslant 1/2\pi f_B Z_{0G} \tag{10.73}$$

Simple considerations of the traveling-wave amplifier suggest that, for maximum output, the phase shift per section along the gate and drain lines must be equal, requiring that

$$f_{CD} = f_{CG} \tag{10.74}$$

i.e.

$$C_D = C_G(Z_{0G}/Z_{0D}) \tag{10.75}$$

There are other considerations relating to the number of stages, N, and to the gain, G, which show that the gain achievable is dependent solely upon the basic FET large-signal parameters and is independent of the output power required. Such conclusions follow from further manipulation of eqns (10.67), (10.69) and (10.75), [10.17].

10.5.3 FET synthesis, an example

The synthesis of FET structure proceeds by inverting (in the mathematical sense) the equations of the preceding two sections. To illustrate the method, we shall take the Raytheon 250 mW, 20 GHz amplifier [10.18] which will be examined by more sophisticated methods in Section 11.6.

The Raytheon amplifier achieved the broad specification given in Table 10.6 from eight FETs grouped in pairs and connected at the drains (see Fig. 11.25). Artificial gate lines, each of 50 Ω, were used to drive the eight FET gates. The two gate lines were combined to a common feed line (see Fig. 11.25), the whole arrangement being equivalent to $N=4$ FETs and $Z_{0G}=25\ \Omega$. We shall insert these values into rearranged equations from Sections 10.5.1 and 10.5.2, starting with eqn (10.69):

$$V_p+V_F=\left(\frac{8P_OZ_{0G}}{G}\right)^{1/2}$$

Taking $V_F=0.5$ V, we then obtain

$$V_P=\left(\frac{8\times 0.25\times 25}{3.16}\right)^{1/2}-0.5=3.5\ \text{V}$$

Equation (10.53) gives

$$C_G\leqslant 1/2\pi(20\times 10^9)25=0.318\ \text{pF}$$

The major contribution to C_G will be the gate-channel capacitance C_{gc}, so we assume

$$C_{gc}\leqslant 0.318\ \text{pF}$$

With reference to Fig. 10.38,

$$\bar{P}_O=(V_{DSB}-V_K)\bar{I}_F/8$$

But, from eqn (10.67),

$$\bar{I}_F=\frac{(V_{DSB}-V_K)}{Z_{OD}}\frac{\bar{P}_O}{P_O}$$

Eliminating $\bar{I}_F$ and $\bar{P}_O$,

$$P_O=\frac{(V_{DSB}-V_K)^2}{8Z_{OD}}$$

Table 10.6 Summary specification of 250 mW, 20 GHz, Raytheon MMIC [10.18]

P_O (mW)	G_{ss} (dB)	f_B (GHz)	Z_{OG} (Ω)	Z_{OD} (Ω)
250	5	20–21	25	50

Note: G_{ss} is the small-signal power gain.

whence

$$V_{DSB} = V_K + (8Z_{OD}P_O)^{1/2}$$
$$V_{DSB} = 2 + (8 \times 50 \times 0.25)^{1/2} = 12 \text{ V}$$

But

$$V_{DGB} = V_{DSB} + V_p = 12 + 3.5 = 15.5 \text{ V}$$

From eqn (6.11), and neglecting substrate current,

$$I_D \approx I_{CH} = qNv_{sat}Z_G(W - d)$$

where N is the channel doping density, and is not to be confused with the number of FET stages (also referred to by N). Writing

$$I = qNv_{sat}Z_G W$$

allows I_D to be written as

$$I_D = I(1 - d/W)$$

But

$$d = \left[\frac{2\epsilon(V_{SG} + V_{BO})}{qN}\right]^{1/2} \quad \text{and} \quad W = \left[\frac{2\epsilon(V_p + V_{BO})}{qN}\right]^{1/2}$$

Substituting, we obtain

$$I_D = I\left[1 - \left(\frac{V_{SG} + V_{BO}}{V_p + V_{BO}}\right)^{1/2}\right] \tag{10.76}$$

A drain current of I_F results when a gate voltage of $V_F = +0.5$ V is applied, i.e. $V_{GS} = +0.5$ V, or $V_{SG} = -0.5$ V. With $V_{BO} = 0.8$ V we then obtain

$$I_F = I\left[1 - \left(\frac{-0.5 + 0.8}{3.5 + 0.8}\right)^{1/2}\right]$$

or

$$I = 1.36 I_F$$

I_{DSS} results when $V_{SG} = 0$ V in eqn (10.76), giving

$$I_{DSS} = 1.36 I_F\left[1 - \left(\frac{0.8}{4.3}\right)^{1/2}\right] = 0.773 I_F$$

From eqn (10.66),

$$n\bar{I}_F = 2(V_{DSB} - V_K)/Z_{OD} = 400 \text{ mA}$$

If $I_{DSS}=0.773 I_F$, then

$$n\overline{I_{DSS}}=309 \text{ mA}$$

From Fig. 6.60, a drain-gate breakdown voltage of approximately 15 V requires, if $L_G=1$ μm,

$$\overline{I_{DSS}}=275 \text{ mA/mm, giving } n=1.12 \text{ mm}$$

or if $L_G=0.5$ μm,

$$\overline{I_{DSS}}=225 \text{ mA/mm, giving } n=1.37 \text{ mm}$$

We shall take $n=1.2$ mm, so each of the four FETs has a gate width equal to the total periphery, n, divided by the number of FETs, N, giving

$$Z_G=300 \text{ μm per FET}$$

So far, we have determined the electrical quantities V_p, C_{gc}, V_{DSB}, V_{DGB} and $\bar{I}_F$, and we have also determined the gatewidth, Z_G. It remains for us to find the gate length, L_G, the channel doping density, N, and the channel layer thickness, W, required to realize these electrical FET quantities. L_G and N may be determined by reverse modeling (see Chapter 8) once the gate voltage bias and the transconductance, g_{m0}, have been found.

To find the bias value of V_{SG}, eqn (10.76) is used. The circuit is biased at $I_D=I_F/2$, so,

$$0.5=1.36\left[1-\left(\frac{V_{SG}+0.8}{4.3}\right)^{1/2}\right]$$

i.e. the gate bias required is $V_{SG}=0.92$ V for $I_D=I_F/2$. At this bias point we find g_m by differentiation of eqn (10.76) to be

$$g_m=\frac{1.36\times I_F}{2}\,\frac{1}{[(V_p+V_{BO})(V_{SG}+V_{BO})]^{1/2}}$$

Now, for each FET,

$$I_F=(n\bar{I}_F)Z_G/n=(400 \text{ mA}/1.2 \text{ mm})0.3 \text{ mm}=100 \text{ mA}$$

Therefore at $V_{SG}=0.92$ V,

$$g_m=\frac{1.36\times 100}{2(4.3\times 1.72)^{1/2}}=25 \text{ mS}$$

As a check, we can use eqn (10.55) to show that the small-signal gain of the amplifier is

$$G = (Ng_m/2)^2 Z_{0G} Z_{0D}$$
$$= (4 \times 0.025/2)^2 \times 25 \times 50 = 3.13 = 4.96 \text{ dB}$$

This value checks with the gain specified, i.e. $G_{ss} = 5$ dB. To determine the gate length, we use eqn (8.6) and assume $g_{m0} \approx g_m = 25$ mS (g_{m0} will actually be a few percent larger when the voltage feedback term $i_d R_s$ across the FET's source resistance is allowed for). It is necessary also to assume typical values for C_{gd} and see what values of L_G result:

$$\text{if } C_{gd} = 0.025 \text{ pF} \quad \text{then} \quad L_G = 0.88\ \mu\text{m}$$

or

$$\text{if } C_{gd} = 0.03 \text{ pF} \quad \text{then} \quad L_G = 0.96\ \mu\text{m}$$

Also, from the equation for reverse modeling of doping density eqn (8.8), we find, with $V_{S'G} \approx V_{SG} = 0.92$ V, $g_{m0} = 25$ mS and $Z_G = 300\ \mu$m,

$$N = N_u = 1.3 \times 10^{17} \text{ cm}^{-3}$$

N_u is the concentration to which a uniformly doped channel would need to be N-type doped.

The final step is to deduce the channel layer thickness, W, and to check that the value obtained gives consistency with the already known value of pinch-off voltage, $V_p = 3.5$ V. Given

$$\overline{I_{DSS}} = qNv_{sat}(W - d_0)$$

we obtain

$$W = \frac{\overline{I_{DSS}}}{qNv_{sat}} + d_0$$

where

$$d_0 = \left(\frac{2\epsilon V_{B0}}{qN}\right)^{1/2}$$

With $\epsilon = 1.15 \times 10^{-10}$ C V^{-1} m^{-1}, $V_{B0} = 0.8$ V, $q = 1.6 \times 10^{-19}$ C and $N = 1.3 \times 10^{23}$ m^{-3},

$$d_0 = 0.094\ \mu\text{m}$$

Also,

$$\overline{I_{DSS}} = n\overline{I_{DSS}}/n = 309 \text{ mA}/1.2 \text{ mm} = 258 \text{ mA/mm}$$

Hence

$$W = \frac{258}{q \times 1.3 \times 10^{23} \times 10^5} + 0.094\ \mu\text{m} = 0.218\ \mu\text{m}$$

As a check, we write

$$V_p + V_{B0} = qNW^2/2\epsilon = 4.3\ \text{V}$$

and therefore

$$V_p = 4.3 - 0.8 = 3.5\ \text{V}$$

which is exactly the result obtained previously. We finally arrive at the comparison given in Table 10.7. This demonstrates just how close to a practical specification it is possible to get using fairly simple equations of device physics. In Chapter 11 it will be shown that a full computer optimization yields very similar values for the technological specification of the FETs required to realize the Raytheon distributed amplifier power MMIC.

Several remarks relating the synthesized FET to practical issues should be made. First, the analysis assumed FET linearity right up to the limits of $(V_p + V_F)$ on the gate line, and $(V_{DSB} - V_K)$ and I_F on the drain-line, all being reached concurrently. In practice, the drain-line limits mark the 1 dB gain compression point as a rough rule of thumb. Secondly, the synthesized value of doping density, N, would be raised by accounting for the fact that $g_{m0} > g_m$; $g_{m0} = 28.8$ mS would raise N to $1.5 \times 10^{17}\ \text{cm}^{-3}$, the Raytheon value. Thirdly, from eqn (10.75) we must have

$$C_D = 0.318\ \text{pF} \times 25\ \Omega/50\ \Omega = 0.159\ \text{pF}$$

which is much larger than the FET output capacitance, $C_o \approx 0.05$ pF. Thus in the practical circuit, the drain capacitance must be increased either by the use of series transformers (as in Fig. 11.25), or by the parallel integration of metal–insulator–metal capacitors.

10.6 Final remarks

This chapter has provided some background on the commonly used amplifier topologies, and has dealt with some examples of how FETs can be designed to suit. What we have tried to advance here is the idea that there are benefits to be had from designing FETs concurrently with amplifier circuits and, therefore, that FET structure synthesis might usefully be put on a formal basis. Even so, there are many aspects to overall MMIC design that have not

Table 10.7 Comparison of Raytheon experimental FETs in power distributed MMIC and FET specification by analytic synthesis

	Amplifier			*FET*					
	P_O (mW)	G_{ss} (dB)	f_B (GHz)	L_G (μm)	N (cm^{-3})	W (μm)	Z_G (μm)	V_p (V)	V_{DGB} (V)
Raytheon [10.18]	250	5	21	1	1.5×10^{17}	–	300	4	15
FET synthesis	250	5	20–21	0.96	1.3×10^{17}	0.22	300	3.5	15.5

been mentioned. For instance, nothing has been said about low-noise amplifiers, despite the fact that such circuits are needed for most systems applications. A few closing remarks on issues of this kind are in order.

Reactively matched amplifiers generally offer lower noise than other topologies; there are no resistors connected to the FET input to add kTB type noise. For low-noise applications, the lossless input network must transform the generator impedance into a special value of impedance to achieve the minimum noise figure, the special value itself being dependent upon the FET input impedance. Because FETs vary from device to device, it is usual to tune the input of practical amplifiers with metal or ceramic disks. MMIC chips are too small to offer any serious prospect of tuning in this way, so low noise amplifiers are, in the main, realized as hybrid MICs. For this reason, added to the low-noise amplifier not being a topology in its own right, such amplifiers have not been explicitly treated in this chapter.

None of the FET synthesis examples given in Sections 10.3–10.5 attempted to design for a specified output conductance, $g_o = 1/r_o$. Substrate current and its associated equivalent circuit elements were ignored altogether. Bearing in mind that g_o depends to first order on technological factors that tend to be poorly controlled, such as the N-layer to buffer–substrate interfacial potential barrier, the gate recess depth and surface-state occupancy, it is inappropriate for the time being to proceed as if FETs having particular values of drain conductance could be designed. At best, all that can be offered are a few qualitative guidelines as to the general type of structure that should be considered if low g_o (i.e. high r_o) is a real requirement. We therefore offer these brief guidelines now. In the gate-drain space, the FET should have a relatively thin, lightly doped channel, no N^+ contact layer, a sharp tail to the channel doping profile, and a shallow gate recess. Low g_o tends to be associated with low values of C_{gd} and high values of τ_{gm}.

In closing this discussion, we observe that much remains to be said about designing and engineering FET elements as part of MMIC design and implementation. The quick, analytic methods examined here should be viewed as a precursor to full computer aided design (CAD) routines; the former offer the highly necessary physical insight, whereas the latter offer the most accurate numerical answers. We consider some CAD developments in Chapter 11.

References

10.1 Podell, A., Moghe, S., Lockie, D., and Ali, F. GaAs real estate—making the most efficient use of semiconductor surface area, *Microwave Journal*, **30** (10), 208–212, November 1987.

10.2 Abrie, P. L. D. *The Design of Impedance-Matching Networks for Radio-Frequency and Microwave Amplifiers*, Artech House, Dedham, 1985.
10.3 Carlin, H. J. and Komiak, J. J. A new method of broad-band equalization applied to microwave amplifiers, *IEEE Trans. Microwave Theory Tech.*, **MTT-27** (2), 93–99, February 1979.
10.4 Mellor, D. J. and Linvill, J. G. Synthesis of interstage networks of prescribed gain versus frequency slopes, *IEEE Trans. Microwave Theory. Tech.*, **MTT-23** (12), 1013–1020, December 1975.
10.5 Hewlett Packard *S-Parameters . . . Circuit Analysis and Design*, application note 95, September 1968.
10.6 Honjo Kazuhiko and Takayama Yoichiro. GaAs FET ultrabroad-band amplifiers for Gbit/s data rate systems, *IEEE Trans. Microwave Theory Tech.*, **MTT-29** (7), 629–636, July 1981.
10.7 Nelson, S., Mitchell, W., Halladay, R., Beall, J., Lehmann, R., Heston, D., Drury, D., Williams, R., Rhine, D., Jones, M., Lerude, G., Palmer, C., Green, D., Nuttall, J. and Sapashe, C. Power amplifier leads MMIC line, *Microwaves and RF*, **25** (13), 125–132, December 1986.
10.8 Halladay, R., Nelson, S. and Anderson, K. Dual MMICs deliver 1 W at Ku band, *Microwave Journal*, **30** (8), 168–178, August 1987.
10.9 Perez, F. and Ortega, V. A graphical method for the design of feedback networks for microwave transistor amplifiers: theory and applications, *IEEE Trans. Microwave Theory Tech.*, **MTT-29** (10), 1019–1027, October 1981.
10.10 Kennan, W., Andrade, T. and Huang, C. C. A 2–18 GHz monolithic distributed amplifier using dual gate GaAs FETs, *IEEE Trans. Electron. Devices*, **ED-31** (12), 1926–1931, December 1984.
10.11 Ayasli, Y., Mozzi, R. L., Vorhaus, J. L., Reynolds, L. D. and Pucel, R. A. A monolithic GaAs 1–13 GHz travelling-wave amplifier, *IEEE Trans. Microwave Theory Tech.*, **MTT-30** (7), 976–981, July 1982.
10.12 Venet, C. and Baudet, P. 2–20 GHz hybrid GaAs MESFET distributed amplifier, *Electron. Lett.*, **21** (9), 376–377, 25 April 1985.
10.13 Ayasli, Y., Reynolds, L. D., Vorhaus, J. L. and Hanes, L. K. 2–20 GHz GaAs travelling-wave amplifier, *IEEE Trans. Microwave Theory Tech.*, **MTT-32** (1), 71–77, January 1984.
10.14 Strid, E. W. and Gleason, K. R. A DC-12 GHz monolithic GaAs FET distributed amplifier, *IEEE Trans. Microwave Theory Tech.*, **MTT-30** (7), 969–975, July 1982.
10.15 Niclas, K. B. and Pereira, R. R. The matrix amplifier: a high gain module for multioctave frequency bands, *IEEE Trans. Microwave Theory Tech.*, **MTT-35** (3), 296–306, March 1987.
10.16 Hoffmann, R. K. *Handbook of Microwave Integrated Circuits*, Artech House, Norwood, 1987, p. 262.
10.17 Ladbrooke, P. H. Large-signal criteria for the design of GaAs FET distributed power amplifiers, *IEEE Trans. Electron Devices*, **ED-32** (9), 1745–1748, September 1985.
10.18 Ayasli, Y., Reynolds, L. D., Mozzi, R. L. and Hanes, L. K. 2–20 GHz GaAs travelling-wave power amplifier, *IEEE Trans. Microwave Theory Tech.*, **MTT-32** (3), 290–295, March 1984.

Chapter 11

Computer-aided design

11.1 Introduction

Having developed the equations which describe the microwave behavior of both active and passive components in terms of their material composition and structural dimensions, the question arises as to how these equations might best be used to design MMICs. As we shall see, at the time of writing a fair part of this process is creative and very much dependent upon the flair of individuals. It is still the task of individual circuit designers to select a certain style, type or topology of network as the first step in designing a circuit to meet a given specification, and to calculate values for all the elements in it. Almost without exception, this practice is conducted around fixed FETs, i.e. FETs which have been well characterized at microwave frequencies by making *S*-parameter measurements upon a sample of perhaps several hundred devices. Having thus arrived at a first design, the next step is the analytic step of simulating the circuit using a proprietary network analysis program such as TOUCHSTONE© or SUPERCOMPACT© to confirm the design, and perhaps to make further adjustment to some elements in the circuit to improve it in some respect or other. At this stage, the circuit may also be analyzed for its sensitivity to variations in some or all of the components, including the FETs. When this is done, typically the elements in the equivalent circuit are varied individually, and variations in the circuit response to, say, variations in gate-drain capacitance are studied. Finally, by allowing all elements to vary simultaneously and randomly between prescribed statistical limits, the yield of the circuit design can be forecast.

The first point to be made about this process of design is that much of it does not involve computer-aided design at all, but computer-aided analysis. Despite this self-evident truth, the term CAD is applied generically, and will be used here to reflect common usage. Secondly, in the foregoing method of designing circuits by trial and analysis there are a number of flaws.

(a) The criteria used in the creative initial phase of selecting a circuit topology and assigning element values tend to be performance-driven criteria aimed at achieving maximum gain-bandwidth, minimum noise, maximum power, or whatever the broad objectives of the circuit are. The performance-driven design approach appears to have been inherited, along with the circuit designers who practise it, from the hybrid MIC fraternity. Unlike hybrids, however, there is next to no scope for tuning MMICs once they are made, with the result that the yield of chips within specification is often very low (units of percent, or even lower). What is wanted from the outset is a change of emphasis in design procedures so that yield becomes a major criterion throughout.

(b) Even the most rudimentary sensitivity and yield studies show that in many circuits it is the GaAs MESFET itself which is the source of out-of-specification behavior of the complete chip. That being so, accurate representation of the FET in computer-aided analysis and design is of paramount importance if meaningful answers are to be obtained.

(c) The practice of varying the FET equivalent circuit elements one-by-one, to study the effect on the chip, is erroneous. As reference to the equations of Table 6.1 in Chapter 6 will show, virtually nothing the technologist can do will change, for example, C_{gd} in isolation. The real variables are N, W, L_G etc. and a change in any one of them will affect many if not all of the equivalent circuit elements. In other words, there are important correlations between the ways in which the equivalent circuit elements vary which must be accounted for.

(d) Acceptance of a pre-characterized or fixed FET may present a severe limitation on the design process. Since the FET exerts a powerful influence over chip performance and reproducibility, it follows logically that careful choice of FET structure should offer much scope for improving chip yield or performance, or both. As will be shown, this reasoning extends even to circuits designed around fixed FETs in the foregoing way; having thus arrived at a design for the passive network, the yield or performance of the overall circuit can often be improved by changing the constructional features of the FET.

(e) The design practice outlined is a consequence of trying to separate the overall task of realizing MMICs into a series of discrete and poorly interacting tasks, each of which has a pre-determined output or result (e.g. a 0.1 μm layer doped to 2.5×10^{17} cm^{-3}; 0.5 μm gate strips), rather than seeing MMIC engineering as an overall process embracing tasks which must of necessity interact. Only in this way can the proper consequences of, say, having a wrong layer doping be identified. In other words, when the interactions at all levels are correctly represented, decisions as to the values of doping density, layer thickness, and gate length can be made for each and every circuit, just as such decisions are made for the inductance, resistance and capacitance values in the passive network. Finally and very importantly, only in this manner can the acceptable tolerances in all these interacting technological factors be determined.

The particular contribution of this book is in these five areas. Progress depends upon providing a set of computer-based tools which allow address of each of the design problems (a) to (e) individually and together, which is what the compact equations of Chapter 6 have attempted to do. We therefore come to regard the FET as an integral, integrable and designable element of a circuit. For the time being, however, we shall persist with the trial design, analysis, and redesign approach to MMIC design, and merely draw attention to the longer-term possibilities for the development of true design-centering computer routines.

11.2 Sensitivity analysis—basis

We start with the assumption that a first design of a circuit has been done around an initial fixed FET. The problem is next to determine the sensitivity of the circuit to variations in all those factors which go to make up the circuit. It is important to address this problem at the technological level and not just the circuit element level, because it is technology which is at the root of all MMIC realization.

Consider, for instance, variations in the layer doping, N. From Table 6.1, Chapter 6, we can expect C_{gc}, C_{gd}, R_i, g_{m0}, τ_{gm} to vary, and from eqns (6.28), (6.29), (6.32), (6.33) and (6.39) we can expect r_o, R_s and R_d to vary too. Hence a change in the single technological variable N leads to correlated changes in all the equivalent circuit elements. The same is true for the other technological variables, W, L_G . . ., including the occupied surface-state density D_S.

There is, however, a second level of correlation which may exist between the technological parameters themselves. A simple example is gate strip sizing and placement errors shown in Fig. 11.1. If the gate is displaced in the source-drain gap, there are correlated changes in L_{SG} and L_{GD}. If the gate length is in error as well, there are correlated differences in the actual values of all three, L_{SG}, L_{GD} and L_G. Consequently, there are correlations at two levels which must be accounted for (Fig. 11.2): the first is vital to both sensitivity and yield analysis; the second is primarily important in yield analysis.

Correlations at level 1 are embodied in the equations given in Chapters 3, 4, 6 and 7. Many of the correlations at level 2 are specific to particular technologies, and so will differ from organization to organization. Thus each organization must determine the nature of the correlated variations which apply in its own case.

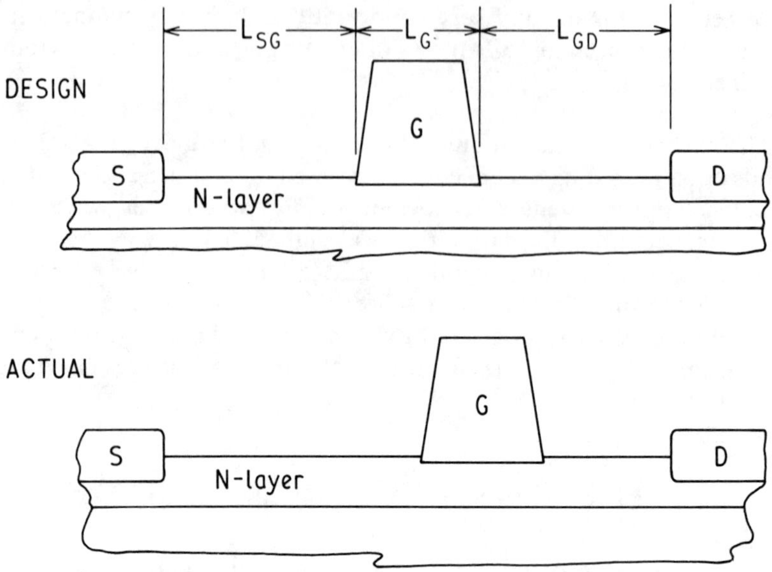

Fig. 11.1 Correlated errors in L_{SG} and L_{GD} due to gate strip misalignment.

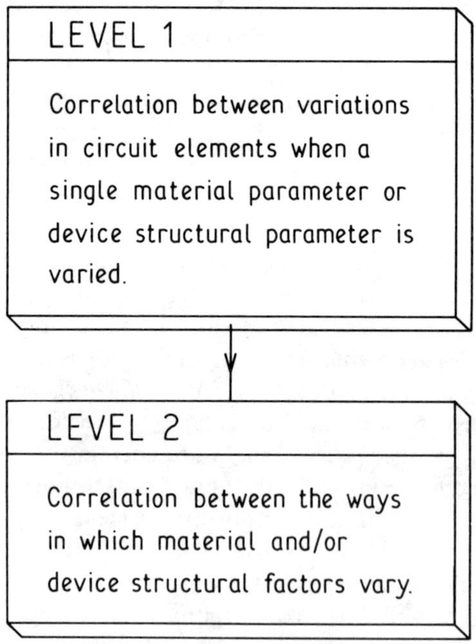

Fig. 11.2 Two levels of correlation in sensitivity and yield analysis.

To illustrate the kind of variations in question, Fig. 11.3 is a sketch of a GaAs wafer with the active layer prepared by VPE: towards the edge of the wafer the doping density and the layer thickness both increase as a consequence of the specific way in which the reactant gases are transported in this particular reactor. Because the number of chip sites at radius r on the wafer is approximately proportional to the circumference $2\pi r$, the statistical distributions of layer thickness W and doping density N are skewed, as well as being correlated, as shown in Fig. 11.4. If ion implantation is used as the method for generating active layers, the distributions of N and W, and the correlations between them, will be affected by the four-fold spatial symmetry of defect distributions which are found in LEC substrates [11.1]. In this case the shape of the profile will be affected, in particular the tail region, so that the N-layer-to-substrate interface potential ϕ_B will vary in a manner correlated with N and W (see Section 6.6).

Returning to the correlations at level 1 in Fig. 11.2, we will consider an example of sensitivity analysis as applied to a 0.5–18 GHz distributed

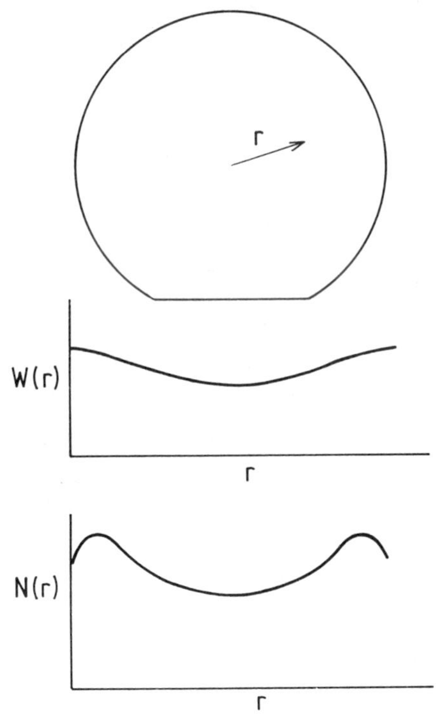

Fig. 11.3 Example of correlated variations in layer thickness (W) and doping density (N) across a VPE layer.

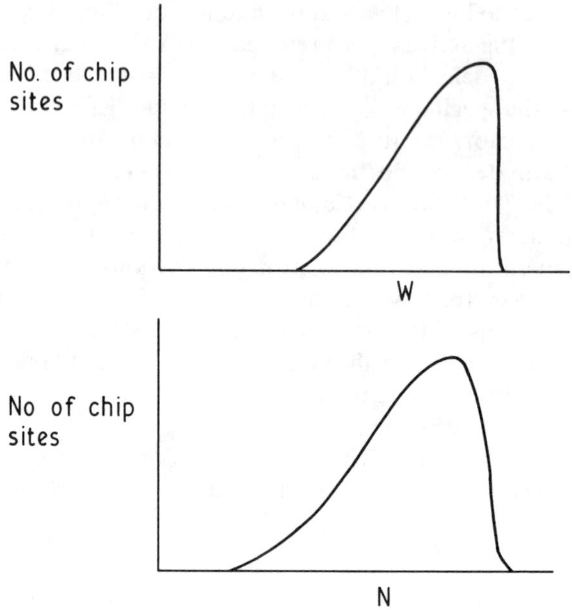

Fig. 11.4 Correlated distributions of W and N from Fig. 11.3.

amplifier chip (Fig. 11.5). This was the first design iteration of a circuit to the following specification:

Frequency band: 0.5–18 GHz
Input and Output match: VSWR < 2.5:1
Feedback: Not critical
Minimum gain: 5 dB
Maximum ripple: 1 dB
Bias: $V_{DS} = 3$ V; $I_D = I_{DSS}/2$

For the purposes of illustration, we will examine the manner in which the forward gain of the amplifier $|s_{21}|$ is affected by FET variations when technological parameters are systematically changed. Fig. 11.6 shows the computational route used, where the FET model is comprised of the equations of Chapter 6†.

Figure 11.7 shows the effect of doping density variations, all other features of the circuit being exactly as designed. Generally speaking, the consequences are not too serious.

Effects of N-layer thickness variations are shown in Fig. 11.8, and are

†In fact, the extended versions of the equations were used, in which a gate recess and surface charge are explicitly included.

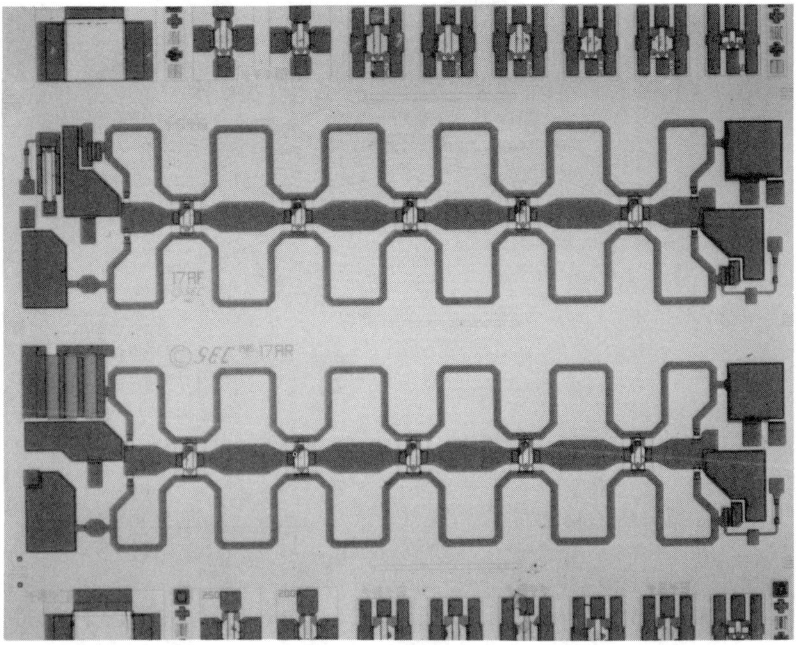

Fig. 11.5 0.5–18 GHz distributed amplifier MMIC (reprinted with permission).

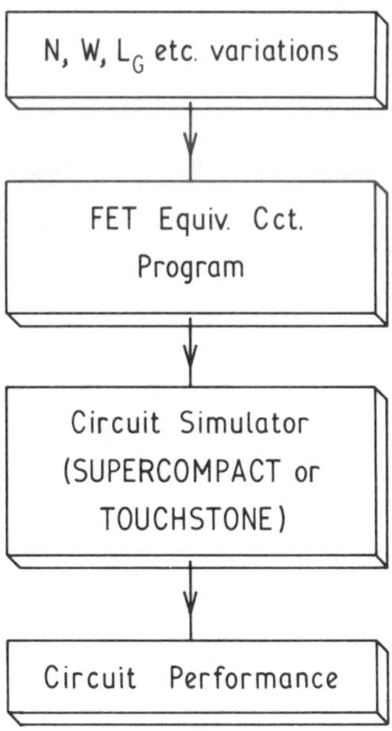

Fig. 11.6 Flow chart for MMIC sensitivity analysis.

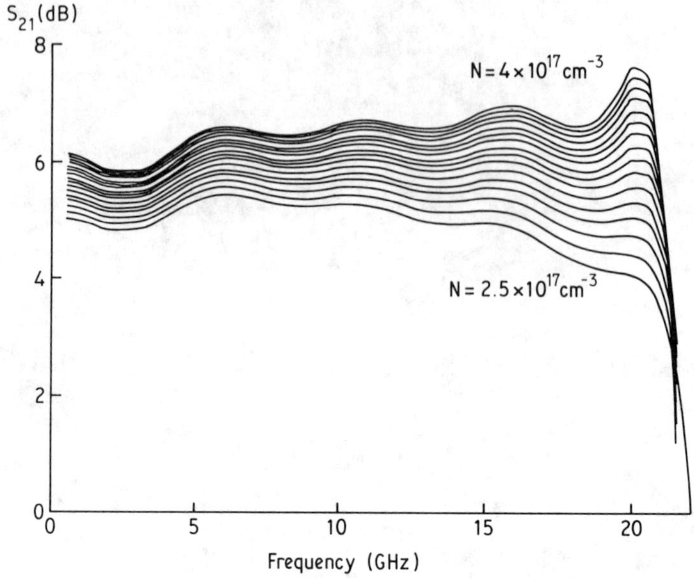

Fig. 11.7 Variation of forward gain of a distributed amplifier with channel doping.

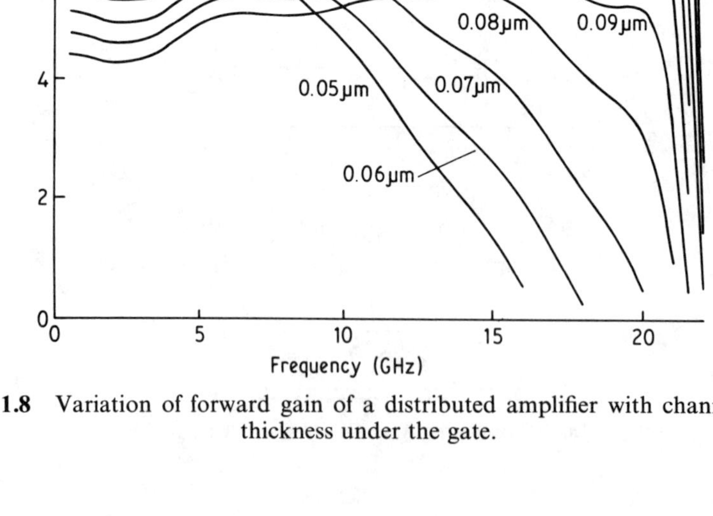

Fig. 11.8 Variation of forward gain of a distributed amplifier with channel layer thickness under the gate.

severe. Part of the reason for the apparent sensitivity to layer thickness errors arises from the bias condition nominated for this particular MMIC. An approximate account of the predominant effects can be given as follows.

When the layer is too thin, I_{DSS} drops. Thus, biasing at $I_D = I_{DSS}/2$ requires a less negative gate bias, V_{GS}. The depletion edge moves closer to the gate metal and, as a direct result, C_{gc} and g_{m0} both increase. Increases in g_{m0} increase the low-frequency gain of a distributed amplifier in accordance with the expression (see eqn (10.55), Section 10.2.5):

$$G_p = (Ng_{m0}/2)^2 Z_{0G} Z_{0D}$$

An increase in C_{gc}, however, increases the gate-line attenuation in accordance with the dependence (Section 10.2.5, eqn (10.50):

$$\alpha_G = f(\omega^2 C_2{}^2 R_2)$$

where C_2 and R_2 are the total series equivalent FET input capacitance and resistance respectively. Thus the gain droops at high frequencies. Combining these two changes gives rise to a gain magnitude which is boosted at low frequencies and which rolls off quickly at the high end of the band. A layer thickness error of only twenty lattice spacings puts the MMIC out of specification.

Figure 11.9 shows the predicted consequences of gate length errors. Here, the predominant effect is on gate-channel capacitance (see Table 6.1). Oversized gates lead once again to increases in gate-line attenuation in accordance with the preceding equation, and gain droop results.

Overall performance errors of the general character forecast in Figs 11.7–11.9 are often seen in practical amplifiers. Figure 11.10, for instance, relates to a development batch in which the FET gates were defined lithographically and were overdeveloped, yielding a gate length found by scanning electron microscopy to be 0.91 μm on a sample chip. The gain curve measured for another chip from the same wafer is overlaid on the computed curves. Bearing in mind that the computed curves assume all other features of the chip are exactly as designed but that some differences are inevitable in the practical amplifier, the agreement evident in Fig. 11.10 is acceptable. (In fact, it was this comparison which led to the initial diagnosis of oversized gates.)

Although the equations of Chapter 6 do not explicitly allow for gate recess depth and surface charge variations, we include here examples of the anticipated effects arising from both (Figs 11.11 and 11.12). Broadly, increases in gate recess depth have an effect similar to decreases in surface potential, or equivalently surface charge; severe ripple appears at the top of the band which can quickly jeopardize the chip performance relative to the ripple specification. Gain peaking is induced by complex changes in the FET equivalent circuit elements, and in the gate- and drain-transmission line cut-off

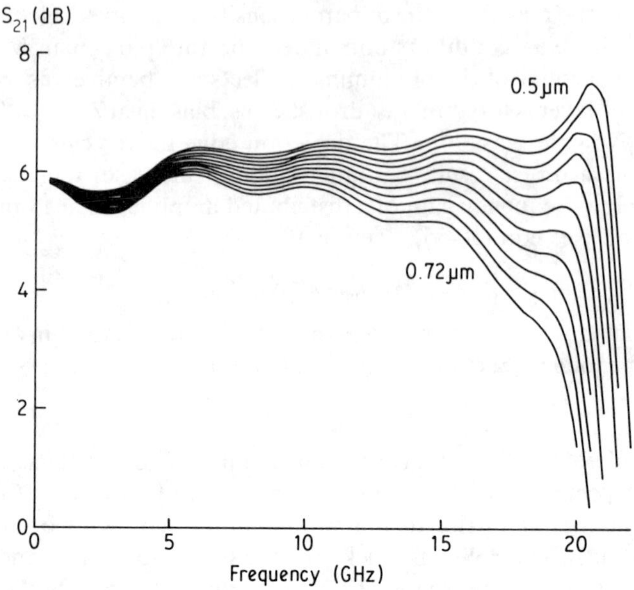

Fig. 11.9 Variation of forward gain of a distributed amplifier with metallurgical gate length.

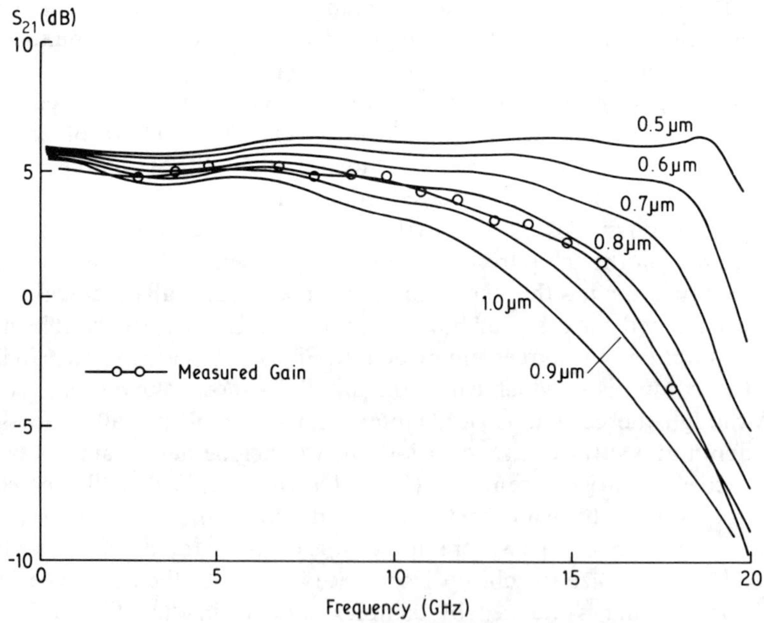

Fig. 11.10 Measured gain of distributed amplifier chip with oversize FET gates compared with forecast from Fig. 11.9.

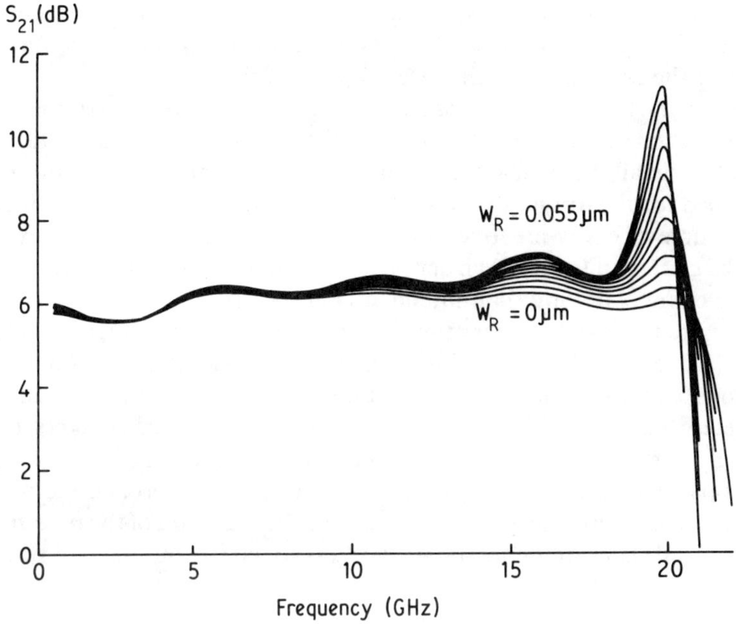

Fig. 11.11 Variation of forward gain of a distributed amplifier with gate recess depth.

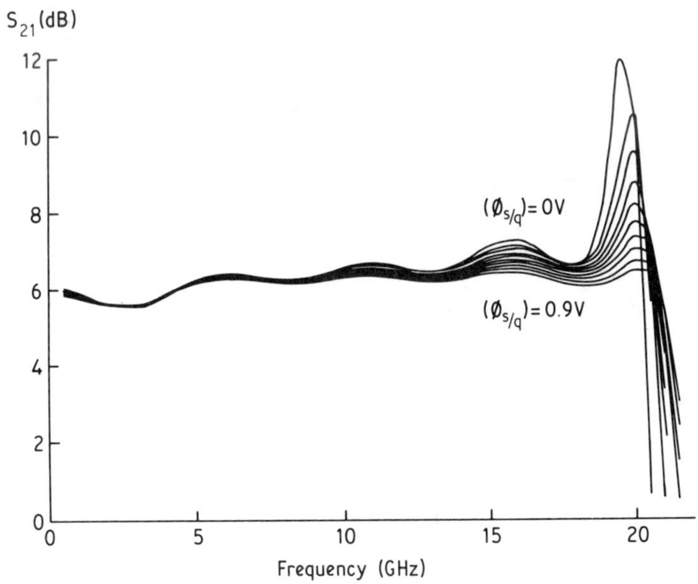

Fig. 11.12 Forecast variation of distributed amplifier gain with surface potential.

frequencies with their associated increases in line impedances as cut-off is approached. Fig. 11.13 shows the response of a practical ion-implanted version of the MMIC exhibiting a gain peak of this kind.

In Fig. 11.14 is shown the gain response of a number of chips from a VPE version of the MMIC in Fig. 11.5. Even though the passive network is now different in detail, the spread in characteristics is highly reminiscent of those determined for the original version, thus suggesting that the FET is indeed a major source of easily identified classes of performance error. For example, when the gain is uniformly high across the band, the most likely single cause is a higher channel doping than intended (Fig. 11.7).

Despite the insights offered by sensitivity analysis as hitherto defined, it ignores all consequences of interactions between effects arising out of variations among technological parameters when several (or all) such parameters vary together. The reality is that process parameters do not obligingly vary one at a time, so that, if the sensitivity of an MMIC to a given process parameter is to be ascertained, it is better to approach the problem with an analysis which progressively restricts the variation of the parameter in question while continuing to allow all others to vary within known statistical

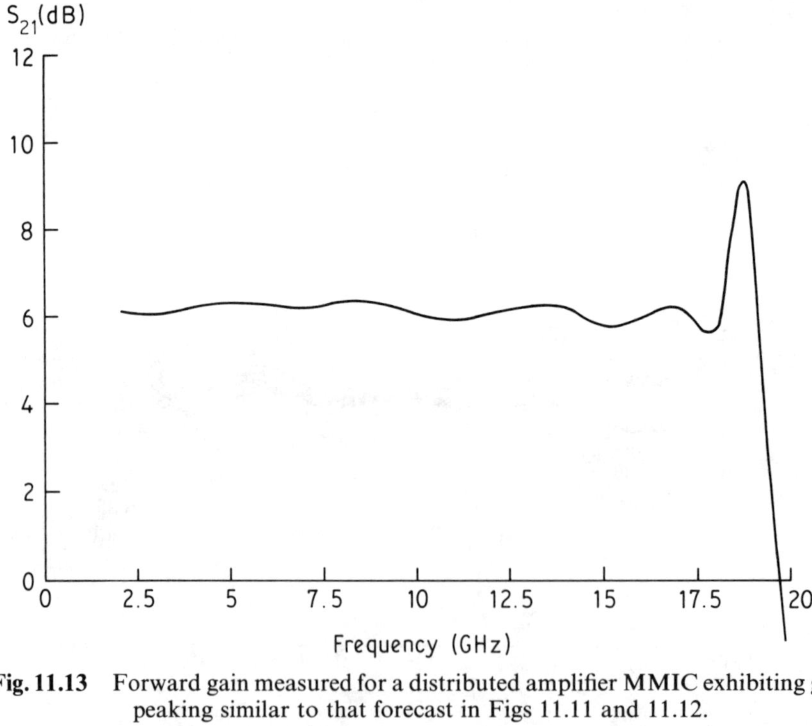

Fig. 11.13 Forward gain measured for a distributed amplifier MMIC exhibiting gain peaking similar to that forecast in Figs 11.11 and 11.12.

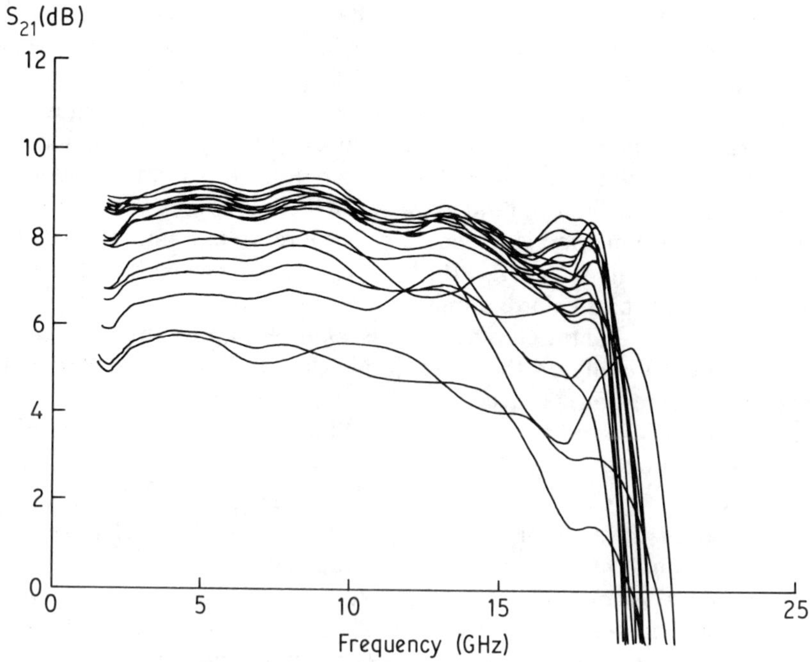

Fig. 11.14 Experimental scatter plot of distributed amplifier responses measured across a VPE wafer without prior screening or selection.

limits, i.e. the deterministic approach of what we have called "sensitivity analysis" should be replaced by a statistical analysis.

To illustrate the point, suppose we wished to establish the likely benefits of controlling more closely the variations in gate length on the circuit of Fig. 11.5. Instead of assuming L_G is the only variable, as was done in Fig. 11.9, we could acknowledge that W, N ... are also variables with statistical distributions as in Fig. 11.4. Analysis then proceeds by first studying the computed responses of the circuit—as all the process parameters are varied within the limits set by their respective distributions—and then looking at how the response improves relative to the circuit specification as the spread of gate length about the mean value is reduced. Such a scheme falls into the general category of yield forecasting.

11.3 Application of the Monte Carlo method to yield forecasting

The Monte Carlo method is a powerful technique which finds widespread application in all kinds of problems requiring solution by statistical trial. In the context of MMICs, consider the distributions of N and L_G shown in Fig. 11.15, which are typical of real distributions obtained by reverse modeling microwave data from test FETs (Chapter 8). The distributions of the other important technological parameters can be obtained similarly. Leaving aside possible correlations among process variables, an amplifier chip could be fabricated with any value of N and, simultaneously, any value of L_G allowed by these distributions (and likewise for W, D_S, W_R . . .). The basic idea of the Monte Carlo method is to generate sets of random numbers such that each set represents values of N, L_G, W, D_S, W_R . . . with which an MMIC might be fabricated. This particular combination is then used as input data to the scheme of Fig. 11.6, and the chip response is computed and compared against the specification. Finally, in the most simple case, a decision is made as to whether the chip response is acceptable.

Figure 11.16 summarizes this sequence of events. Several points need to be made concerning the method.

(a) The random values of N, L_G . . . must be generated in such a way that over a large enough number of trials they duplicate exactly the empirically determined distributions (Fig. 11.15). Because computers generate random numbers R on the interval $0 \leqslant R \leqslant 1$ only, transformations must be employed to map the normalized distributions initially generated by the computer onto the actual distribution for N (and similarly for L_G, W, D_S . . .).

(b) Where two (or more) tightly correlated distributions are involved, as for example with N and W in Fig. 11.4, just one computer-generated random number is used to generate the two correlated values of N and W.

(c) Because the FET model software is compact, it executes rapidly enough to make many hundred simulations of a typical MMIC possible in a realistic timescale, even on a moderately-sized computing facility.

(d) In place of the simple accept or reject performance criterion in Fig. 11.16, likely distributions of return loss, transmission gain and phase can be ascertained. Figure 11.17 shows a density plot of s_{21} magnitude for the amplifier of Fig. 11.5 with, for the purposes of illustration, gaussian distributions assumed for all technological parameters with 10% standard deviation in every case.

(e) As mentioned at the end of Section 11.2, changing the spread or deviation of any one parameter about its mean value is the preferred way of carrying out an analysis of the sensitivity of a circuit to that particular parameter. The term "process robustness" is sometimes applied to this activity.

The manner of transforming between computer-generated random

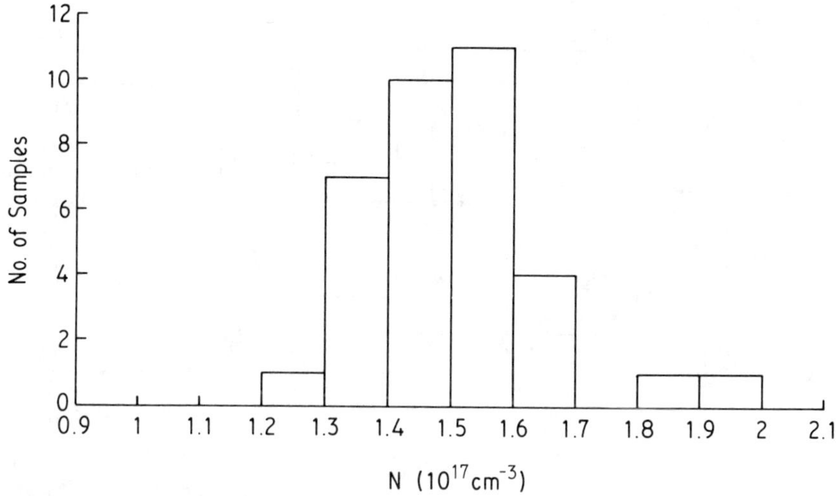

(a)

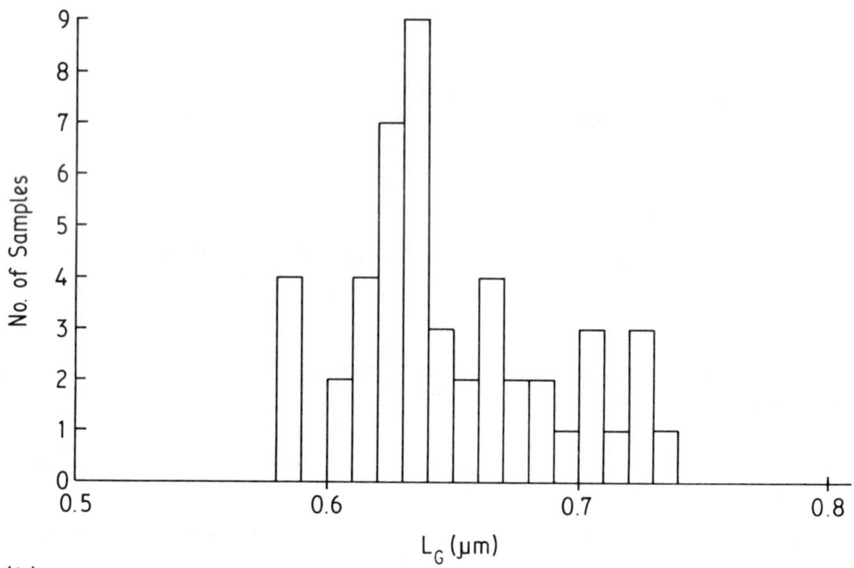

(b)

Fig. 11.15 (a) Typical histogram of doping density and (b) typical histogram of metallurgical gate length obtained by reverse modeling of microwave test FETs from process control monitors on an ion-implanted MMIC wafer.

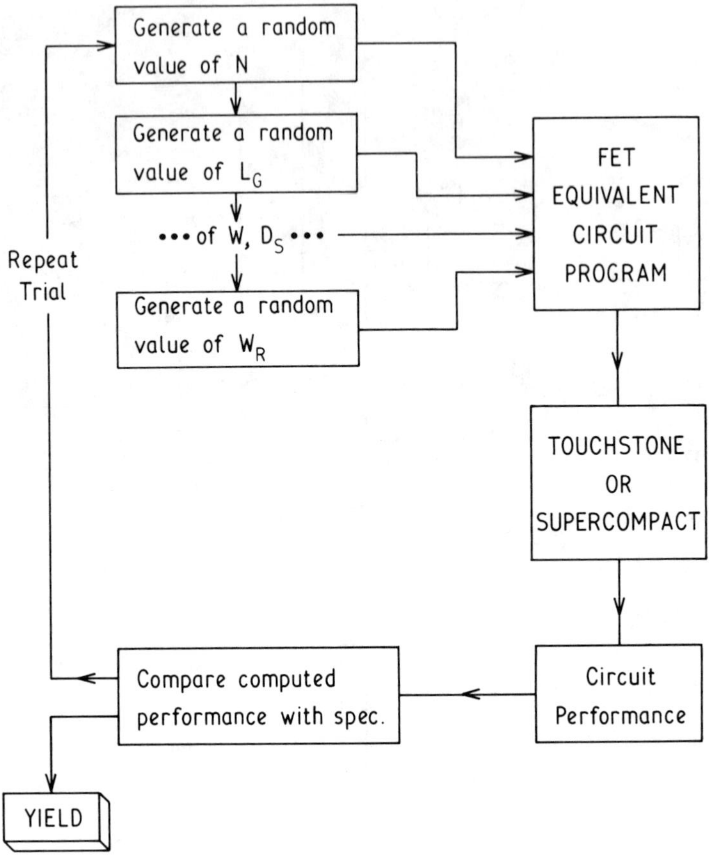

Fig. 11.16 Application of the Monte Carlo method to MMIC yield forecasting.

numbers, $0 \leqslant R \leqslant 1$, and the actual distribution relating to any particular process parameter is as follows.

(i) Let $f(x)$, where $x = N$ or L_G or . . ., represent the actual distribution (Fig. 11.18(a)).

(ii) Normalize the distribution such that the area under the normalized distribution $\bar{f}(x)$ is unity (Fig. 11.18(b)):

$$\bar{f}(x) = \frac{f(x)}{\int_0^{\infty} f(x)\,dx}$$

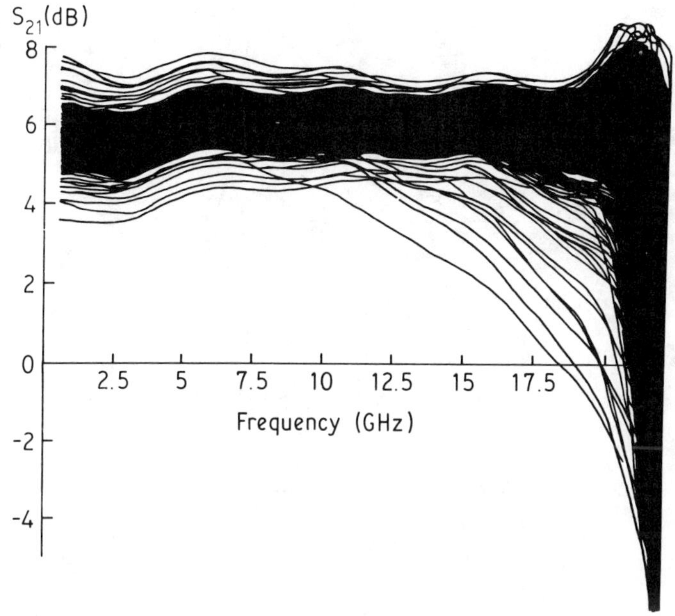

Fig. 11.17 Density plot of gain magnitude from Monte Carlo investigation of a distributed amplifier MMIC.

We thus have,

$$\int_0^x \bar{f}(x')\,\mathrm{d}x' \leqslant 1$$

(iii) If random numbers R are now computer generated on the interval $0 \leqslant R \leqslant 1$, random values of x (i.e. N, L_G . . .) can be generated from R by solving the simple integral relationship

$$R = \int_0^x \bar{f}(x')\,\mathrm{d}x'$$

for x. Graphically, x is simply read off the horizontal axis as in Fig. 11.18(c).

Using this procedure, all the features of the actual distributions are duplicated in the computer-generated numbers, including the mean value, deviation, skew and so on.

Detailed information regarding technological parameter distributions tends to be regarded as proprietary. Hence, for illustrative purposes, we quote the yield forecast for a circuit of the type shown in Fig. 11.5 for assumed gaussian distributions as in Table 11.1.

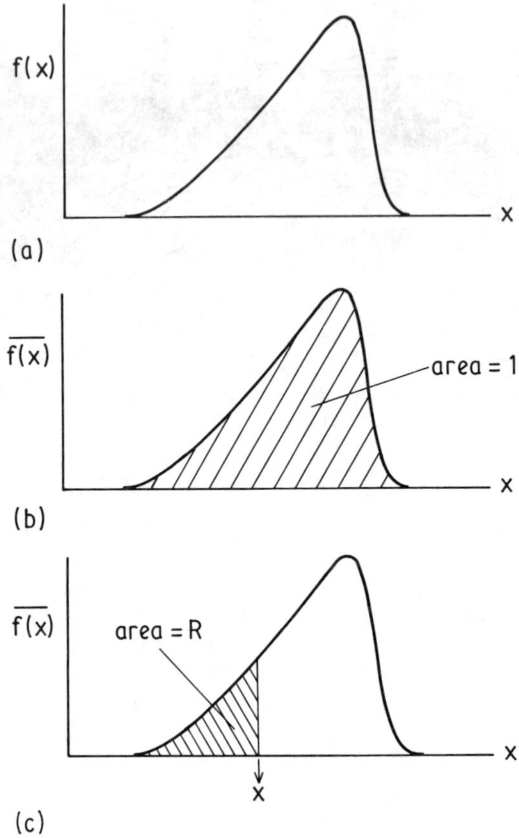

Fig. 11.18 Generation of variable values conforming to known statistical distributions using a computer-generated random number lying on the interval $0 \leq R \leq 1$.

Table 11.1 Intrinsic yield forecast for the circuit of Fig. 11.5 with assumed gaussian distributions on all process parameters

Standard deviation on all parameters	*Intrinsic yield (approximate)*
10%	32%
2%	80%

By redesign, it has proved possible to double, approximately, the yield for 10% deviations [11.2]. The yield calculated by Monte Carlo is what we have chosen to call the intrinsic yield, termed the design yield by Pucel [11.3]. The overall yield of chips is the product of the intrinsic yield and the mechanical yield, so

$$\text{overall yield} = \text{mechanical yield} \times \text{intrinsic yield}$$

where

Mechanical yield is the fraction of chips remaining after failures due to such factors as wafer breakage, poor photolithographic lift off, poor alloying and the like.

Intrinsic yield is the fraction of chips which, having been successfully fabricated in the mechanical sense, are within RF specification.

A practical measure of intrinsic yield is

$$\text{intrinsic yield} \approx \frac{\text{number of chips within RF specification}}{\text{number of chips that are dc functional}}$$

Similarly, mechanical yield is usefully regarded as

$$\text{mechanical yield} \approx \frac{\text{number of dc functional chips}}{\text{chip sites on wafer}}$$

Hypothetically, for a mechanical yield of 20%, and with 10% deviations leading to an intrinsic yield of 32%, the overall yield of good RF chips would be $0.2 \times 0.32 = 0.064$ or 6.4%. For high performance chips, it is quite likely that, for a number of organizations, yield figures such as these came as an unwelcome surprise.

11.4 Uses of yield forecasting

Monte Carlo analysis has uses which are almost self-evident, but for completeness they are listed here

(a) forecasting the intrinsic yield of a given circuit fabricated by a fixed process with known variations;

(b) assessment of the robustness of alternative circuit designs to the uncontrolled variations in a given process;

(c) calculation of the degree of materials and process control needed to achieve any desired yield target;

(d) from (a), and knowing the wafer processing cost and mechanical yield, determination of the chip selling price required for economic viability;

(e) from (b), formulation of a set of design rules for circuit robustness; and

(f) from (c), forecasting the investment in capital equipment, time, and human resources needed to achieve economic viability in GaAs IC ventures.

There may well be further applications of Monte Carlo analysis which will become evident as experience is gained using it. At all events, it offers an alternative to the vastly expensive and time-consuming approach of practical trial and iteration to MMIC chip realization.

11.5 FET centering

So far in this chapter we have considered the design of passive networks around FETs whose specification is fixed, and we have discussed tools for assessing the sensitivity or robustness of the MMIC to declared process variations, together with the intrinsic (or design) yield. We have also canvassed the idea that further design iterations of the passive network may offer improvements in yield while retaining the same microwave specification. This procedure is summarized in Fig. 11.19. We may call it the "trial design" method.

A fundamental weakness of design by trial is that it fails to capitalize on the generally strong influence exerted by FETs on the overall chip response. A common disposition among adherents to this most widespread of methods is to blame the FET technologist when the designer's circuit does not work. The FET's failure to do what it was supposed to is seen as a costly annoyance. This is not a helpful reaction.

Rather than label the FET as the source of failure, we prefer to regard the strong influence it exerts over circuit behavior as an opportunity to be exploited; this opportunity rests on the view that the fixed FET around which the circuit designer created his passive network is unlikely to be the best FET, in any sense, to extract the best compromise between performance and yield from that passive network. So, having designed a circuit, improvements can probably be made simply by changing the cross-sectional parameters of the FET, N, W, L_G, W_R, D_S, ϕ_B . . . (see Fig. 6.2 of Chapter 6). We refer to this process as "reactive design centering" or "FET centering", as summarized in Fig. 11.20.

One advantage of FET centering is that useful improvements in overall MMIC performance can sometimes be made without requiring any changes to the mask set, and that includes gate length changes when gate strips are direct E-beam written.

The point can be illustrated by reference to Fig. 11.21, which is a repeat of

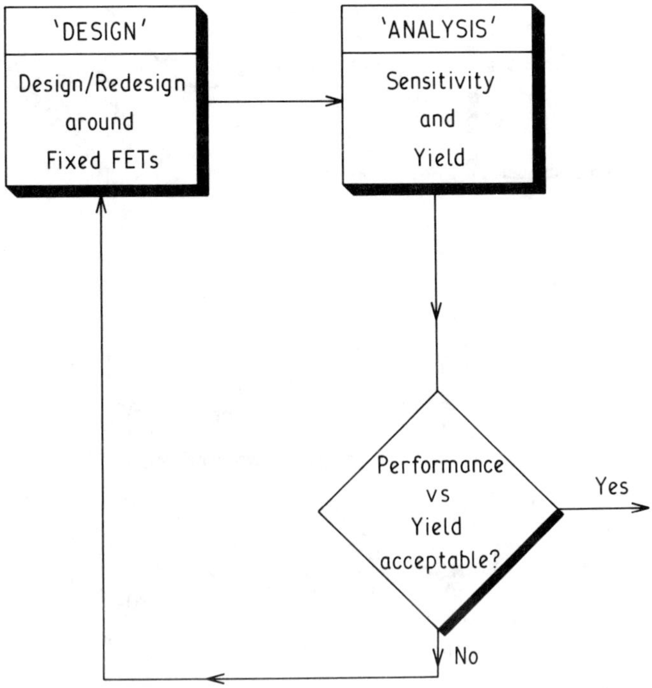

Fig. 11.19 MMIC design by the method of trial design and analysis.

Fig. 11.7. To keep the argument clear we will assume that, in a particular process technology, very tight control over all parameters has been achieved except the layer doping, N. Consequently, when a chip fails to meet the specification, it does so because of doping variations. We further assume a simplified specification for the forward gain magnitude:

$$|s_{21}| > 5 \text{ dB over 0.5–18 GHz}$$

Lastly, we assume the distribution of N to be rectangular, with limits of $\pm 0.5 \times 10^{17}$ cm^{-3} either side of the mean value, $\langle N \rangle$. Thus if the nominal value of N is $\langle N \rangle = 3 \times 10^{17}$ cm^{-3}, values between 2.5×10^{17} cm^{-3} and 3.5×10^{17} cm^{-3} could occur in the finished chip with equal probability.

The way in which the yield improves can now be determined, in this very much simplified case, by inspection of Fig. 11.21, wherein the curves for $N = 2.5$, 2.6, 2.7 and 2.8×10^{17} cm^{-3} all fall on or below the 5 dB gain line. Table 11.2 gives the analysis of yield improvement as the nominal doping is increased.

Thus, if the amplifier had originally been designed around a FET with a nominal channel doping of 3×10^{17} cm^{-3}, useful yield improvement could be

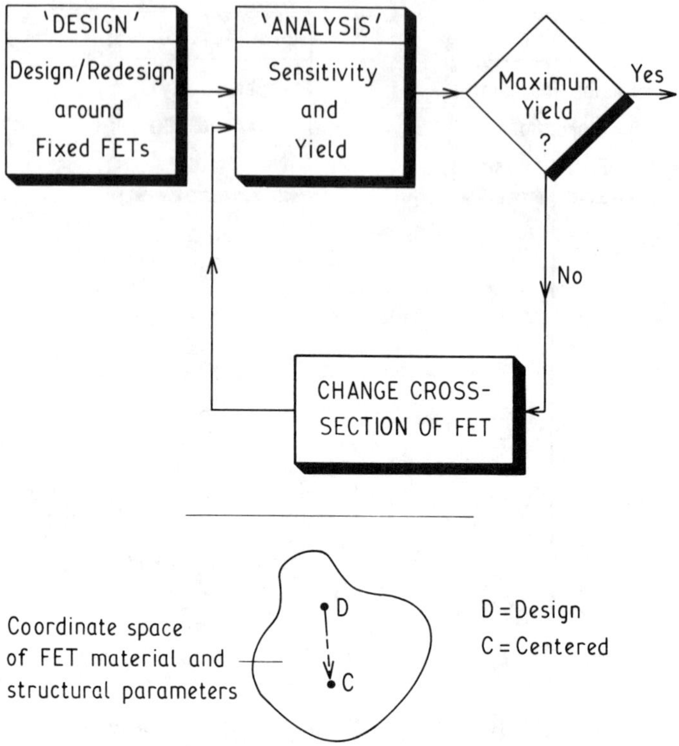

Fig. 11.20 Scheme for improvement in yield versus performance compromise by FET centering.

expected from now centering the FET on a channel doping of at least 3.4×10^{17} cm^{-3}, say 3.5×10^{17} cm^{-3}. In all other respects the FET in this example would remain the same.

In practice, of course, the full Monte Carlo program is used to study the yield as the mean values of the FET's technological parameters are systematically varied, with the widths of the distributions held constant. This procedure of varying the positions of the statistical distributions in search of an optimum should be contrasted with the earlier robustness calculation of point (e) in Section 11.3 in which the spread of the distribution was altered around fixed means.

11.6 Longer-term developments

The technique of FET centering considered in the preceding section is clearly a partial implementation of a more general technique in which every

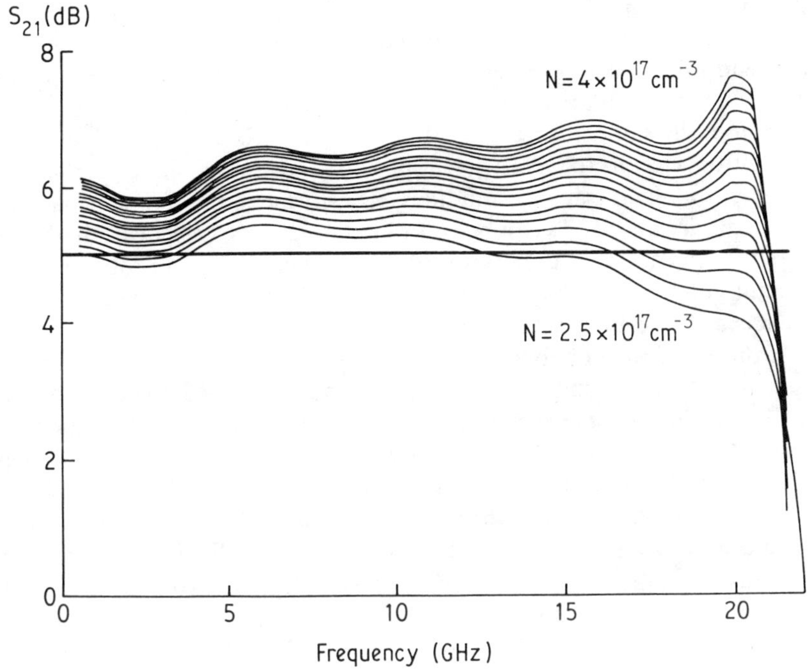

Fig. 11.21 Illustrating potential for yield improvement by FET centering.

Table 11.2 Simple example to show how yield can improve with FET centering

$\langle N \rangle$ (10^{17} cm^{-3})	*Range of N* (10^{17} cm^{-3})	*No. of samples of N in range*	*No. meeting spec.*	*Yield* (%)
2.5	2.0–3.0	11	2	18
2.6	2.1–3.1	11	3	27
2.7	2.2–3.2	11	4	36
2.8	2.3–3.3	11	5	45
2.9	2.4–3.4	11	6	55
3.0	2.5–3.5	11	7	64
3.1	2.6–3.6	11	8	73
3.2	2.7–3.7	11	9	82
3.3	2.8–3.8	11	10	91
3.4	2.9–3.9	11	11	100
3.5	3.0–4.0	11	11	100

single component in the circuit is adjusted, or centered, so as best to achieve yield and performance objectives. When this is done around a passive network of pre-determined configuration, so that only the values of the passive elements in the circuit are changed but not their interconnection, the two are near enough the same in principle. Whereas a FET is always unchanging in configuration, circuits to meet a given specification are not; there are normally many ways in which transmission lines, transformers, inductors, capacitors, resistors, diodes and FETs can be connected together to meet certain desired gain, return loss and output power objectives. There is then a problem of how the circuit configuration might best be chosen so as both to maximize yield and achieve the desired performance.

At present, it would appear that there are few guidelines for configuring circuits in this way. A simple example has been given by Rummelhard and Le Brun [11.4], who considered simple inductor-capacitor matching sections for a single FET amplifier. Four possible input matching networks were considered, and four output matching networks, and the best of each was identified on the basis of the sensitivity of the input and output return losses to variations in the network elements themselves (but not the FET). A detached view of this example, however, shows it still to be design by trial in that various circuit options were identified, and a particular configuration ultimately chosen on the basis of its predicted performance. As experience is gained with trial design and circuit centering methods, it would be reasonable to expect some guidelines to emerge to help constrain the *ab initio* choice of circuit configuration.

Leaving aside the detailed question of rules for defining the circuit configuration to meet concurrent yield and performance objectives, enough is now known to suggest that a distinction can be drawn between classes, or types, of circuit: coarsely, categories which might be termed high yield with low cost, and high performance with low yield, appear to exist. The difference between them involves, in part, the type of FET employed.

Figure 11.22 gives examples from each category. Normally, we would expect a 0.5 μm GaAs FET to offer useful gain up to 20 GHz at least, so its use in an amplifier designed for operation at 2–6 GHz at first sight seems extravagant. The trick, however, is that it has performance potential which can be sacrificed by loading the device with passive components in the interests of achieving reproducibility at the "new" FET terminals, and hence yield of the overall circuit. Simple resistive feedback between the drain and gate is one possibility, and the resistor can be neatly integrated between the gate and drain pads (Fig. 11.23). The example of a high-yield, low-cost circuit given in Fig. 11.22 is typical of the kind of MMIC offered by foundry operations as a merchant market device.

The high performance category, on the other hand, relates to circuits

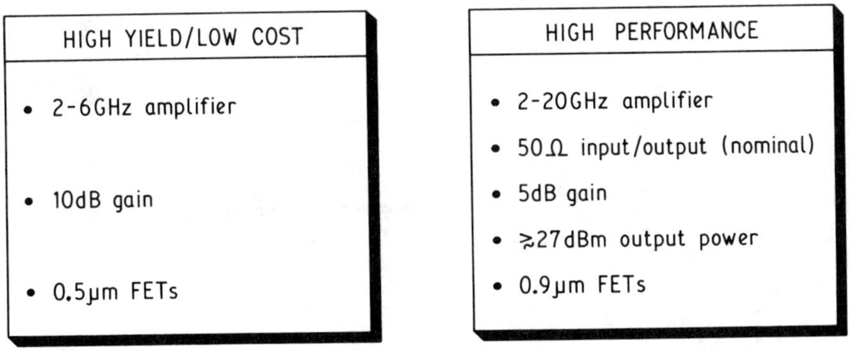

Fig. 11.22 Examples of high yield low cost, and high performance high cost MMIC specifications.

which seek to extract the limits of performance the FET can offer. Of necessity, there is a high degree of particularization, or matching in the general sense, between the FET and the passive network. If either deviates in practice from its design value, performance is quickly lost. The example shown in Fig. 11.22 is of a power amplifier: a low-noise amplifier example would have served equally as well. We shall return to the example of this power amplifier shortly.

Broadly, the aims of longer-term developments in MMIC design might be stated as concurrent design of FETs and passive networks for the best compromise between yield, cost and performance, as well as for out-and-out performance.

Figure 11.24 attempts to show the hierarchy of approaches we have

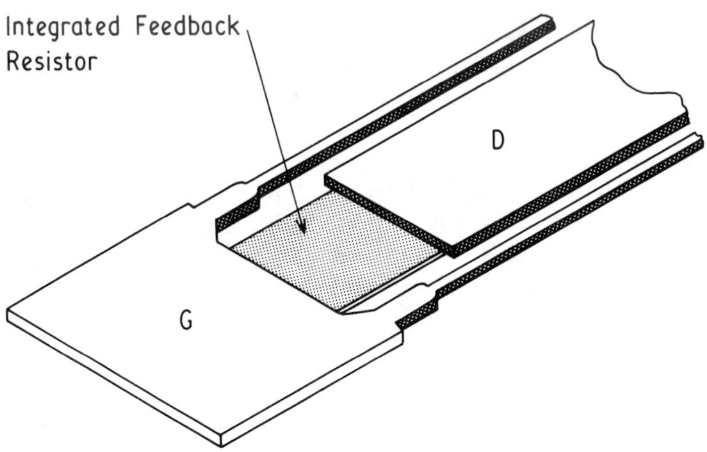

Fig. 11.23 Integral drain-gate shunt feedback.

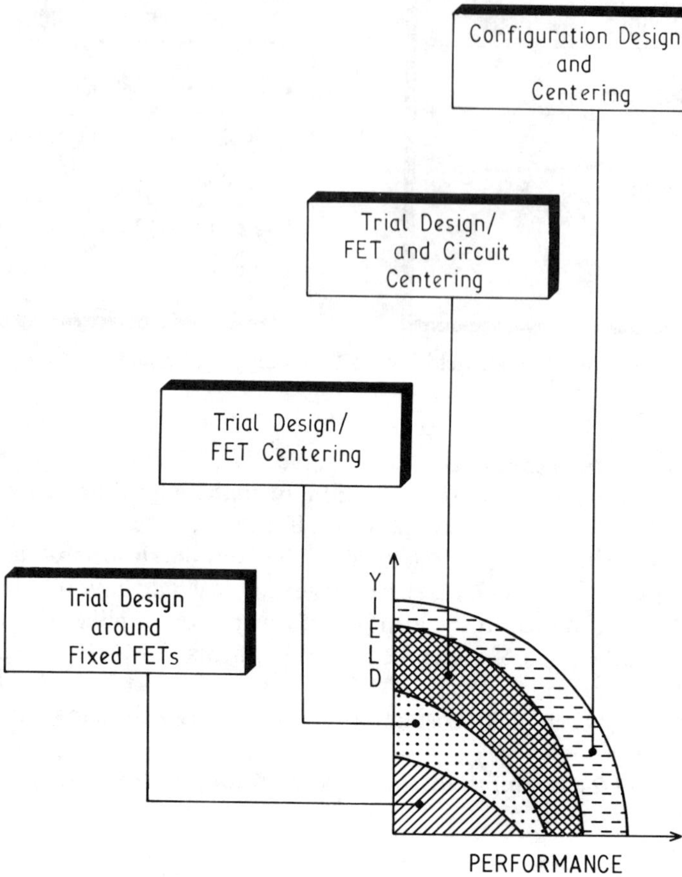

Fig. 11.24 Perceived yield versus performance merits of the hierarchy of design and centering methods.

discussed, and to indicate the relative benefits to yield versus performance of which each might be capable. Each shaded area is meant to convey an impression of how yield might be traded for performance, and vice versa. Ascending levels of sophistication in design methods should thus allow a higher yield to be achieved while maintaining a given microwave specification or, conversely, an improved performance to be realized while maintaining a constant yield.

As an exploratory exercise in FET and circuit centering we have taken the power amplifier example of Fig. 11.22 [11.5]. A number of potential applications requiring power levels of up to 1 W over a broad band, with the upper band edge at about 20 GHz, prompted this study. The task set was to design an MMIC by centering the network and the FET together, in the first instance to give the best possible performance irrespective of yield. By this means a performance reference would be established against which to judge manufacturable MMIC designs. A brief description of this exploratory investigation is of value here because it serves to demonstrate what might be done in the future†.

The type of circuit considered is a distributed amplifier, or traveling-wave amplifier, shown in Fig. 11.25. The particular MMIC shown is among the foremost examples of distributed amplifier design and realization in monolithic form, and is one of a number of such circuits that Raytheon have produced [11.6]. The special characteristics of distributed amplifiers are their comparatively constant gain and constant input and output impedances over a very broad band. Because the amplifier shown in Fig. 11.25 has a band specified as 2–20 GHz [11.6], it comfortably embraces the band of current interest and, consequently, it is a very convenient study case with which to check the computer program written for traveling-wave amplifier synthesis. A brief account of this work has been given before [11.5].

Under small-signal conditions the amplifier power gain is governed by

$$G_p(\omega) = \frac{g_{m0}{}^2 Z_{0G} Z_{0D}}{4} \left[\frac{\exp(-\alpha_G N l_g) - \exp(-\alpha_D N l_d)}{\alpha_D l_d - \alpha_G l_g} \right]^2$$

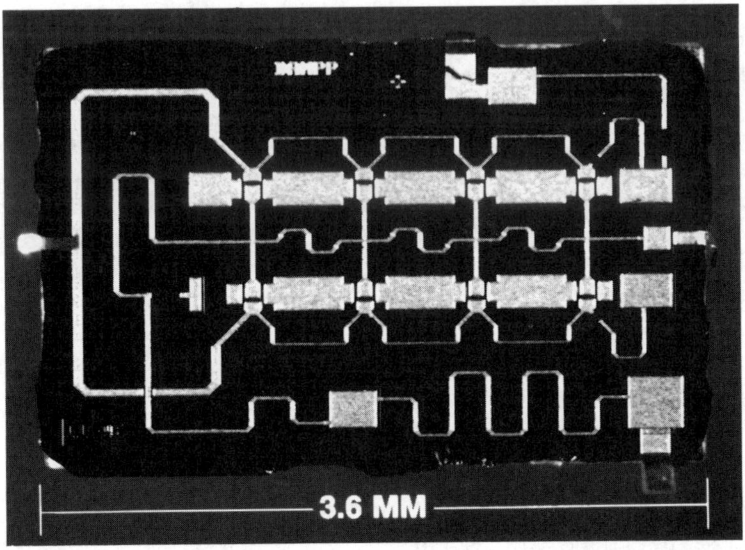

Fig. 11.25 Raytheon 250 mW, 2–20 GHz distributed power MMIC (photograph courtesy of Raytheon).

†An analytic treatment of the same problem was given in Section 10.5.

where the gate-line and drain-line attenuation coefficients are respectively given by eqns (10.50) and (10.47). Using also eqns (10.49) and (10.52), we obtain

$$\alpha_G = \frac{\omega^2 C_2{}^2 R_2}{2l_g} Z_{0G}$$

$$\alpha_D = \frac{Z_{0D}}{2r_o l_d}$$

Z_{0G} and Z_{0D} are the characteristic impedances of the artificial gate and drain lines. C_2 and R_2 are the equivalent series input capacitance and resistance of the FETs when loaded at the drain with the effective microwave impedance presented by the drain line†, i.e. C_2 and R_2 are the effective series capacitance and resistance seen looking into the gate-source terminals of the FET equivalent circuit of Fig. 6.3. All the small-signal equivalent circuit elements are calculated in the program using the equations of Chapter 6.

Under large signal conditions, the largest conceivable peak-to-peak input signal that will still give rise to current modulation in a FET is

$$V_{\text{in peak-peak}} = V_p + V_{BO} + I_F R_S + E_S L_G$$

The output signal swing is either I_F limited or V_{DSB} limited, in accordance with Fig. 6.54; these limits are given by eqns (6.54) and (6.55). Up to the point where any one of these three limits is reached, the FETs are assumed to behave linearly. Note, however, that generally the FETs in the amplifier will not saturate (either at the input or the output) at the same time but will saturate progressively, one by one, as the drive level is increased. Thus despite the "linear-up-to-a-limit" representation of the FETs, the overall amplifier will exhibit a progressively saturating (or compressed) power gain characteristic.

All the foregoing equations have been combined in a program consisting of about 2000 lines of source code‡, with the following additional restrictions and assumptions.

(a) The topology of the circuit is predetermined (in terms of number of FETs and all component inter-connections).

(b) The gate- and drain artificial transmission lines are uniform.

†To state the obvious, the effective load impedance presented to the drain of any FET is given by the ratio v_d/i_d. This ratio clearly depends upon the current on the drain line originating from other FETs in the distributed array, so the impedance is not simply $Z_{0D}/2$ (that is, Z_{0D} looking to the left, and Z_{0D} looking to the right, in parallel with each other).

‡The program was written by P.L.D. Abrie.

(c) The FET active layer is uniformly doped, with a sharp profile tail so that ϕ_B is large enough to suppress injected substrate current effectively.

(d) The output power capability of the amplifier is calculated on the basis of $V_{\text{in peak-peak}}$, V_{DSB}, and I_F limits.

(e) All FETs in the distributed amplifier have the same characteristics.

(f) The surface charge density remains at the equilibrium value even when the FETs are operating under large-signal microwave conditions.

The data input to the program defines the small-signal response of the distributed amplifier, principally (Fig. 11.26):

G_{ss} ($\omega \approx 0$): small signal gain (at zero frequency)
ΔG: gain droop over the band
f_B: upper edge of the frequency band
f_C: transmission line cut-off frequency
Z_{0G}: gate line impedance
Z_{0D}: drain line impedance

The program then calculates the parameters of a whole series of amplifiers with FETs which will satisfy the specification. For each FET, P_O is calculated,

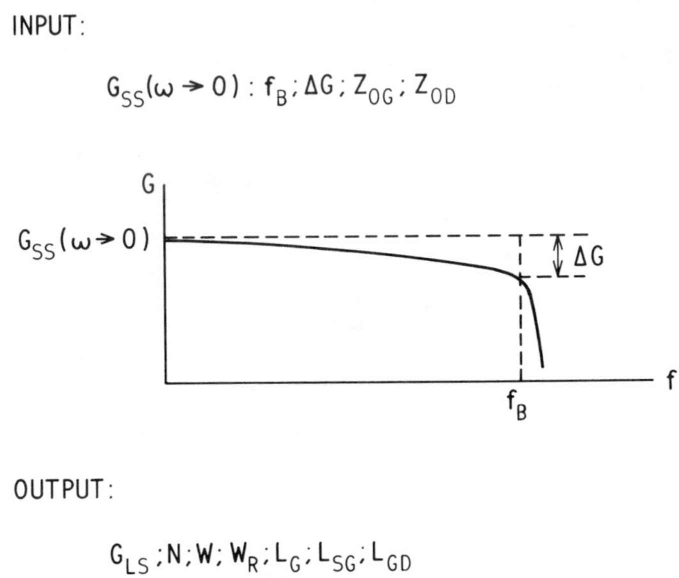

Fig. 11.26 Definition of input and output variables for distributed power amplifier design program.

and the FET and passive network which maximizes P_O is identified. The large-signal gain, G_{LS}, is also calculated. Distributed (or traveling-wave) amplifiers consisting of up to 16 elements can be handled (4–7 is typical), and high-power designs incorporating a capacitor, C_{EXT}, external to and in series with the FET gates, can be synthesized. Amplifiers without such an external capacitor are referred to as directly coupled.

As an independent test of the method, the Raytheon 2–20 GHz traveling-wave amplifier [11.6] has been taken as a study case. Table 11.3 gives a detailed comparison of the Raytheon amplifier and FET parameters, where given in [11.6] and [11.7], and those calculated by the computer routine. There are two points to be made in connection with Table 11.3.

(a) The practical amplifier [11.6] was designed around 0.8 μm, 4 V pinch-off voltage FET models. The actual gates on-wafer turned out to be 1 μm long. It is not known how accurately the actual gate length was determined.

(b) The calculated value of V_p (=3.53 V) applies at low V_{DS} ($\leqslant$100 mV). It is well known that, at a V_{DS} of several volts, GaAs FETs almost always require an increased V_{SG} to pinch them off, and that the number is often quoted to the nearest sub-unit of voltage as applied by a curve tracer.

With these facts in mind, Table 11.3 constitutes a satisfying comparison—certainly sufficient to give confidence that the routine can be applied to real-world design tasks.

The Raytheon circuit is essentially a directly-coupled amplifier with four FET elements. We find that it is just one of a range of amplifiers, albeit rather similar in detail, which will give the same small-signal gain, the same bandwidth and the same output power at approximately 1 dB gain compression. By further particularization, the theoretical output power can be

Table 11.3 Comparison of predicted FET technological parameters with actual (Raytheon) FET to realize a given amplifier specification

	Circuit					*FET*				
	P_O (mW)	G_{ss} (dB)	G_{LS} (dB)	ΔG (dB)	f_B (GHz)	L_G (μm)	N (10^{17}) (cm^{-3})	Z_G (μm)	V_p (V)	V_{DGB} (V)
Raytheon	250	5	3.7–4	2	21	1	1.5	300	4	15
FET synthesis program	250	5	3.9	2.1	21	0.92	1.51	303	3.53	15

increased as in Table 11.4, which gives only a bare outline of the FET structure required.

As already demonstrated practically [11.8], further improvements are possible using a series gate capacitor. Depending upon the acceptable values of G_{ss}, G_{LS}, ΔG and Z_{OD}, values of P_O in excess of 1 W should be possible.

Although yield forecasts were not made for any of the above performance-centered amplifier designs, it would be reasonable to expect the yield to diminish with performance, roughly in accordance with the trends depicted in Fig. 11.24. The real value lies in having established the limit of what should be possible from a given network configuration after design centering.

Table 11.4 Maximum (theoretical) output power for directly coupled distributed amplifiers having 4, 5 and 6 FETs

	P_O (mW)	G_{ss} (dB)	G_{LS} (dB)	ΔG (dB)	f_B (GHz)	Z_{OD} (Ω)	Z_{OG} (Ω)	N (cm^{-3})	L_G (μm)
4 cell	547	5	4.2	2.1	21	50	25	8×10^{16}*	0.96
5 cell	601	5	4.5	2.1	21	50	25	7.6×10^{16}	1.00
6 cell	650	5	4.6	2.1	21	50	25	7.5×10^{16}	1.02

*Approximate value.

There is also value in seeing what technological lessons might be learned. Above all, to capitalize on design-centering methods will require a change in technological thinking away from the idea of fixed processes, to one where flexibility is offered with accuracy and precision in the complete range of technological services. It will be necessary, for example, to integrate FETs with 1×10^{17} cm^{-3}, 0.2 μm active layers, even for the restricted example of 4-element, directly-coupled traveling-wave amplifiers in Table 11.4. More generally, a complete range of layer doping, layer thickness, profile tail, gate length, gate recess depth and so forth, will all need to be available with extremely tight tolerances. Contemporary technology shows this promise.

References

11.1 Di Lorenzo, J. V., Jordan, A. S., Von Neida, A. R. and O'Connor, P. Substrate quality: how important for future GaAs ICs? *Semi-Insulating III–V Materials Conf.*, 308–334, Kah-ne-ta, Oregon, 24–26 April 1984.

11.2 Riches, M. J., Bridge, J. P., Hill, A. J. and Ladbrooke, P. H. Unpublished work carried out at GEC Hirst Research Centre, 1987.

11.3 Pucel, R. A. MMICs, modelling, and CAD—where do we go from here?, *Proc. 16th European Microwave Conf.*, 61–70, Dublin, 8–12 September 1986.
11.4 Rummelhard, C. and LeBrun, M. in Soares, R., Graffeuil, J. and Obrégon, J. (eds), *Applications of GaAs MESFETs*, 420–421, Artech House, Dedham, 1983.
11.5 Abrie, P. L. D., Bridge, J. P. and Ladbrooke, P. H. A total active device and passive network design capability for optimum travelling-wave power amplifiers, *IEEE Electronics Division Colloquium Dig.* 5/1–5/5, Savoy Place, London, 30 May 1986.
11.6 Ayasli, Y., Reynolds, L. D., Mozzi, R. L. and Hanes, L. K. 2–20 GHz GaAs traveling-wave power amplifier, *IEEE Trans. Microwave Theory Tech.*, **MTT-32** (3), 290–295, March 1984.
11.7 Ayasli, Y., Reynolds, L. D., Vorhaus, J. L. and Hanes, L. K. 2–20 GHz GaAs traveling-wave amplifier, *IEEE Trans. Microwave Theory Tech.*, **MTT-32** (1), 71–77, January 1984.
11.8 Kim, B. and Tserng, H. Q. 0.5 W 2–21 GHz monolithic GaAs distributed amplifier, *Electron. Lett.*, **20** (7), 288–289, 29 March 1984.

Chapter 12

Future developments

The underlying theme throughout this book has been that the physical processes occurring within the GaAs MESFET have a major influence on the microwave characteristics of MMIC amplifiers, and that there is a need for physics-based computer-aided tools to design MMICs for high yield with acceptable performance. More precisely, there is a need for techniques which allow the compromises possible between performance and yield to be systematically explored long before a design is committed to masks and fabrication. Integrated circuit yield is principally determined by the effect of process variations on the microwave performance of the active elements, be they GaAs FETs or HEMTs. There is a need, therefore, for computer-aided methods that allow the scattering parameters of an overall MMIC to be forecast and analyzed in terms of the technological make-up of the chip, particularly, insofar as they affect the transistor and hence the whole circuit.

This idea is not new; it was done about fifteen years ago for the silicon analog integrated circuit industry, much of it by the University of California at Berkeley, as summarized in the excellent book by Getreu [12.1]. All that is required, therefore, is for the same principles to be applied to GaAs FETs in MMICs, and to a lesser extent to HEMTs. For the analysis of a complete circuit to be feasible using the computing power available from personal computers or minicomputers, the code relating the microwave characteristics of a FET to its material and physical structure must execute extremely rapidly, typically in a few CPU seconds. It is hard to think of a better word than "compact" to describe this FET code and the mathematical-physical model from which it is derived.

In Chapters 6 and 7 we have given some compact, closed-form expressions for the small-signal properties of GaAs FETs and HEMTs, more completely for the first device than the second. Limits to the class A output power of a MESFET were also given. These expressions are simplifications of more complete descriptions which have been developed through many years of device physics research supported by purpose-designed experiments to test theoretical propositions, and to provide fresh experimental facts to assist with analytical developments. There is, therefore, no question that these methods work in practice. If the practical structure is known accurately enough, it will be found that they do. Even though the FET and HEMT equations presented in this book are of a lower level than the original and more complete descriptions, they nevertheless allow a useful demonstration of the basic computer tools under discussion.

Having established the device models in Chapters 6 and 7, Chapter 9 considered some design limits that apply to the gain or power available from a device in a circuit at a particular frequency or band of interest. Chapter 10 covered the principal amplifier configurations used in MMICs, giving equations which, taken together with the FET equations, allow some preliminary study of the ways in which FET variations affect circuits, and of the design of FET elements to meet specific performance targets for an overall MMIC. In Chapter 11, the use of computer methods such as sensitivity analysis, Monte Carlo analysis, and concurrent FET-and-circuit design were considered at the initial level of development at which they currently exist.

Chapter 8 offered a different kind of tool: an algorithm for working backwards from measured *S*-parameters to determine the metallurgical gate length of the FET or HEMT from which those parameters came. In other words, Chapter 8 considered a tool to help the process technologist analyze and gain control of FET gate length, which the experience gained during this work shows to be an important factor to MMIC reproducibility.

Against this background, what of the future? What software developments can be anticipated in the light of known physics? What might be possible in the way of developments in these device-physical models, and what software capability might follow from them? What are the implications for process technology? And how will all this serve the needs of MMIC designers and foundries?

In the first place, it should be possible to augment existing commercial network analysis packages such as SUPERCOMPACT© and TOUCHSTONE© with small-signal FET tools of the kind described here. Many levels of improved sophistication in the compact FET code are possible, allowing for different recess depths and shapes, non-uniform channel doping profiles, and different surface-state charging rates during FET operation (where this is known in practice). Equations and code to cover these cases can be written on the basis of currently-known FET physics, thereby making FET-based sensitivity and yield analysis readily available.

Any technological change that results in improved FET reproducibility should result also in improved MMIC reproducibility. Where ion implantation is employed as the means of preparing the channel layer, it is possible to achieve improved FET reproducibility by careful choice of the profile when other known process variations are taken into account. Some organizations obviously practise these tricks of the trade already, but for those who do not a systematic exploration using FET predictors that accommodate non-uniform doping profiles should allow implant conditions for reproducibility to be identified.

One of the important realizations to emerge from device physics is that there are many correlations among the equivalent circuit elements of a FET that, certainly in the writer's case, were unsuspected. An example is the way in which τ_{gm} and C_{gd} change in response to a bias change or a change in structure. Such correlations represent a wealth of knowledge when it comes to fitting equivalent circuits to measured S-parameters. By measuring the S-parameters of the FET at a number of selected bias points, it should be possible to use the physical correlation criteria to carry out equivalent circuit fits of improved quality. Thus improvements in commercially available equivalent circuit fitting routines can be foreseen. As an illustration as to why this should matter, a straightforward S-parameter-to-circuit fitting algorithm tends to leave some uncertainty in the way the real part of y_{11} (obtained after conversion to Y parameters) should be divided-up into R_g, R_i and R_s. Yet there are some amplifier circuits where the detailed distribution of a given total resistance among R_g, R_i and R_s makes a clear difference to the MMIC gain response.

Allied to the foregoing development, we can envisage the evolution of higher-level reverse modeling routines that would allow not only the metallurgical gate length to be determined, but also the channel doping profile, gate recess depth, surface-state charge, and electron velocity in the channel. Such a capability would be a powerful aid to process technologists, particularly if it included statistical distributions of gate length, doping and other technological data in the form of a wafer map at the same time. All the mathematical-physical descriptions of GaAs FET operation required for the development of such software tools are already known. With them, a process technologist could analyze the results of his work, and develop a technology appropriate to whatever performance versus yield requirements for MMICs he may have. It is only the advent of on-wafer microwave probing, afforded by the CASCADE prober for example, that makes such tools possible.

Present-day process tolerances may be too wide for acceptable yield on some circuits operating up to 20 GHz. Standard deviations of around 10% on doping, gate length and layer thickness may not be good enough for high-performance MMICs at these frequencies. Instead, tolerances of a few percent may be required, with some fairly radical implications for substrate growth, buffer layer and N-layer preparation, passivation film growth and gate lithography. One may ask whether vertical gradient freeze (VGF) substrates offer improved device reproducibility compared with LEC substrates, whether in-situ growth of passivating dielectrics is feasible, and whether high-energy electron beam lithography will become a necessity. When capital expenditure and development costs of this magnitude are at stake, it makes sense to base the decision on structured argument and calculation. Advanced CAD tools have a clear role to play. Furthermore, the whole question of process tolerances seems to become more acute at higher frequencies. For instance, what are the chances of holding a $\pm 5\%$ tolerance on a 0.1 μm gate for a 94 GHz millimetre-wave monolithic integrated circuit?

To take proper advantage of cost and performance trade-offs, fixed process schedules may have to be abandoned in favor of process flexibility while preserving extremely tight process tolerances and control. In other words, fixed FETs, as typically offered by some foundry operators, will prove too much of a constraint on the circuit design process. To say again what has already been said in earlier chapters, FETs should be configured to meet the needs of circuits rather than circuit design having to accommodate the limitations of FETs. The implications of such an idea are fairly extreme, almost requiring "dial-up" materials growth and process technologies. Evidence abounds already that dry processing will displace wet chemical processing in all the critical steps of MMIC fabrication. Wet etching, for example, offers little hope of maintaining tight enough control over the recess depth and shape for adequate yield of some high-performance circuits operating up to, say, 26.5 GHz. Again, such process control becomes progressively more important at higher frequencies. It appears that the gain of a 94 GHz HEMT, for instance, depends crucially on the shape of the gate recess relative to the gate (metal) cross-section.

The combination of forward-modeling and reverse-modeling software would be valuable in helping circuit designers specify the material and structure of a FET, in conjunction with some chosen circuit topology, needed to satisfy MMIC performance criteria. A need exists for designers to be able to explore the benefits of moving away from a standard foundry type FET to a FET of different doping, gate length, and so on. There is little point in making such a study in terms of simply varying the equivalent circuit elements, unless the physical structure required to realize those element values simultaneously can be specified. Active encouragement might be given to such practices to try and develop a better understanding between circuit designers and process technologists.

Turning now to power MMICs, we can expect to see the development of compact FET models and code for large-signal analysis and design. It is likely that the most successful MMIC enterprises will be those that can design and fabricate power chips. In no other case is it so important to have precise process control, together with an ability to design FETs and the rest of the network concurrently. Power MMICs may typically require a specification to be satisfied embracing power gain, output power at 1 dB compression, power-added efficiency and phase tracking, all over a wide range of temperatures. In the long run, it is in this area that the methods and techniques advanced in this book will prove the most valuable, because there may be no other way of getting the job done, except by extended trial, insight, and a little luck. There is every indication that device physics is equal to this task.

To conclude this book, we focus on one of the most challenging and potentially rewarding applications of monolithic integrated circuits: 94 GHz M^3ICs, shown as "TGSMs" in Fig. 2.3. As reference to the professional literature will show, to date there have been few reports of useful circuit gain (meaning 5 dB or greater) from FETs or HEMTs at 94 GHz. The reason is simple: 5 dB gain at 94 GHz is hard to achieve. Slightly oversized gates or gate recesses, or MBE layers with a doping-thickness product that is slightly wrong, cause a loss in gain of a few dB so that all useful device performance is lost. Herein lies a challenge to the depth of understanding of device physicists and engineers. The problem suggests yet another question: are there any opportunities for manipulating the various material and structural features of the device so as to enhance the gain at frequencies approaching 100 GHz?

The final paragraph below suggests an affirmative answer to this question. The paragraph, and the two figures which accompany it, were provided by Professor R. J. Trew of North Carolina State University, Raleigh, North Carolina†. Particular note should be taken of the last two sentences.

The unilateral gains calculated for four transistors representative of the state of the art are shown in Fig. 12.1. The four transistors consist of a 0.1 micron gate length GaAs MESFET, a 0.25 gate length AlGaAs–GaAs HEMT, a 0.25 micron gate length pseudomorphic HEMT, and a mm-wave permeable base transistor (PBT). The unilateral gains are calculated from physically based RF equivalent circuits which were determined by parameter extraction from measured data. The frequency responses of the equivalent circuits were extrapolated to the unity gain frequency in order to investigate the expected frequency limits for each of the devices. The flat responses of the MESFET and PBT at the lower frequencies are due to forward conduction of the control electrode. The gains for all four devices fall at a -6 dB per octave rate up to approximately 20–30 GHz, indicating a single pole characteristic response. Above this frequency and up to f_{max} the gain falls at a -12 dB per octave rate. The additional pole is due to feedback effects in the device. The AlGaAs–GaAs HEMT shows an anomalous response beginning at about 50 GHz. The unilateral gain shows an enhancement before falling at the -12 dB per octave rate. The enhancement is due to the time delay in the current source. This is illustrated in Fig. 12.2. By varying the time delay, the unilateral gain near f_{max} can be tuned to obtain optimized performance. This indicates that a minimized time delay (i.e. a maximized carrier velocity under

†The text is reproduced as received, and describes work done and conclusions reached independently by Professor Trew. The author is indebted to Bob Trew for permission to include his work.

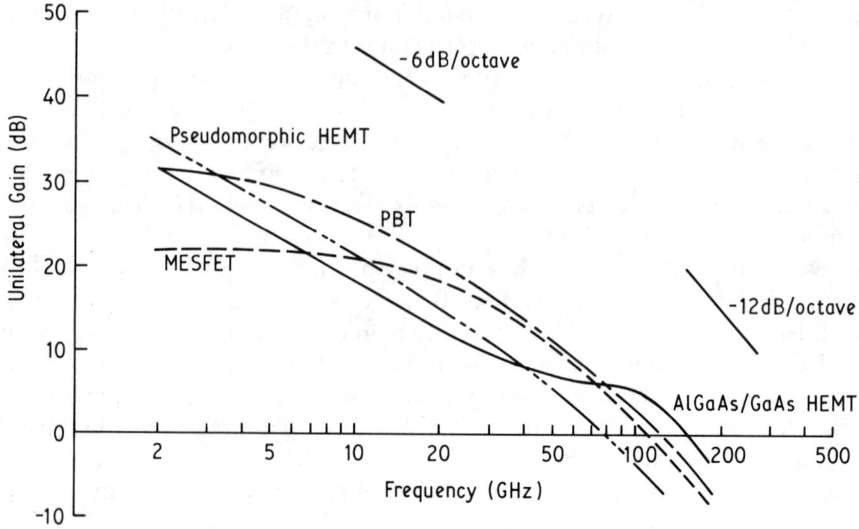

Fig. 12.1 Unilateral gains calculated for four state-of-the-art transistors.

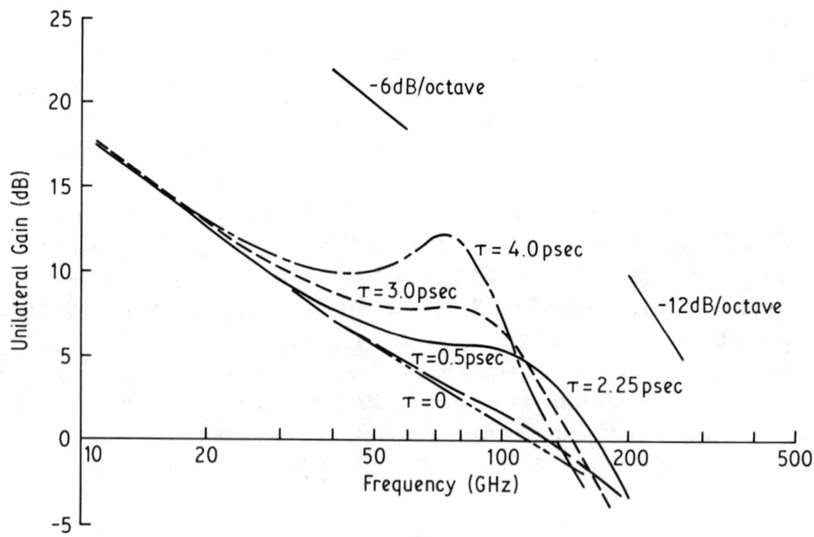

Fig. 12.2 Unilateral gain of 0.25 μm gate AlGaAs–GaAs HEMT showing gain enhancement possibility by varying the time delay in the current generator ($\tau \equiv \tau_{gm}$).

the gate) may not be sufficient to obtain a maximized f_{max}. The frequency performance of the device is a complicated function of the device design, and the device structure can be optimized to obtain a tuned high frequency response.

Reference

12.1 Getreu, I. E. *Modeling the Bipolar Transistor*, Tektronix, Beaverton, 1976.

Index